*An Anthology of Nineteer
American Science Writing*

An Anthology of Nineteenth-Century American Science Writing

Edited by
C. R. RESETARITS

ANTHEM PRESS
LONDON · NEW YORK · DELHI

Anthem Press
An imprint of Wimbledon Publishing Company
www.anthempress.com

This edition first published in UK and USA 2013
by ANTHEM PRESS
75-76 Blackfriars Road, London SE1 8HA, UK
or PO Box 9779, London SW19 7ZG, UK
and
244 Madison Ave. #116, New York, NY 10016, USA

First published in hardback by Anthem Press in 2012

British Library Cataloguing-in-Publication Data
A catalogue record for this book is available from the British Library.

Library of Congress Cataloging-in-Publication Data
The Library of Congress has cataloged the hardcover edition as follows:
An anthology of nineteenth-century American science writing / edited
by C. R. Resetarits.
p. cm.
Includes bibliographical references.
ISBN 978-0-85728-974-2 (hbk : alk. paper) – ISBN 0-85728-974-8 (hbk
: alk. paper)
1. Science–United States–History–19th century. 2.
Scientists–United States–History–19th century. I. Resetarits, C. R.
II. Title: Anthology of 19th-century American science writing.
Q127.U6A77 2012
509.73–dc23
2011048161

ISBN-13: 978 1 78308 062 5 (Pbk)
ISBN-10: 1 78308 062 0 (Pbk)

This title is also available as an ebook.

CONTENTS

Part Two: 1846–1876   Warriors

Part Three: 1876–1900   Scientists

PREFACE

THIS anthology of nineteenth-century American science writing began with my pursuit of American studies. Over the years I became intrigued by the scientists of the period, especially the first half of the century. They often wrote about science and natural history, literature and exploration. When I queried my science friends on their familiarity with scientists of the nineteenth century, they recognized names, knew who represented their own fields, but had rarely read any of the original work. I soon realized one of the reasons why when I tried to find ready sources for the primary works of the early naturalists (Say, Nuttall, Wilson). Not only were there no anthologies of early American science, but even the compendiums of nineteenth-century science writing, of which there are scores, rarely bothered with the Americans. If by chance they should, the same three or four names appeared: Joseph Henry, Oliver Wendell Holmes Sr., a smattering of William James as philosophy of science or Thoreau as natural history. I came away from these anthologies with the uncomfortable feeling that American science in the nineteenth century was inferior at best, non-existent at worst. Of course, there might be good reason for this omission from the British and European point of view; simply put, they were far ahead of the Americans in both productivity and scientific infrastructure, but that did not mean science in the United States had not made contributions to the century or that it would not be of interest to students of the American nineteenth century.

After reading in the subject for a few years, I discovered many excellent studies on a whole range of issues in American science through the nineteenth century, from specialized studies of individual scientists to synthetic studies of movements and controversies to sweeping overviews of the evolution of scientific thought. Missing, however, were collections of primary materials. There was Nathan Reingold's *Science in Nineteenth-Century America: A Documentary History*, which is still an excellent source of private documents between scientists of the era, but its private nature only underscored the need for a volume on the public writings.

My goal here, then, is to provide an anthology that stands alone but also serves as a supplement, source, and inspiration for the studies of American science already in play or under construction.

The Selection Process

The selections in this volume are the result of reading widely in the history of science, philosophy of science, American studies and in the specific scientific fields. My method for arriving at this list was simple in delineation although slightly more complex in fine-tuning. Firstly, I established the parameters of exclusion; this is, admittedly, a narrowly focused science anthology. I do not include the social sciences or science education, both of which might be better served with anthologies of their own, nor the literature of American exploration, which has already been brilliantly and broadly edited by Rueben Gold Thwaites. Considering the lack of any anthology of nineteenth-century American science, I prefer to keep my focus narrow in order to include the fullest measure of the natural/physical sciences and their evolution through the century. Secondly, I established the criteria for inclusion within this focus: seminal, published, nineteenth-century American science writing. By seminal I mean published essays of lasting worth that were ground-breaking at the time *and* still remain pertinent to science in some manner today. I was not looking to represent all disciplines or factions; this is more a distillation than a survey. Once I had my parameters, I began to read. I made lists of all scientists I encountered and noted the frequency with which they were mentioned. I then began reading and rereading with an eye to the seminal writings of these scientists, paying particular attention to who referenced who and why. I read in the specialized histories and anthologies of the individual disciplines (physics, chemistry, biology, etc.) to identify which writings present-day scientists hold dear in their own fields, and, lastly, I stepped back to look at the controversies that defined the science of the time – from the early neglect of the Lewis and Clark collections, the evolution debate, the Taconic/geological naming controversy, to the "Bones War" – in order to gather a sense of what was most influential then. This was the method used to determine the following selections.

Women and Science

I must make note here of the limits of this anthology with regard to women and suggest two studies on the subject, Margaret W. Rossiter's *Women Scientists in America: Struggles and Strategies to 1940* and Nina Baym's *American Women of Letters and the Nineteenth Century Sciences: Style of Affiliation*. There are also fine essays and collections of specific aspects of this topic. For instance, I must recommend Abir-Am and Outram's *Uneasy Careers and Intimate Lives: Women in Science, 1789–1979*, although its view goes far beyond the nineteenth-century American woman of science. Rossiter and Baym are particularly helpful in understanding the dearth of women writing science of the sort specified in this volume. Rossiter traces an odd reaction against women in science just at the time that educational opportunities were, supposedly, opening up. She suggests that in their goal to professionalize, the male science community actively tried to masculinize the profession by defeminizing its higher ranks: "The coming of professionalism in the 1880s and 1890s had contained and circumscribed the women and restricted them to the fringes of science" (p. 99). Nina Baym's book discusses a variety of ways in which women worked at the fringe, with their limited affiliation to the larger science world.

Part Divisions and Introductions

The part introductions are intended to give general background and address pivotal issues and figures not featured in the anthology. Further information on the anthologized scientists themselves is reserved for the biographies that front each section. These biographies were crafted in an attempt to present in condensed form the relevance of the scientist and the science, as well as to provide a humanizing aspect. Finally, the chronologies ending each part introduction help identify the milestones of the period with the anthologized selections in boldface to orient them in time. All told, I hope this multifaceted approach will prove helpful to the greatest variety of readers.

This anthology is divided into three parts in an attempt to address the movement of American science in the nineteenth century from its beginnings in exploration and classification, with scientists who were largely self-taught (Part One), to college-educated scientists with generalist inclinations and several fields of study (Part Two), to the university-educated scientists who were by necessity specialists (Part Three). Part One covers the turn of the century up to 1846–7 when a number of pivotal events took place: the opening of Yale's School of Applied Chemistry, which would become the Sheffield Scientific School; the arrival of Louis Agassiz to lecture for the Lowell Institute and his subsequent hire at the new Lawrence Scientific School at Harvard in 1847; the founding of the Smithsonian Institution with Joseph Henry as its first secretary; and the decision of the Association of American Geologists and Naturalists to take on a more national and general scientific scope, renaming themselves the American Association for the Advancement of Science. (I must, of course, thank Robert V. Bruce for these delineations; see his *The Launching of Modern American Science: 1846–1876*.) Part Two: 1846–1876 focuses on the turbulent years surrounding the Civil War, the reception of Darwin's *On the Origin of Species*, and the movement of the machinery of scientific exploration into state and federal agencies. Part Three: 1877–1900 offers the writings of a more modern, specialized, and academic scientific community. Johns Hopkins University opened in 1876 with half of its departments dedicated to science, the American Chemical Society was established the same year, J. W. Gibbs began publishing his major scientific contributions, and T. H. Huxley's visit and lecture series on evolution in 1876 stood as both a parallel and a striking contrast to Agassiz's 1846 visit and lectures. I have edited the essays as little as possible, retaining the original spelling, punctuation, and word usage, correcting only if I felt sure that there was a misprint in the original or if the punctuation was arbitrary and superfluous as to hinder understanding. All footnotes in this volume, unless indicated, belong to the scientists being anthologised, and are presented as originally published.

I have enjoyed every bit of this journey sifting through the written record of American science. I hope you enjoy the results.

C. R. Resetarits

Part One: 1800–1846

Naturals and Naturalists

NATURALS AND NATURALISTS

THREE themes dominated the science of this period: exploration, classification, and utilization. Jefferson, in his famous letter of instruction to Meriwether Lewis, illustrated the tone: "The object of your mission is to explore the Missouri river, & such principal stream of it, as, by it's course & communication with the waters of the Pacific Ocean, whether the Columbia, Oregon, Colorado or any other river, may offer the most direct and practicable water communication across this continent, for the purposes of commerce." He also listed among the "objects worthy of notice" animals, plants, minerals, soil, topography, and climate. Embedded in this letter – the main focus of which was trade, commerce, and navigation – were Jefferson's own scientific interests, kept subordinate because of the reluctance of Jefferson's largely mercantile oriented congress to pay for science that did not promise a commercial payoff. The pro-British mercantile elite found many of the enlightenment tendencies of the Francophile Jefferson suspect, superfluous, ungodly, but this judgment was more a matter of politics, balance, economy, and utility than any true anti-science bias. The outstanding American scientists of the previous century, Franklin, Bartram, Rittenhouse, all had roots in the mercantile class and were themselves mindful of the young country's need for practical and useful knowledge. Jefferson too thought of science in terms of utility, for though he might not view commerce as the sole guide for scientific activity, he always tended toward a measure of utility. In his *Notes on the State of Virginia* (1787) he classified plants not only by their scientific name but by their utility: medicinal, esculent, ornamental, and "useful for fabrication."

Philosophical Considerations and Classification

The philosophical underpinnings of science at the start of the nineteenth century were much the same both sides of the Atlantic: Baconian philosophy ruled, and Scottish "common-sense" eased the ambiguities that a larger Baconian tendency might reveal, as did the influential works of Humboldt, who was widely read in America and Britain, as well as Europe; and Natural Theology was there to offer a sense of solace and harmony in the midst of change. Eighteenth-century and early nineteenth-century philosophers of natural history substituted classification of species for the Great Chain of Being and assumed that all hierarchies lead naturally to God. Throughout the first third of the nineteenth century, the majority of those who pondered such things at all would have agreed with Samuel Tyler's claim in *The Discourse of Baconian Philosophy* (1844): "philosophy, revelation, natural theology, and physical science are united in perfect harmony" (Daniels, *Age of Jackson*, p. 200). For Tyler the whole of Bacon's scientific method was embodied in the classification of species, and his book was a summary, and a defense, of science at the time. It is important to bear in mind, however, that there would have been less need for such a defense if there were not already an undercurrent of resistance to the status quo. Issues, questions and inconsistencies were arising that could not be easily subsumed or classified away. As early as 1818 Amos Eaton was beginning work that would lead to the controversy over the naming of geological strata in America. A similar controversy was paralleled in England. In American meteorology there was controversy over the interpretation of storm data and theories of storm movement. Fossils and new species were turning up everywhere as inspiration, incentive, and obstacle to classification systems. Two of the youngest and brightest of America's fledgling science community, Asa Gray and James Dwight Dana (see Part Two), were each in their own fields turning to a more advanced and organic method of classification, what was becoming known as the natural classification system, which employed the use of type specimens in Gray's case and chemical analysis in Dana's. Classification may have seemed the whole of science for amateur and casual observers early in the nineteenth century, but that whole was already raising as many questions as answers for those few scientists of a more critical bent.

The Lewis and Clark Expedition

Even with hindsight it is difficult to judge the full effects of the scientific pursuits of the Lewis and Clark Expedition. Its position at the beginning of the century established it as both a pivotal event and a window on the growth of American science. Too often the science of the Lewis and Clark Expedition is glossed over with easy praise, citing Jefferson's letter, his foresight in sending Lewis to study under Benjamin Smith Barton, or the lengthy lists of plants and animals cited in the journals. However, the complications that accompanied the actual specimens are an excellent measure of the difficulties of this exploratory and fledgling phase of American science. The first crates of specimens that arrived in the east were later destroyed en route to storage. Another grouping of specimens stored at Great Falls was lost when the Missouri flooded. Nevertheless, several hundred plant and animal specimens did arrive safely. Jefferson sent the botanical specimens to the American Philosophical Society and the zoological specimens to the Peale Museum, which served as a de facto national repository before the Smithsonian Institute. Both Jefferson and Lewis had hoped that Benjamin Smith Barton would prepare a report on the botanicals, but Barton was over committed and failing in health, so Lewis turned to Barton's German protégé Frederick Pursh. Pursh worked on preparing and drawing the plants until Lewis's death, a presumed suicide, in 1809 when Pursh abruptly sent a small subset of specimens to Clark and left for London. The specimens sent to Clark were returned to Barton and then passed to the American Philosophical Society on Barton's death in 1815, where they were protectively ignored until Thomas Meehan rediscovered them in

1896. Meanwhile, Pursh had actually kept the majority of specimens as well as his notes on all the rest. His *Flora Americae Septentrionalis: Or, A Systematic Arrangement and Description of the Plants of North America* was published in London in 1814 and offered American botanists a contemporary look at the plants, although the details of publication were a source of some ire. Pursh's specimens went into the collection of his English employer, A. B. Lambert. Upon Lambert's death in 1842, all but nine of the Lewis and Clark specimens were purchased by Edward Tuckerman, an American artist and naturalist, and given to the Academy of Natural Sciences of Philadelphia in 1856. The errant nine remaining in England ended up in the collection of the Royal Botanic Garden, Kew, where they can still be found today. In the 1920s the American Philosophical Society specimens were transferred to the Academy of Natural Sciences where the remaining 226 extant sheets were finally reunited.

The fate of the zoological specimens at the Peale museum is less breathtaking and more tragic. Over time the specimens were dispersed or loaned to museums and collections and subsequently lost or destroyed. None of the specimens remain. Luckily while at the Peale much of the collection was described and published by Wilson, Ord, Say, and Rafinesque.

Benjamin Silliman and *The American Journal of Science and Arts*

In 1802 Timothy Dwight, president of Yale College, asked a 22-year-old Yale law student, Benjamin Silliman, to consider the new chair of chemistry and natural history at Yale, part of plans to expand the college curriculum. After accepting the professorship, Silliman left for two winters to study at the Medical College of the University of Pennsylvania and then spent 1805–1806 studying abroad. Even with the University of Pennsylvania's accreditations, Silliman's principal qualification for his new position remained unchanged: his orthodoxy. Silliman was a man that Yale could trust, as the college's trustees were more than a little hesitant to expand beyond the core curriculum of divinity and classics texts for fear of ungodly influences. Silliman proved to be the right man not only for Yale but for American science too. As Nathan Reingold summarized, Silliman "combined many of the virtues of both the amateurs and the emerging professionals of science. He knew more than the former but not enough to keep up far with the latter" (*Science in Nineteenth Century America*, p. 2). Silliman may not have left a legacy of groundbreaking science, but he did leave an important legacy as exemplar, promoter, and patron of science in the first half of the nineteenth century. He not only bridged the gap between the years of the amateur and the professional but actively set many of the following generations into place. He brought science not just to Yale but to the public at large. He was the opening act for the Lowell Institute, giving twelve sold out lectures on geology; he wrote textbooks and travel books, all of which sold very well; and he helped found the Yale Medical School. He was a well-respected man of science in the academe at a time when science in the academe was not particularly respected. His most lasting contribution, however, is undoubtedly the *American Journal of Science and Arts*, known as the *American Journal of Science* after 1880. Silliman started the publication in 1818, editing and financing it himself, and in the early years it was often known simply as Silliman's journal. The editorship of the journal remained in the Silliman family until 1926, and this lineage is itself quite impressive: Silliman to son-in-law James Dwight Dana to grandson Edward Salisbury Dana. Today the journal is housed in the Department of Geology and Geophysics at Yale and is a leading periodical in earth sciences. One of the most interesting measures of the early import of Silliman's journal might be Hawthorne's reaction after noticing a copy on a desk in the Radcliffe Library at Oxford. It was, Hawthorne remarked, "the only trace of American science, or American learning or ability in any department, which I discovered in the University at Oxford" (Daniels, *Science in American Society*, p. 151).

The Useful Arts

One of the mainstays of the history of nineteenth-century American science is the premium placed on usefulness. Certainly by both inclination and necessity, usefulness in the New World was paramount, but the Baconian idea that science would be employed to elevate the condition of man and give him more control over Nature was near and dear to men of science throughout Britain and the Continent. In America, however, there was the additional practical consideration of a pioneering community rich in natural resources, sparse in population, and dedicated from the start to more democratic ideals and opportunities. The title of Benjamin Franklin's journal *The American Philosophical Society for Promoting Useful Knowledge* nicely summarized both the need and the plan for promotion and dissemination to the masses. A hundred years later, Tocqueville's comment in *Democracy in America* that Americans were addicted to practical rather than to theoretical science was more an observation, query, and caution than an admonishment. Tocqueville displayed the roundedness of his critique later in that same chapter:

> I do not contend that the democratic nations of our time are destined to witness the extinction of the great luminaries of man's intelligence, or even that they will never bring new lights into existence. At the age at which the world has now arrived, and among so many cultivated nations perpetually excited by the fever of productive industry, the bonds that connect the different parts of science cannot fail to strike the observer; and the taste for practical science itself, if it is enlightened, ought to lead men not to neglect theory. In the midst of so many attempted applications of so many experiments repeated every day, it is almost impossible that general laws should not frequently be brought to light; so that great discoveries would be frequent, though great inventors may be few. (p. 53)

Tocqueville was attempting an extended, honest look at a changing world. The democratic experiments of the United States and France were epoch-making. Like the advent of the industrial age that was already redefining Tocqueville's world or the technological age that has defined our own, contemporary commentators and social sages were destined to get as much wrong as right. How true it would prove that the great chaotic experiment of educating the common man would make it almost impossible that general laws not come to light. However, as to the paucity of great inventors, the century was destined for a few more than Tocqueville might have expected: Bell, Morse, Edison, Tesla. What I would like readers to remember as they come to the contributions in this section and throughout the anthology is that there was this double helix of pure and applied science in America from the beginning, sometimes attracting and fortifying, sometimes repelling and adversarial. During the Enlightenment the two seemed naturally woven together. As the pro-British mercantile class gained power after the Revolution and clashed with the pro-French Jeffersonians, applied and pure scientific pursuits sometimes seemed at odds. Yet if you look a little deeper into this period you always see a complementary action and reaction, for at the very height of mercantile and nation-building motivations, we have movements like New Harmony and the Transcendentalists, both of which viewed science and nature in new, nontraditional ways and wealthy benefactors like Maclure, Lowell, Sheffield, and Smithson stepping up to provide the sort of patronage of science that had previously been the purview of the Crown. I also hope that the reader will reconsider some of the simpler assumptions about both the science and scientists of the time. Consider the image of Audubon, perhaps the most recognizable naturalist of America: foreign-born, vagabond, artist, tramp, entrepreneur, businessman, con-artist, national treasure; American science blackballed him until British science made him our quintessential image maker. Consider the common-man, self-reliant origins of the quiet, stellar contributors, Bowditch, Henry, Loomis, or the Brahmin backgrounds of such contrasting contributors as Say and Holmes. Consider the loose cannons, Rafinesque and Ord; consider the artist/scientist/vagabond explorers Nuttall, Wilson, Audubon. These figures made significant contributions to science, and not just American science. Hopefully the summary below will help place them in context.

1801   Charles Willson Peale and his son Rembrandt excavate and assemble two nearly complete mastodons in New York State.

Robert Hare gives "Memoir of the Supply and Application of the Blow-Pipe," to Chemical Society of Philadelphia. His oxyhydrogen blow-pipe is progenitor of the welding torch.

Alexander Wilson, *American Ornithology*.

1802   **Nathaniel Bowditch, "Currents,"** *The New American Practical Navigator.*

1803   Benjamin Smith Barton, *Elements of Botany: Or Outlines of the Natural History of Vegetables* (Philadelphia).

Andre Michaux, *North American Flora* (Paris).

Count de Volney, *Tableau du climat et du sol des Etats-Unis*, trans. Charles Brockden Brown (Philadelphia 1804).

1804   Dunbar scientific expedition to lower Red and Ouachita Rivers.

Alexander von Humboldt visits America.

1804–6   Lewis and Clark Expedition.

1811   Caspar Wistar, *A System of Anatomy for the Use of Students of Medicine* (Philadelphia).

1812   Benjamin Rush, *Medical Inquiries and Observations Upon the Diseases of the Mind* (Philadelphia).

1812–14   *Emporium of Arts and Sciences* (Philadelphia). The serial stresses application of scientific knowledge to technological interests.

1814   Frederick Pursh, *Flora Americae Septentrionalis: Or, A Systematic Arrangement and Description of the Plants of North America* (London).

1818   Amos Eaton, *Index to the Geology of the Northern States, with a Transverse Section form Catskill Mountain to the Atlantic* (Leicester, Mass.).

Thomas Say, "On the Genus Ocythoe," *Philosophical Transactions of the Royal Society of London,* **vol. 2.**

Thomas Nuttall, "Preface," *The Genera of North American Plants and a Catalogue of the Species, to the Year 1817* **(Philadelphia)**.

1819–20   Long Expedition to Rocky Mountains. Military expedition to carry out map survey work, includes naturalists Edwin James, Thomas Say, and Titian Peale. Torrey will publish many of James's specimens.

1820   **C. S. Rafinesque, "Introduction,"** *Ichthyologia ohiensis, or Natural History of the Fishes Inhabiting the River Ohio and Its Tributary Streams* **(1820)**.

C. S. Rafinesque, "Notices of Materia Medica, or New Medical Properties of Some American Plants," *Western Minerva or, American Annals of Knowledge and Literature* **(1820)**.

1823   First state sponsored survey conducted by Denison Olmsted in North Carolina.

1824   **Thomas Say, "Preface,"** *American Entomology, or Descriptions of the Insects of North America* **(Philadelphia)**.

John Torrey, *Flora of the Northern and Middle Sections of the United States, Or a Systematic Arrangement and Description of All the Plants Hitherto Discovered in the United States North of Virginia* (New York).

Rensselaer School, renamed Rensselaer Polytechnic Institute in 1851, founded for study of science and engineering, "farm and factory," at Troy, N.Y.

1825   Settlement at New Harmony, Indiana, of utopian model community started by the Welsh social reformer Robert Dale Owens and backed by William Maclure. Maclure, a Scotsman who had settled in the states as early at 1778, was a prosperous business man who turned his attentions to geology and American science. He was president of the Academy of Natural Sciences of Philadelphia and a financial supporter of many science and educational ventures until his death in Mexico in 1840. He is often referred to as the "father of American geology."

Dr. William Beaumont conducts digestive experiments on the open stomach wound of Alex St. Martin.

1826   Ebenezer Emmons, *Manual of Mineralogy and Geology* (Albany).

1829   Almira Hart Lincoln Phelps, *Familiar Lectures on Botany* (Hartford). The most popular botanical work of the nineteenth century.

Jacob Bigelow, *Elements of Technology* (Boston). Introduces the word "technology" to modern usage. Bigelow had used the term for years in his lectures as Rumford Professor on the Application of Science to the Useful Arts at Harvard.

James Smithson, a British scientist, left money in his will to the U.S. government to establish an institution for the "increase and diffusion" of knowledge if his nephew/heir should die without heirs, which he did in 1835.

1831   **John James Audubon, "Passenger Pigeon,"** *Ornithological Biography* **(1831).**

Chloroform discovered by Samuel Guthrie.

William C. Redfield, "Remarks on the Prevailing Storms of the Atlantic Coast of the North American States," *American Journal of Science* 20: 17–51.

1832   **Thomas Nuttall, "Pileated Woodpecker,"** *A Manual of the Ornithology of the United States and Canada* **(Boston)**.

Joseph Henry, "On the Production of Currents and Sparks of Electricity from Magnetism," *American Journal of Science and Arts* **(July)**.

1833   Edward Hitchcock, *Geology of Massachusetts* (Amherst).

1834   Samuel George Morton, *Synopsis of the Organic Remains of the cretaceous Group of the United States* (Philadelphia).

1835   Timothy Abbott Conrad, "Observations of the Tertiary Strata of the Atlantic Coast," *American Journal of Science* 28: 104–11, 280–82. Early American case for the use of organic remains for geological dating.

1836   Natural History Survey of New York established; it would employ such scientists as Conrad, Emmons, Hall, Mather, and Vanuxem.

1837   Samuel F. B. Morse patents telegraph.

Indiana Geological Survey established.

State survey of Massachusetts authorized by its legislature.

1837–40   Earliest Ohio geological survey.

1838   American Philosophical Society begins publishing its Proceedings.

1838–42   Wilkes U.S. Exploring Expedition carried out scientific study and collecting in Latin America, Antarctica, Central Pacific islands, and North American northwest coast.

1838–43   Asa Gray and John Torrey, *Flora of North America* (New York), which modified Linnaean's system with natural classification for American botany.

1839   Lowell Institute lectures begin in Boston with lectures on geology by Benjamin Silliman, Sr.

Samuel George Morton, *Crania Americana* (Philadelphia and London). Morton argues that different races are not related.

1840   John William Draper produces first photograph of the moon.

Augustus Addison Gould, "Results of an Examination of the Species of Shells of Massachusetts and Their Geographical Distribution," *Boston Journal of Natural History* 3: 483–94.

1842   Henry Darwin Rogers and William Barton Rogers, "On the Physical Structure of the Appalachian Chain, as Exemplifying the Laws Which Have Regulated the Elevation of Great Mountain Chains Generally," *Reports of the Meetings of the Association of American Geologists and Naturalists* (1843): 474–531.

Fremont Expedition to Rocky Mountains.

1843   **Oliver Wendell Holmes, "The Contagiousness of Puerperal Fever,"**
         ***New England Quarterly Journal of Medicine* 1.**

1844   John William Draper produces first photograph of diffraction spectrum.

1845   **Elias Loomis, "On the Two Storms Which Were Experienced**
         throughout the United States, in the Month of February, 1842,"
         Transactions of the American Philosophical Society.

         Humboldt publishes *Cosmos* and reviews begin to appear in the United
         States.

1846–47 Yale establishes professorship of agricultural chemistry and practical
         (applied) chemistry for graduates and other non-undergraduate
         students. School of Applied Chemistry allowed within the newly created
         Department of Philosophy and Arts. Science is still not considered part
         of traditional undergraduate curriculum.

         Louis Agassiz comes to lecture at Lowell Institute and is persuaded to
         accept a professorship at Harvard.

         Lawrence Scientific School established at Harvard.

         Association of American Geologist and Naturalists becomes the American
         Association for the Advancement of Science.

         Smithsonian Institution founded by congressional act; Joseph Henry
         chosen as first secretary.

         American Medical Association established.

THOMAS JEFFERSON (1743–1826)

Politician, naturalist, agriculturalist.
Born in Albermarle County, Virginia. Died in Monticello, Virginia.

JEFFERSON had a traditional gentleman's course of classical study at the College of William and Mary before reading law. A more effective writer than speaker, Jefferson drafted the Declaration of Independence as a revolutionary and then continued to apply his pen to the shaping of the new United States government as a leader of the Democratic-Republicans. He served as U.S. vice president under his opponent Adams in 1796 and then in 1801 was elected president serving until 1809. During his tenure as president he completed the Louisiana Purchase of territory from Napoleon and was a strong advocate of the Lewis and Clark Expedition. At the age of 76 he worked to found the University of Virginia.

Throughout his life, Jefferson read and wrote about science, from formal works like *Notes on the State of Virginia*, where he first refutes Comte de Buffon's notion that American animal life was inferior – that is, degenerative – to similar forms in Europe, to a lifetime of fascinating private letters in which he often discussed his latest reading in natural philosophy. The selection provided here, however, was his only writing on science to be published in a science-oriented journal. The fossil bones he described would prove to be the first giant sloth discovered in North America, later named the *Megalonyx jeffersoni*. Jefferson's interest in paleontology was a constant throughout his life, even to the point of his laying out newly arrived fossil bones in the unfinished rooms of the White House. The early twentieth-century paleontologist, Henry Fairfield Osborn, wrote often of Jefferson as a pioneer in American paleontology.

Thomas Jefferson, "A Memoir on the Discovery
of Certain Bones of a Quadruped of the Clawed Kind
in the Western Parts of Virginia," *Transactions of the
American Philosophical Society* (1799)

IN a letter of July 3d, I informed our late most worthy president that some bones of a very large animal of the clawed kind had been recently discovered within this state, and promised a communication on the subject as soon as we could recover what were still recoverable of them. It is well known that the substratum of the country beyond the Blue Ridge is a limestone, abounding with large caverns, the earthy floors of which are highly impregnated with nitre; and that the inhabitants are in the habit of extracting the nitre from them. In digging the floor of one of these eaves, belonging to Frederic Cromer in the country of Greenbriar, the labourers at the depth of two or three feet, came to some bones, the size and form of which bespoke an animal unknown to them. The nitrous impregnation of the earth together with a small degree of petrification had probably been the means of their preservation. The importance of the discovery was not known to those who made it, yet it excited conversation in the neighbourhood, and led persons of vague curiosity to seek and take away the bones. It was fortunate for science that one of its zealous and well-informed friends, Colonel John Stewart of that neighbourhood, heard of the discovery, and, sensible from their description, that they were of an animal not known, took measure without delay for saving those which still remained. He was kind enough to inform me of the incident, and to forward me the bones from time to time as they were recovered. To these I was enabled accidentally to add some others by the kindness of a Mr. Hopkins of New York, who had visited the cave. These bones are,

1st. A small fragment of the femur or thigh bone; being in fact only its lower extremity, separated from the main bone at its epiphysis, so as to give us only the two condyles, but these are nearly entire.
2d. A radius, perfect.
3d. An ulna, or fore-arm, perfect, except that it is broken in two.
4th. Three claws, and half a dozen other bones of the foot; but whether of a fore or hinder food, is not evident.

About a foot in length of the residue of the femur was found, it was split through the middle, and in that state was used as a support for one of the salt petre vats, this piece was afterwards loft, but it measures had been first taken as will be stated hereafter.

These bones only enable us to class the animal with the unguiculated quadrupeds; and of these the lion being nearest to him in size, we will compare him with that animal, of whose anatomy Monsieur Daubenton has furnished very accurate measures in his tables at the end of Buffon's Natural History of the lion. These measures were taken as he[1] informs us from "a large lion of Africa," in which quarter the largest[2] are said to be produced. I shall select from his measures only those where we have the corresponding bones, converting them into our own inch and its fractions, that the comparison may be more obvious: and to avoid the embarrassment of designating our animal always by circumlocution and description, I will venture to refer to him by the name of the Great-Claw or Megalonyx, to which he seems sufficiently entitled by the distinguished size of that member.

[1] Buffon, XVIII. 38. Paris edition in 31 vols. 12mo.
[2] De Manet, 117.

	Megalonyx (inches)	Lion (inches)
Length of the ulna, or fore-arm	20.1	13.7
Height of the olecranum	3.5	1.85
Breadth of the ulna, from point of coronoide apophysis to the extremity of the olecranum	9.55	
Breadth of the ulna at its middle	3.8	
Thickness at the same place	1.14	
Circumference at the same place	6.7	
Length of the radius	17.75	12.37
Breadth of the radius at its head	2.65	1.38
Circumberence at its middle	7.4	3.62
Breadth at its lower extremity	4.05	1.18
Diameter of the lower extremity of the femur at the base of the two condyles	4.2	2.65
Transverse diameter of the larger condyle at its base	3	
Circumference of both condyles at their base	11.65	
Diameter of the middle of the femur	4.25	1.15
Hollow of the femur at the same place	1.25	
Thickness of the bone surrounding the hollow	1.5	
Length of the longest claw	7.5	1.41
Length of the second phalanx of the same	3.2	1.41

The dimensions of the largest of the foot bones are as follow (inches),

Its greatest diameter, or breadth at the joint	2.45
Its smallest diameter, or thickness at the same place	2.28
Its circumference at the same place	7.1
Its circumference at the middle	5.3

	Longest toe	Middle-sized toe	Shortest toe
2d. Phalanx. Its length	3.2	2.95	
Greatest diameter at its head or upper joint	1.84	2.05	

	Longest toe	Middle-sized toe	Shortest toe
Smallest diameter at the same place	1.4	1.54	
Circumference at the same place	5.25	5.8	
3d. Phalanx. Its length	7.5[3]	5.9[4]	3.5
Greatest diameter at its head or upper joint	2.7	2	1.45
Smallest diameter at the same place	.95	.9	.55
Circumference at the same place	6.45	4.8	

Were we to estimate the size of our animal by a comparison with that of the lion on the principle of *ex pede Herculem,* by taking the longest claw of each as the module of their measure, it would give us a being out of the limits of nature. It is fortunate therefore that we have some of the larger bones of the limbs which may furnish a more certain estimate of his stature. Let us suppose then that his dimensions of height, length and thickness, and of the principal members composing these, were of the fame proportions with those of the lion. In the table of M. Daubenton an ulna of 13.78 inches belonged to a lion 42 ½ inches high over the shoulders: then an ulna of 20.1 inches bespeaks a megalonyx of 5 feet 1.75 inches height, and as animals who have the same proportions of height, length, and thickness have their bulk or weights proportioned to the cubes[5] of any one of their dimensions, the cube of 42.5 inches is to 262 lb. the height and weight of M. Daubenton's lion as the cube of 61.75 inches to 803 lb. the height and weight of the megaloynx; which would prove him a little more than three times the size of the lion. I suppose that we should be safe in considering, on the authority of M. Daubenton, his lion as a large one. But let it pass as one only of the ordinary size, and that the megalonyx whose bones happen to have been found was also of the ordinary size. It does[6] appear that there was dissected for the academy of sciences at Paris, a lion of 4 feet 9 3/8 inches height. This individual would weigh 644 lb. and would be in his species, what a man of eight feet height would be in ours. Such men have existed. A megalonyx equally monstrous would be 7 feet high, and would weigh 2000 lb. but the ordinary race, and not the monsters of it, are the object of our present enquiry.

    I have used the height alone of this animal to deduce his bulk, on the supposition that he might have been formed in the proportions of the lion. But these were not his proportions, he was much thicker than the lion in proportion to his height, in

---

3 It is actually 6 ¾ inches long, but about ¾ inch appear to have been broken off.
4 Actually 5.65 but about ¼ inch is broken off.
5 Buffon xxii. 121.
6 Buffon xviii. 15.

his limbs certainly, and probably therefore in his body. The diameter of his radius, at its upper end, is near twice as great as that of the lion, and, at its lower end, more than thrice as great which gives a mean proportion of 2 ½ for one. The femur of the lion was less than 1 ¼ inch diameter. That of the megalonyx is 4 ¼ inches, which is more than three for one. And as bodies of the same length and substance have their weights proportioned to the squares of their diameters, this excess of caliber compounded with the height, would greatly aggravate the bulk of this animal. But when our subject has already carried us beyond the limits of nature hitherto known, it is safest to stop at the most moderate conclusions, and not to follow appearances through all the conjectures they would furnish, but leave these to be corroborated or corrected by future discoveries. Let us only say then, what we may safely say, that he was *more* than three times as large as the lion: that he stood as pre-eminently at the head of the column of clawed animals as the mammoth stood at that of the elephant, rhinoceros, and hippopotamus: and that he may have been as formidable an antagonist to the mammoth as the lion to the elephant.

A difficult question now presents itself. What is become of the great-claw? Some light may be thrown on this by asking another question. Do the wild animals of the first magnitude in any instance fix their dwellings in a thickly inhabited country? Such, I mean, as the elephant, the rhinoceros, the lion, the tyger? As far as my reading and recollection serve me, I think they do not: but I hazard the opinion doubtingly, because it is not the result of full enquiry. Africa is chiefly inhabited along the margin of its seas and rivers. The interior desert is the domain of the elephant, the rhinoceros, the lion, the tyger. Such individuals as have their haunts nearest the inhabited frontier, enter it occasionally, and commit depredations when pressed by hunger: but the mass of their nation (if I may use the term) never approach the habitation of man, nor are within reach of it. When our ancestors arrived here, the Indian population, below the falls of the rivers, was about the twentieth part of what it now is. In this state of things, an animal resembling the lion seems to have been known even in the lower country. Most of the accounts given by the earlier adventurers to this part of America make a lion one of the animals of our forests. Sir John Hawkins[7] mentions this in 1564. Thomas Harriot, a man of learning, and of distinguished candor, who resided in Virginia in 1587[8] does the same, so also does Bullock in his account of Virginia,[9] written about 1627, he says he drew his information from Pierce, Willoughby, Claiborne, and others who had been here, and from his own father who had lived here twelve years. It does not appear whether the fact is stated on their own view, or on information from the Indians, probably the latter. The progress of the new population would soon drive off the larger animals, and the largest first. In the present interior of our continent there

[7] Hakluyt, 541. Edition of 1589.
[8] Ibid. 757, and Smith's History of Virginia, 10.
[9] Bullock, page 5.

is surely space and range enough for elephants and lions, if in that climate they could subsist; and for mammoths and megaloynxes who may subsist there. Our entire ignorance of the immense country to the West and North-West, and of its contents, does not authorise us to say what it does not contain.

Moreover it is a fact well known, and always susceptible of verification, that on a rock on the bank of the Kanhawa, near its confluence with the Ohio, there are carvings of many animals of that country, and among these one which has always been considered as a perfect figure of a lion. And these are so rudely done as to leave no room to suspect a foreign hand. This could not have been of the smaller and maneless lion of Mexico and Peru, known also in Africa both in[10] ancient and[11] modern times, though denied by [12]M. de Buffon: because like the greater African lion, he is a tropical animal; and his want of a mane would not satisfy the figure. This figure then must have been taken from some other prototype, and that prototype must have resembled the lion sufficiently to satisfy the figure, and was probably the animal the description of which by the Indians made Hawkins, Harriot, and others conclude there were lions here. May we not presume that prototype to have been the great-claw?

Many traditions are in possession of our upper inhabitants, which themselves have heretofore considered as fables, but which have regained credit since the discovery of these bones. There has always been a story current that the first company of adventurers who went to seek an establishment in the county of Greenbriar, the night of their arrival were alarmed at their camp by the terrible roarings of some animals unknown to them: that he went round and round their camp, that at times they saw his eyes like two balls of fire, that their horses were so agonised with fear that they couched down on the earth, and their dogs crept in among them, not daring to bark. Their fires, it was thought, protected them, and the next morning they abandoned the country. This was little more than 30 years ago. – In the year 1765, George Wilson and John Davies, having gone to hunt on Cheat river, a branch of the Monongahela, heard one night, at a distance from their camp, a tremendous roaring, which became louder and louder as it approached till they thought it resembled thunder, and even made the earth tremble under them. The animal prowled round their camp a considerable time, during which their dogs, though on all other occasions fierce, crept to their feet, could not be excited from their camp, nor even encouraged to bark. About day light they heard the same sound repeated from the knob of a mountain about a mile off, and within a minute it was answered by a similar voice from a neighbouring knob. Colonel John Stewart had this account from Wilson in the year 1769, who was afterwards Lieutenant Colonel of a Pennsylvania regiment in the revolution-war; and some years after from Davies, who is now living in Kentucky.

[10] Aristot. Animal, 9. 4. Pliny, 8. 16.
[11] Kolbe.
[12] Buffon, xviii. 18.

These circumstances multiply the points of resemblance between this animal and the lion. M. de la Harpe of the French Academy, in his abridgment of the General History of Voyages, speaking of the Moors, says[13] "it is remarkable that when, during their huntings, they meet with lions, their horses, though famous for swiftness, are seized with such a terror that they become motionless, and their dogs equally frightened, creep to the feet of their master, or of his horse." Mr. Sparrman in his voyage to the Cape of Good Hope, chap. 11, says, "we could plainly discover by our animals when the lions, whether they roared or not, were observing us at a small distance. For in that case the hounds did not venture to bark, but crept quite close to the Hottentots; and our oxen and horses sighed deeply, frequently hanging back, and pulling slowly with all their might at the strong straps with which they were tied to the wagon. They also laid themselves down on the ground, and stood up alternately, as if they did not know what to do with themselves, and even as if they were in the agonies of death." He adds that "when the lion roars, he puts his mouth to the ground, so that the sound is equally diffused to every quarter." M. de Buffon (xviii. 31) describes the roaring of the lion as, by its echoes resembling thunder: and Sparrman c. 12, mentions that the eyes of the lion can be seen a considerable distance in the dark, and that the Hottentots watch for his eyes for their government. The phosphoric appearance of the eye in the dark seems common to all animals of the cat kind.

The terror excited by these animals is not confined to brutes alone. A person of the name of Draper had gone in the year 1770, to hunt on the Kanhawa. He had turned his horse loose with a bell on, and had not yet got out of hearing when his attention was recalled by the rapid ringing of the bell. Suspecting that Indians might be attempting to take off his horse, he immediately returned to him, but before he arrived he was half eaten up. His dog scenting the trace of a wild beast, he followed him on it, and soon came in sight of an animal of such enormous size, that though one of our most daring hunters and best marksmen, he withdrew instantly, and as silently as possible, checking and bringing off his dog. He could recollect no more of the animal than this terrific bulk, and that his general outlines were those of the cat kind. He was familiar with our animal miscalled the panther, with our wolves and wild beasts generally, and would not have mistaken nor shrunk from them.

In fine, the bones exist: therefore the animal has existed. The movements of nature are in a never ending circle. The animal species which has once been put into a train of motion, is still probably moving in that train. For if one link in nature's chain might be lost, another and another might be lost, till this whole system of things should evanish by piece-meal; a conclusion not warranted by the local disappearance of one or two species of animals, and opposed by the thousands and thousands of instances of the renovating power constantly exercised by nature for the reproduction of all her subjects, animals, vegetable, and mineral. If this animal then has once existed, it is probable on this general view of the

---

[13] Gentleman's, and London Magazines, for 1783.

movements of nature that he still exists, and rendered still more probable by the relations of honest men applicable to him and to him alone. It would indeed be but conformable to the ordinary economy of nature to conjecture that she had opposed sufficient barriers to the too great multiplication of so powerful a destroyer. If lions and tygers multiplied as rabbits do, or eagles as pigeons, all other animal nature would have been long ago destroyed, and themselves would have ultimately extinguished after eating out their pasture. It is probable then that the great-claw has at times been the rarest of animals. Hence so little is known, and so little remains of him. His existence however being at length discovered, enquiry will be excited, and further information of him will probably be obtained.

The Cosmogony of M. de Buffon supposes that the earth and all the other planets primary and secondary, have been masses of melted matter struck off from the sun by the incidence of a comet on it: that these have been cooling by degree, first at the poles, and afterwards more and more towards their Equators: consequently that on our earth there has been a time when the temperature of the poles suited the constitution of the elephant, the rhinoceros, and hippopotamus: and in proportion as the remoter zones became successively too cold, these animals have retired more and more towards the Equatorial regions, till now that they are reduced to the torrid zone as the ultimate stage of their existence. To support this theory, he[14] assumes the tusks of the mammoth to have been those of an elephant, some of his teeth to have belonged to the hippopotamus, and this largest grinders to an animal much greater than either, and to have been deposited on the Missouri, the Ohio, the Holston, when those latitudes were not yet too cold for the constitutions of these animals. Should the bones of our animals, which may hereafter be found, differ only in size from those of the lion, they may on this hypothesis be claimed for the lion, now also reduced to the torrid zone, and its vicinities, and may be considered as an additional proof of this system; and that there has been a time when our latitudes suited the lion as well as the other animals of that temperament. This is not the place to discuss theories of the earth, nor to question the gratuitous allotment to different animals of teeth not differing in any circumstance. But let us for a moment grant this with his former postulata, and ask how they will consist with another theory of this "qu'il y a dans la combinaison des elemens et des autres causes physiques, quelque chose de contraire a l'aggrandisement de la nature vivante *dans ce nouveau monde*; qu'il y a des obstacles au developpement et peutetre a la formation des grands germes."[15] He says that the mammoth was an elephant, yet[16] two or three times as large as the elephants of Asia and Africa: that some of his teeth were those of a hippopotamus, yet of a hippopotamus[17] four times as large as those of Africa: that the mammoth himself, for he still considers him as a distinct

[14] Buffon, Epoq. 2. 233, 234.
[15] Buffon, xviii. 145.
[16] 2. Epoq. 223.
[17] 1. Epoq. 246. 2. Epoq. 232.

animal,[18] "was of a size superior to that of the largest elephants. That he was the primary and greatest of all terrestrial animals." If the bones of the megalonyx be ascribed to the lion, they must certainly have been of a lion of more than three times the volume of the African. I delivered to M. de Buffon the skeleton of our palmated elk, called orignal or moose, 7 feet high over the shoulders, he is often considerably higher. I cannot find that the European elk is more than two thirds of that height: consequently not one third of the bulk of the American. He[19] acknowledges the palmated deer (daim) of America to be larger and stronger than that of the Old World. He[20] considers the round horned deer of these States and of Louisiana as the roe, and admits they are of three times his size. Are we then from all this to draw a conclusion, the reverse of that of M. de Buffon. That nature, has formed the larger animals of America, like its lakes, its rivers, and mountains, on a greater and prouder scale than in the other hemisphere? Not at all, we are to conclude that she has formed some things large and some things small, on both sides of the earth for reasons which she has not enabled us to penetrate; and that we ought not to shut our eyes upon one half of her facts, and build systems on the other half.

To return to our great-claw; I deposit his bones with the Philosophical Society, as well in evidence of their existence and of their dimensions, as for their safe-keeping; and I shall think it my duty to do the same by such others as I may be fortunate enough to obtain the recovery of hereafter.

Th: Jefferson.

Monticello, Feb. 10th, 1797.

P.S. *March 10th,* 1797. After the preceding communication was ready to be delivered in to the Society, in a periodical publication[21] from London I met with an account and drawing of the skeleton of an animal dug up near the river La Plata in Paraguay, and now mounted in the cabinet of Natural History of Madrid. The figure is not so done as to be relied on, and the account in only an abstract from that of Cuvier and Roume. This skeleton is also of the clawed-kind, and having only four teeth on each side above and below, all grinders, is in this account classed in the family of unguiculated quadrupeds destitute of cutting teeth, and receives the new denomination of megatherium, having nothing of our animal but the leg and foot bones, we have few points for a comparison between them. They resemble in their stature, that being 12 feet 9 inches long, and 6 feet 4 ½ inches high, and ours by computation 5 feet 1.75 inches high: they are alike in the colossal thickness of the thigh and leg bones also. They resemble too in having claws but those of the figure appear very small, and the verbal description does not

---

[18] 2. Epoq. 234, 235.
[19] Buffon, xxix. 245.
[20] Ibid. xii. 91. 92. xxix. 245. Vide Suppl. 201.
[21] Monthly Magazine, Sep. 1796.

satisfy us whether the claw-bone, or only its horny cover be large. They agree too in the circumstance of the two bones of the fore-arm being distinct and moveable on each other; which however is believed to be so usual as to form no mark of distinction. They differ in the following circumstances, if our relations are to be trusted. The megatherium is not of the cat form, as are the lion, tyger, and panther, but is said to have striking relations in all parts of its body with the bradypus, dasypus, pangolin, &c. According to analogy then, it probably was not carnivorous, had not the phosphoric eye, nor leonine roar. But to solve satisfactorily the question of identity, the discovery of fore-teeth, or of a jaw bone shewing it had, or had not, such teeth, must be waited for, and hoped with patience. It may be better, in the mean time, to keep up the difference of name.

ALEXANDER WILSON (1766–1813)

Ornithologist.
Born in Paisley, Renfrewshire, Scotland. Died in Philadelphia, Pennsylvania.

A Scottish weaver, itinerant peddler, and poet until the age of 20, Alexander Wilson's main interests as a young man centered on poetry and the unfair treatment of weavers. His poetry was popular with the people but not with the authorities and landed him in jail for libel. Following his release, he immigrated to America in 1794 and became a schoolmaster, first in New Jersey and then Pennsylvania. It was at the latter location that Wilson met a neighbor, the famous eighteenth-century naturalist William Bartram, who encouraged him to collect, draw, and paint birds. Wilson took the encouragement to heart and began serious ornithological work around 1802, traveling widely in search of birds to paint, as well as collecting subscriptions for a book illustrating all the birds of North America. The result was the nine volumes of *American Ornithology*, which illustrated 268 species of birds, 26 of which had never been previously described. Wilson was considered the greatest American ornithologist and wildlife illustrator prior to Audubon, and much has been made of their meeting in 1810 in Kentucky as the inspiration for Audubon's own work. Wilson's collection was based on all his own observations and the plates produced by engravings from his own drawings. Wilson even colored a sample proof for use as model by the hand-colorists. Seven of the volumes (1808–13) were published under his direction and in his lifetime, but volumes eight and nine were brought out under the direction of his protégé, George Ord, after his death. Wilson died in the pursuit of his continued study, drowning in a river while pursuing a bird. Wilson's story of a penniless, self-educated emigrant who achieved success through diligence and natural talent appealed to the already growing American legend of the philosopher-backwoodsman, much as Audubon's story would two decades later. Charles Darwin wrote in his *Second Notebook on Transmutation of Species* that he found Wilson's *American Ornithology* "a mine of valuable facts, regarding habits, range and all kinds of information [and] instinct" (Darwin, "Second Notebook", p. 69).

Alexander Wilson, "Pileated Woodpecker," *American Ornithology; or the Natural History of the Birds of the United States* (1801)

THIS American species is the second in size among his tribe, and may be styled the great northern chief of the woodpeckers, though, in fact, his range extends over the whole of the United States from the interior of Canada to the Gulf of Mexico. He is very numerous in the Gennesee country, and in all the tracts of high timbered forests, particularly in the neighbourhood of our large rivers, where he is noted for making a loud and almost incessant cackling before wet weather; flying at such times in a restless uneasy manner from tree to tree, making the woods echo to his outcry. In Pennsylvania and the northern states he is called the black woodcock; in the southern states, the logcock. Almost every old trunk in the forest where he resides bears the marks of his chisel. Wherever he perceives a tree beginning to decay, he examines it round and round with great skill and dexterity, strips off the bark in sheets of five or six feet in length, to get at the hidden cause of the disease, and labours with a gaiety and activity really surprising. I have seen him separate the greatest part of the bark from a large dead pine tree, for twenty or thirty feet, in less than a quarter of an hour. Whether engaged in flying from tree to tree, in digging, climbing, or barking, he seems perpetually in a hurry. He is extremely hard to kill, clinging close to the tree even after he has received his mortal wound; nor yielding up his hold but with his expiring breath. If slightly wounded in the wing, and dropt while flying, he instantly makes for the nearest tree, and strikes with great bitterness at the hand stretched out to seize him; and can rarely be reconciled to confinement. He is sometimes observed among the hills of Indian corn, and it is said by some that he frequently feeds on it. Complaints of this kind are however, not general; many farmers doubting the fact, and conceiving that at these times he is in search of insects which lie concealed in the husk. I will not be positive that they never occasionally taste maize; yet I have opened and examined great numbers of these birds, killed in various parts of the United States, from Lake Ontario to the Alatamaha river, but never found a grain of Indian corn in their stomachs.

The pileated woodpecker is not migratory, but braves the extremes of both the arctic and torrid regions. Neither is he gregarious, for it is rare to see more than one or two, or at the most three, in company. Formerly they were numerous in the neighbourhood of Philadelphia; but gradually, as the old timber fell, and the country became better cleared, they retreated to the forest. At present few of those birds are to be found within ten or fifteen miles of the city.

Their nest is built, or rather the eggs are deposited, in the hole of a tree, dug out by themselves, no other materials being used but the soft chips of rotten wood. The female lays six large eggs of a snowy whiteness; and, it is said, they generally raise two broods in the same season.

This species is eighteen inches long, and twenty-eight in extent; the general colour is a dusky brownish black; the head is ornamented with a conical cap of bright scarlet; two scarlet mustaches proceed from the lower mandible; the chin is white; the nostrils are covered with brownish white hair-like feathers, and this stripe of white passes from thence down the side of the neck to the sides, spreading under the wings; the upper half of the wings are white, but concealed by the black coverts; the lower extremities of the wings are black, so that the white on the wing is not seen but when the bird is flying, at which time it is very prominent; the tail is tapering, the feathers being very convex above, and strong; the legs are of a leaden gray colour, very short, scarcely half an inch; the toes very long; the claws strong and semicircular, and of a pale blue; the bill is fluted, sharply ridged very broad at the base, bluish black above, below and at the point bluish white; the eye is of a bright golden colour, the pupil black; the tongue, like those of its tribe, is worm-shaped, except near the tip, where for one-eighth of an inch it is horny, pointed, and beset with barbs.

The female has the forehead, and nearly to the crown, of a light brown colour, and the mustaches are dusky, instead of red. In both a fine line of white separates the red crest from the dusky line that passes over the eye.

NATHANIEL BOWDITCH (1773–1838)

Mathematician and astronomer.
Born in Salem, Massachusetts. Died in Boston, Massachusetts.

NATHANIEL Bowditch left school at ten to work in his father's cooperage. At twelve he became an apprentice clerk in a ship's chandlery. While there, Bowditch was encouraged to train himself thoroughly in mathematics and navigation in order to pursue a career on the sea. At 14 he began to study algebra and at 16 calculus. By 17 he had taught himself Latin and French in order to read mathematical works. By the time he went to sea in 1795 as a supercargo, he had acquired knowledge of mathematics to rival any of his contemporaries in the United States. Between the years 1795–1803, Bowditch worked as supercargo and eventually became captain of his own ship. Because of Bowditch's analytical skills and his seafaring experience, the publisher Edmund M. Blunt urged him to edit an American edition of John Hamilton Moore's *The Practical Navigator*. Bowditch found the book rife with mistakes and decided he would attempt a book literally free from error and written so that any sailor could use it. He not only redid all the mathematics in Moore's edition but expanded and greatly improved the accuracy and basic information. The 1802 edition included several solutions to spherical triangle problems as well as recalculated formulae and tables for navigation. From 1802 on, all editions carried his name. In 1866 the U.S. hydrographic office bought the copyright and has kept it in continuous publication with revisions to keep it current. In 1799 Bowditch was elected a member of the American Academy of Arts and Sciences and in 1802 received the honorary degree of Master of Arts from Harvard University. His career on the high seas ended in 1803 with a spectacularly navigated landfall into Salem during a raging snow storm, a classical example of "dead reckoning." That same year, at 30, he took up the rather prosaic field of maritime insurance, becoming the president of the Essex Fire and Marine Insurance Company (1804–23) and then actuary of the Massachusetts Hospital Life Insurance Company (1823–38). He was offered chairs of mathematics and physics at Harvard, West Point, and the University of Virginia, all of which he turned down. He did, however, publish articles on meteors, the orbits of comets, and Lissajous figures in studies of pendulum motions in American, British and continental journals. His passion and avocation for the rest of his life was translating Laplace's monumental *Mecanique Celeste*, adding many explanatory notes to the original. He seems to have been a man more of analytical than creative gifts but he so perfected the mathematics used in the methods of navigation that he greatly reduced the costly errors of navigation to both purse and person. This excerpt from *The Practical Navigator* should illustrate the clarity and enduring aspect of his prose.

Nathaniel Bowditch, "Currents,"
The New American Practical Navigator (1802)

A current is a progressive motion of the water, causing all floating bodies to move that way towards which the stream is directed. The *set of a current* is that point of the compass towards which the waters run, and its *drift* is the rate it runs per hour. The most usual way of discovering the set and drift of an unknown current, is the following, supposing the current at the surface to be much more powerful than at a great distance below the surface:—

Take a boat a short distance from the ship, and, by a rope fastened to the boat's stern, lower down a heavy iron pot or loaded kettle to the depth of 80 or 100 fathoms; then heave the log, and the number of knots run out in half a minute will be the miles the current sets per hour, and the bearing of the log will show the set of it.

There is a very remarkable current, called the *Gulf Stream*, which sets in a north-east direction along the coast of America, from Cape Florida towards the Isle of Sables, at unequal distances from the land, being about 75 miles from the short of the southern States, but more distant from the shore of the northern States. The width of the stream is about 40 or 50 miles, widening towards the north. The velocity is various, from one to three knots per hour, or more, being greatest in the channel between Florida and the Bahamas, and gradually decreasing in passing to the northward, but is greatly influenced by the winds, both in drift and set.

We are chiefly indebted to Doctor Franklin, Commodore Truxton, and Mr. Jonathan Williams, for the knowledge we possess of the direction and velocity of this stream. Its general course, as given by them, is marked on the chart affixed to this work. They all concur in recommending the use of the thermometer, as the best means of discovering when in, or near, the stream; for it appears, by their observations, that the water is warmer than the air when in the stream; and that at leaving it, and approaching towards the land, the water will be found six or eight degrees colder than in the stream, and six or eight degrees colder still when on soundings. Vessels coming from Europe to America, by the northern passage, should keep a little to the northward of the stream, where they may probably be assisted by a counter current. When bound from any southern port in the United States of America to Europe, a ship may generally shorten her passage by keeping in the Gulf Stream. By steering N.W. you will generally cross it in the shortest time, as its direction is nearly N.E.

In other parts of the Atlantic Ocean, the currents are variable, but are generally south-easterly along the coast of Spain, Portugal, and Africa, from the Bay of Biscay towards Madeira and the Cape de Verds. Between the tropics, there is generally a current setting to the westward.

There is also a remarkable current which sets through the Mozambique Channel, between the Island of Madagascar and the main continent of Africa, in a south-westerly direction. In proceeding towards Cape Lagullas, the current takes a more westerly course, and then trends round the cape towards St. Helena. Ships bound to the westward from India, may generally shorten their passage by taking advantage of this current. On the contrary, when bound to the eastward, round the Cape of Good Hope, they ought to keep far to the southward of it. However, there appears to be a great difference in the velocity of this current at different times; for some ships have been off this cape several days endeavoring to get to the westward, and have found no current; others have experienced it setting constantly to the westward, during their passage from the cape towards St. Helena, Ascension, and the West India Islands. Instances have however occurred, where an easterly current was experienced off the Cape of Good Hope.

All cases of sailing in a current are calculated upon the principle that the ship is affected by it in the same manner as if she had sailed in still water, with an additional course and distance exactly equal to its set and drift. On this principle the projection and calculation of any problem of this kind may be easily made.

Example

If a ship sail 98 miles N.E. by N., in a current which sets S. by W. 27 miles, in the same time, required her true course and distance.

By Projection

Describe the compass NESW; through the centre A draw the N.E. by N. line AC equal to 98 miles; through C draw the line BC parallel to the S. by W. line, make BC

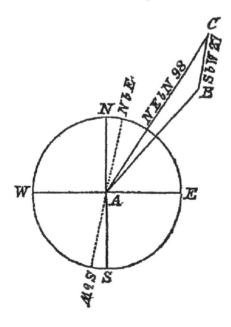

equal to 27 miles, and join AB. Then AB will be the course and distance made good; and by measuring we find the course to be N.E. ¼ N., the distance 74 miles.

By Calculation

The shortest method of calculating this problem is by means of Table I., as in the adjoined Traverse Table; putting in it the course sailed by the ship and the set of the current; then finding the difference of latitude and departure by the table. The course and distance made good is then found as in Case VI of Plane Sailing. In the present example, the course is N.E. ¼ N., and the distance 74 miles nearly.

TRAVERSE TABLE.

Courses.	Dist.	N.	S.	E.	W.
N. E. by N.	98	81.5		54.4	
S. by W.	27		26.5		5.3
		81.5	26.5	54.4	5.3
		26.5		5.3	
Diff. Lat...55.0				Dep. 49.1	

GEORGE ORD (1781–1866)

Ornithologist and naturalist.
Born and died in Philadelphia, Pennsylvania.

A sea captain and rope maker by profession, Ord befriended Alexander Wilson in the early 1800's, shared with him a great interest in the natural world, and accompanied him on many of his collecting trips for *American Ornithology*. He was one of the first members of the Academy of Natural Sciences of Philadelphia (serving as its president 1851–58) as well as the American Philosophical Society (where he also held office). He contributed numerous descriptive articles to the premier scientific periodicals of the day, both American and British, and described several of the Lewis and Clark specimens, including the grizzly bear and the pronghorn antelope. After Wilson's untimely death, Ord helped finish the eighth and ninth volumes of *American Ornithology*. Ord may, however, be most remembered for his vitriolic opposition to John James Audubon. He was certainly in part responsible for the initial American rejection of Audubon's work. Ord also took excessive and near libelous offense to Nuttall and his work. Both responses seem to be connected to the strength of his attachment to Wilson and his investment in Wilson's work. At the time of Audubon's initial introduction to the Academy of Natural Sciences of Philadelphia, Ord and others were preparing to bring out another edition of Wilson's work. In his later years, Ord became more and more the recluse. Fittingly, he was buried in Philadelphia next to his mentor and friend Wilson. George Ord did not make any groundbreaking contributions to American science, but I wanted to include his description of the rocky-mountain sheep because it is a quintessential example of what was prevalent throughout the first part of the nineteenth century. Ord was also the sort of gentleman naturalist who contributed his own time, money and resources to understanding the flora and fauna of this amazing, unexplored land. He was a man of his time, but his time, the time of the nonprofessional, self-taught science devotee, was drawing to a close.

George Ord, "Account of a North American Quadruped, supposed to belong to the Genus *Ovis*, Rocky-Mountain Sheep, *Ovis Montana*," *Journal of the Academy of Natural Sciences of Philadelphia* (1817)

IN the Journal of Lewis and Clark, there is an account of a quadruped which appears to have not excited that attention which it merits. The following extracts are made from the above mentioned work: "Saw the skin of a mountain sheep, which the

Indians say lives among the rocks in the mountains: the skin was covered with white hair, the wool long, thick and coarse, with long coarse hair on the top of the neck and the back, resembling somewhat the bristles of a goat." Vol. II. p. 49.

"The sheep is found in many places, but mostly in the timbered parts of the rocky mountains. They live in greater numbers on the chain of mountains forming the commencement of the woody country on the coast, and passing the Columbia between the falls and rapids." Vol. II. p. 169.

The latter passage was written while our travelers wintered at the mouth of the Columbia river. But on their return, at Brant Island, an Indian "offered two sheep skins for sale: one, which was the skin of a full-grown sheep, was as large as that of a common deer; the second was smaller, and the head part, with the horns remaining, was made into a cap, and highly prized as an ornament by the owner. The Clahelellahs informed us that the sheep was very abundant on the heights, and among the cliffs, of the adjacent mountains; and that these two had been lately killed out of a herd of thirty-six, at no great distance from the village." Vol. II. p. 233.

"The Indians assert, that there are great numbers of the white buffalo or mountain sheep, on the snowy heights of the mountains, west of Clark's river. They generally inhabit the rocky and most inaccessible parts of the mountain, but as they are not fleet, are easily killed by the hunters." Vol. II. p. 331.

In the above passages we are made acquainted with the important fact, that, besides the Argali or Big-horned sheep, we have another species in North America of the genus Ovis. The smaller of the two skins, which the Indian offered to sale at Brant Island, was purchased by captain Lewis, and was presented by him to the museum of Philadelphia. It is undoubtedly the skin of a young animal: it measures three feet from the insertion of the tail to the neck, its breadth is twenty-six inches; the tail is short, but it was probably not skinned to the end; along the back there runs a ridge of coarse hair, about three inches in length, and bristled up in the manner of that of the common goat, this ridge is continued up the neck, forming a kind of mane, and is thicker, coarser, and longer there than that of the back; the whole of the skin is closely covered with short wool, of an extreme fineness, surpassing in this quality that of any breed with which I am acquainted, not excepting the wool of the Merino lamb – a coat of hair conceals this wool, but on dividing the former with the hands, the latter lies so thick that the hairs are scarcely visible; the ears are narrow, and taper to a point, they are nearly four inches long; the whole is white; the horns appear to have stood on the top of the head, somewhat in the manner of those of a goat, or of those on the figure of Shaw's Pigmy Antelope, Gen. Zool. vol. ii, plate 188, and vignette on the title-page. But one[1] horn is now attached to the skin, and that measures three inches and three quarters in length, on the fore part; it is slightly recurved, cylindrical and acuminated, its base is somewhat tumid, and, with its lower half, is scabrous, its upper part smooth, obsoletely striated, and of a black colour.

---

[1] The other horn is in Peale's Museum.

A cut of this horn, of the size of nature, accompanies this account, by which figure it will be evident to the naturalist, that the above described sheep is a distinct species. It is true that the animal was young, and we have no positive evidence that when full-grown or old the horns do not increase in size, so as to resemble those of some well-known species or varieties of the genus. One of Lewis and Clark's men informed them that he had seen the animal in the Black Hills, and that the horns were *lunated* like those of the domestic sheep. The Indians asserted that the horns were *erect* and *pointed*. The latter account is the more probable, as it has been remarked by travellers, that, in describing those natural productions with which they are conversant, our Indians seldom deviate from the truth.

We would incite the attention of our citizens to this important discovery, for although the Spanish missionaries, in 1697, made mention of this sheep, and it is again noticed in Venegas' History of California,[2] yet these accounts were discredited. It is to captain Lewis to whom belongs the honour of having been the first to assure his countrymen, by the exhibition of a genuine specimen, that the animal does exist. How subservient to the wants and pleasures of mankind it may be rendered by domestication, we cannot at present declare; but there is room for conjecture, that the introduction of this new species of a race of quadrupeds immemorially ranked among the most valuable of the gifts of the Creator, will confer a lasting benefit upon the agricultural and manufacturing interests of the community.

Since writing the foregoing, I have seen the three first volumes of the *Nouveau Dictionnaire d'Histoire Naturelle*, which work is now publishing in Paris; and in the article *Antelope* I find a description of an American quadruped, which is in the collection of the Linnean society of London. This description appears to have been extracted from a *memoire*, read before the Philomatique Society of Paris, by M. de Blainville, wherein the author proposes a new arrangement of the ruminants with hollow and persistent horns, and a subdivision of the *Genus Antilope*; and classes the above animal under the name of *Rupicapra Americana* (Bulletin de la Societe' Philomatique, 1816, p. 80). As I have not the satisfaction of seeing the Bulletin, I must be content with the information conveyed in the article in the *Nouveau Dictionnaire*. The specimen is said to be of the bigness of a middling sized goat; the body is entirely covered with long *pendent* hair, silky and totally white, but not curled; the head is elongated, without a muzzle or naked part, the ears of a middling size; the forehead not protuberant; the horns are short, tolerably thick, black, slightly annulated, they are round, almost straight, bent backwards, and terminated in a blunt point (*pointe mousse*); the legs are short, stout, and supported on short and thick hoofs; the tail is hardly perceptible, perhaps on account of the length of the hair. M. de Blainville inclined to the opinion that this animal is the same as the Pudu of Molina, Shaw's Gen. Zool. vol. ii. p. 392.

It is probable that the specimen belonging to the Linnean Society is of the same species as that brought by captain Lewis; and it is further probable that M. de Blainville

---

[2] Vol. 1. p. 36. English translation, London, 1759.

was not permitted to examine his subject as closely as was requisite, otherwise the important circumstance of the thick coat of wool, beneath the outer covering of straight hair, would not have escaped his attention. As to the horns being obtuse, this may have arisen from an accident, or some other cause.

It is much to be wished that some traveler would bring a living specimen of this singular quadruped, or at least a dead specimen in such a state as should enable the naturalist to determine, with decision, its characters. From the information derived from Captain Lewis, and from the descriptions above, we cannot, with propriety, arrange this animal with the Antelopes; and if it should not prove to be a true *Ovis*, it will, probably, constitute a new genus, and take its station, in the systems, between the sheep and the goat.

THOMAS SAY (1787–1834)

Entomologist, zoologist, paleontologist, conchologist and naturalist.
Born in Philadelphia, Pennsylvania. Died in New Harmony, Indiana.

THOMAS Say was born into a prominent Quaker family that included his great grandfather John Bartram (whom Linnaeus named the greatest contemporary "natural botanist") and his great uncle William Bartram (naturalist and author of *Travels* [1791]). He was originally trained as an apothecary but, considering his family, could hardly help but take a keen interest in nature. In 1812 the self-taught Say became one of the founding members of the Academy of Natural Sciences of Philadelphia. Say was so invested in his study of nature that he actually lived in the Academy building and took care of its growing museum. He was the chief zoologist on the 1819 Long expedition to the tributaries of the Missouri River (where he first described coyote, swift fox, western kingbird, band-tailed pigeon among many others) as well as on the 1823 expedition to the headwaters of the Mississippi. Between 1822 and 1828 he was Professor of Natural History at the University of Pennsylvania. Throughout this time he also worked on identifying the insect life he encountered. In 1824 the first of three volumes (1824–28) of the monumental *American Entomology, or Descriptions of the Insects of North American* was published. This would prove to be an important and classic work which would garner Say the title of the Father of American Descriptive Entomology. In 1825 Say was part of the so-called Boatload of Knowledge, a group of likeminded scientists, educators, and utopians who settled in New Harmony, Indiana, in order to start a utopian colony. This experiment in social living only lasted two years, but even after its acknowledged leaders had moved on, Say and his wife Lucy (an artist, illustrator and fellow New Harmony traveler) would stay. While there, Say began his other classic work, *American Conchology*. Volumes 1–6 were completed by Say in New Harmony and Volume 7 was overseen by his wife Lucy and published in 1836 two years after he died of typhoid fever. Lucy Say would contribute all but two of the amazing 68 plates featured in *American Conchology*. In 1841, Lucy Say turned her husband's materials over to the Academy of Natural Sciences and in gratitude the members voted her an Associate Member, the first American woman to be so honored. Thomas Say's complete entomological papers were collected and edited by J. L. LeConte in 1859 and his writings on conchology edited by W. G. Binney in 1858.

Thomas Say, "A Monograph of North American insects, of the genus Cincindela" (excerpt), *Transactions of the American Philosophical Society* (1818)

IT will perhaps be thought necessary, previous to entering into a technical detail of the characters of the genus *Cicindela*, and of the indigenous individuals which are comprehended by it, that some account of the manners of this sprightly tribe should be given, and of such circumstances, relating to them, as may serve to present them to the recollections of the general observer. I shall accordingly proceed to state, that these insects usually frequent arid, denudated soils; are very agile, run with greater celerity than the majority of the vast order to which they belong; and rise upon the wing, almost with the facility of the common fly. They are always to be seen, during the warm season, in roads or pathways, open to the sun, where the earth is beaten firm and level. At the approach of the traveler, they fly up suddenly to the height of a few feet, pursuing then a horizontal course, and alighting again at a short distance in advance, as suddenly as they arose. The same individual may be roused again and again, but when he perceives himself the object of a particular pursuit, he evades the danger by a distant and circuitous flight, usually directed towards his original station. It is worthy of observation, as a peculiarity common to the species, that when they alight, after having been driven from their previous position, they usually perform an evolution in the air near the earth, so as to bring the head in the direction of the advancing danger, in order to be the more certainly warned of its too near approach.

They lead a predatory life, and as it would appear, are well adapted to it, by their swiftness, and powerful weapons of attack. The beaten path, or open sandy plain, is preferred, that the operations of the insects may not be impeded by the stems and leaves of vegetables, through which, owing to their elongated feet, they pass with evident difficulty and embarrassment. They prey voraciously upon the smaller and weaker insects, upon larvae and worms, preferring those whose bodies are furnished with a membranaceous cuticle, more readily permeable to their *instrumenta cibaria*.

The same rapacity is observable in the larva, or imperfect stage of existence, of these insects, that we have occasion to remark in the parent; but not having been endowed by nature with the same light and active frame of body, they are under the necessity of resorting to stratagem and ambuscade for the acquisition of the prey, which is denied to their sluggish gait. The remark is, I believe, generally correct, though liable to many signal exceptions, that carnivorous animals display more cunning, industry, and intelligence, than those whose food is herbs, for the acquisition of which, fewer of the mental attributes are requisite; we see throughout the animated creation, that the development of these qualities, as well as of the corporeal functions, are in exact correspondence with their necessities; and that where a portion of the one is withheld, an additional proportion of the other is imparted. This larva has a very large head, elongated abdomen, and six short feet placed near the head; when walking, the body rests upon the earth, and is dragged forward slowly by the feet. Notwithstanding these disadvantages they contrive means to administer plentifully to an appetite, sharpened by a rapid increase of size. A cylindrical hole is dug in the ground to a considerable depth, by means of the feet and mandibles, and the earth transported from it, on the concave surface of the head; this cell is enlarged and deepened, as the inhabitant increases in size, so that its diameter is always nearly equal to that of the head. At the surface of the earth they lay in wait for their prey, nicely closing the orifice of the hole by the depressed head, that the plain may appear uninterrupted; when an incautious or unsuspecting insect approaches sufficiently near, it is seized by a sudden effort of the larva, and hurried to the bottom of the dwelling, to be devoured at leisure. These holes we sometimes remark, dug in a footpath; they draw the eye by the motion of the inhabitant retreating from the surface, alarmed at the approach of danger.

I shall now proceed to offer some remarks on the affinities of this genus, and endeavor to point out the differential traits, by which it may be distinguished from its congeners. *Cicindela*, according to Linnaeus, included not only all the insects, which would at this day be referred to it, but many others, which, however closely allied by habit, are widely distinct in the formation of their oral organs. These were separated by the celebrated systematists, Fabricius and Latreille, into several new genera, to which well defined essential characters have been affixed. These separations have been made upon the best possible grounds; the convenience of the student, and the approximation to natural method. So circumscribed, *Cicindela* presents a natural group, in which each individual so perfectly corresponds with the others, as well in its internal organization and parts of the mouth, as in habit, or general form of the body, that the entomologist finds no difficulty in distinguishing it from insects of neighboring genera, and referring it to its relative situation.

The genera to which allusion is here made, as having affinity with the one under consideration, are principally *Colliuris*, *Therates*, *Megacephala*, *Manticora*, *Elaphrus*, and *Notiophilus*. In constructing the essential character, I have endeavored to ascertain such traits as will at once, invariably, distinguish *Cicindela* from all other known genera of the pentamerous Coleoptera, and prevent the occurrence of error in the reference of species to it.

Thomas Say, "On the Genus Ocythoe,"
Philosophical Transactions of the Royal Society of London (1819)

Read February 4, 1819

I have before me a specimen of *ocythoe* in an *argonauta*, forming part of the collection of the Acad. of Nat. Sciences. It was taken from the stomach of a dolphin, which was caught in soundings on our Atlantic coast, and is in the most perfect state of preservation, not having suffered the slightest decomposition from gastric action.

It is sufficiently distinct from your *O. cranchii*, as well as from the animal of *Nautilus sulcatus* of KLEIN; and if the figure given by SHAW of the animal of *Argonauta argo* has any pretensions to accuracy, it is most probably an unknown species.

I here attempt a description of it, and also submit a few remarks on the genus.

Ocythoe punctata

Body pale, punctured with purplish; abdomen conic-compressed, vertical, semifasciate near the summit, with a profoundly indented transverse line; arms much longer than the body, attenuated, filiform at their tips, alated; membranes rounded.

Inhabits the Atlantic ocean near the North American coast.

Descrip. *Abdomen* conical, slightly compressed, nearly vertical with respect to the disk of the head, with a profoundly indented transverse line, which extends half round, near the summit. *Arms* attenuated, much longer than the body, filiform towards the tip, slightly varied with brassy, inferior ones, when extended double the length of the body; *suckers* alternate, becoming gradually smaller towards the extremities of the arms, where they are very minute; *membranes* of the anterior arms rounded or suborbicular, extending half way to the base of the arms; periphery occupied by the attenuated portion of the arms, which near its extremity passes upon the disk of the membrane, and terminates abruptly near the base of the expansion; the membrane is carinately decurrent on the inferior surface of the arm, near the base of which it terminates; the inferior surface of the membrane is brassy, and more numerously maculated than the superior surface, which is pale.

Length from the disk to the tip of the abdomen				2 inches.
--- of the abdomen	--	--	--	1 ½
Greatest breadth ditto	--	--	--	1 1/10
Length of the alated arms	--	--	--	2 ¾
--- of those of the opposite side	--	--		5

Eggs subovate, attached to a delicate pedicle by a small basilar tubercle. These fill the involuted spire in the specimen, besides a considerable portion of the body of the shell.

The suckers are very like those of the *O. cranchii*, but the arms are much more elongated, and the abdomen longitudinal with respect to the head. This animal seems to be not unfrequently the prey of some of the larger fishes, for in addition to the instance above mentioned, Bosc informs us that in his passage between Europe and America, he found a specimen in the stomach of a *Coryphoena equiselis*, GMEL. but very much decomposed: and in the Museum of Mr. PEALE, in this city, a fine *argonauta* occurs, which was taken from the stomach of a shark.

With respect to the contested question relative to the parasitic nature of the animals of this genus, I believe the remark will hold good generally, if not absolutely, that those mulluscous animals that form the shell in which they reside, are more or less connected with it by muscular or membranaceous attachment, or by the permanent spiral form of the posterior part of the body; and that the body of the animal complies with the inequalities of the chamber of the shell or rather that the shell is moulded upon the body, so as to be in contact with it in every part. So careful are they to fill the cavity to its very summit, that when from their increase of growth the apex of the shell is vacated in consequence of its straightness, either the part is removed by the animal, and additional calcareous matter is secreted to close the aperture thus formed, or it is permitted to remain, and the cavity is filled up by the same secretion; of the former process we have an instance in *Bulimus decollata*; and of the latter many instances occur, familiar to the knowledge of conchologists. The *Ocythoe* offers to our consideration a remote deviation from these ordinary laws which apply to the testaceous mollusca, inasmuch as it only resides in the last volution or body of the shell. In the specimen above described, the sides of the abdomen are slightly canaliculated, in conformity with the sculpture of the inner lateral surface of the shell; but it is worthy of remark, that the portion which corresponds with the most unequal part of the chamber, the carina, is not at all indented; which fact induces the supposition that the shell does not fit the body, and of course, was not made for it, otherwise it does not seem probably that the body would be remote from the shell in one part, and impressed by its asperities in another.

Such also is the form of the inferior part of the abdomen, that it never could have revolved in the cavity of the involuted spire; yet we have never been informed that the vacated spire has been either broken or solidified. Neither is there any attachment whatever between any part of the body and the including shell, by an organ appropriated to the office. In consequence of this organization, the *Ocythoe* cannot adapt itself to the form of the cavity in which it rests, or secure itself there so completely as the well known parasitic paguri are enabled to do, in consequence of the pliability of their vesicular abdomen, and by the agency of their terminal hooks or holder. Such observations seem to afford presumptive evidence of the parasitic nature of these animals.

It does not appear to me probable that the *Ocythoe* ascends to the surface of the water by exhausting its shell of the included water; for if this were the fact, those females whose shell is in great part filled with eggs, could not visit the surface. But the change of specific gravity is doubtless effected in its own body, by which it is enabled to sustain itself on the surface at will, or to descend to the bottom promptly at the approach of danger.

The shells which in structure and appearance approach nearest to *argonauta*, are unquestionably to be found in the PTEROPODA; and the examination of *Carinaria*, *Atlanta* and *Spiratella*, would almost lead us to suppose, that the artificer of *argonauta* is in reality of that division; but if this supposition be indicated by the conformation of the shell, it does not seem to be corroborated by the probable habits of the animal. All those hitherto discovered of that group, are known to swim at the surface of the ocean, and not being furnished with other organs of locomotion than fins, they cannot glide upon the bottom; we must therefore (analogically) suppose this to have been the habit of the animal; and yet it is hardly admissible that it should, in that case, have eluded the observation of voyagers, since the shell has not unfrequently been found in a state of occupancy of the parasite.

Thomas Say, "Preface," *American Entomology, or Descriptions of the Insects of North America* (1824–28)

THE present number is intended as the integral portion of a publication of no inconsiderable magnitude on the insects of North America.

But little, I might almost say nothing, has yet been done in the United States, in relation to the very interesting and important science upon which this work is intended to treat; while, in other departments of natural history, we have publications honourable to the republic, there is not, as far as I know, in the archives of American science, the record of an indigenous work on this subject.

In Europe, a celebrated writer informs us, the insects, so numerous, so diversified in their characters, in their colouring so elegant and varied, and so singular in their manner, have so much interest under these different relations, that of all the animals, they have been the most observed, the most studied, and are those upon which the labours of naturalists have been the most exercised. But in the United States, entomology, of all the sciences, has been regarded with the least attention by the learned. The attractive charms of natural history have, indeed, with us, allured many votaries; but these, in general, choose those departments for their study of which the knowledge is more readily acquired, and where the labour of initiation is not arduous or protracted. Hence the higher departments of zoology, botany, &c. are more frequently selected, offering as they do more prominent and obvious characters, easily detected by the investigator. Entomology, on the other hand, although at a distance captivating to the beholder, yet, when arrived at its threshold, he is opposed by so many difficulties, that he is deterred from prosecuting further his researches. The variety of systems, the obscurity of the distinctive characteristics, and often the great requisite nicety of discrimination upon which some of those systems are founded, the want of a guide such as would be afforded by good books of plates, – all conspire to retard the progress of the student. To these obstacles we may also add, the difficulty of procuring the many splendid and costly works of European authors – our booksellers being unwilling to incur the risk of importing them unless expressly ordered. Attributable to these causes is the absence of knowledge of this science and of taste for its cultivation. Indeed, there are not wanting among the uninformed, individuals who harbour the almost impious opinion that insects are despicable because they are minute, and that the study of them is little better than contemptible trifling, and prodigality of time. This opinion is too unphilosophical to deserve notice, or serious reply; it is impious, inasmuch as it assumes that a portion of the labours of the Creator, which we are informed he contemplated with pleasure, and in his wisdom pronounced good, are altogether futile and of a nature too trifling for the serious attention of man.

Much might be said in opposition to this absurd notion, the offspring of ignorance, and enough has been said by numerous authors of the first authority, to establish the claims of these minute, but most formidable of all animals to an exalted rank in our respect and consideration.

Hitherto the American insects have been collected and sent to England, France, and Germany, by intelligent foreigner's resident amongst us. These were described by their entomologists, figured and published in their various works, and through that medium were familiarized to the knowledge of the scientific in those countries. In short, they are infinitely better acquainted with them than we are.

Having, for a considerable length of time, attached myself particularly to this department of zoology, I saw with regret that no one arose among us to investigate and describe such of the individuals belonging to it as were unknown.

I was anxious in vain that some one of my countrymen, whose talents and scientific acquirements were greater than my own, and more adequate to the task, would make known to the world this Protean people.

At length, urged exclusively by a love for the pursuit, and not, as my friends will unhesitatingly admit, by any expectation or desire of mere pecuniary acquisition, I have, though not without considerable hesitation, undertaken this work.

THOMAS NUTTALL (1786–1859)

Botanist and zoologist.
Born in Settle, Yorkshire, England. Died in St. Helens, Lancashire, England.

Nuttall worked briefly as a journeyman printer until he immigrated to the United States in 1808. Very soon after arriving, he sought out Benjamin Smith Barton's help identifying a plant and quickly became Barton's new protégé (replacing Frederick Pursh who had left for England with some of the Lewis and Clark plant collection in his possession). Within months Nuttall was collecting plants for Barton. In 1810 he was collecting for Barton around the Great Lakes when he heard of plans for the Astor expedition (fur-trading interests) up the Missouri. He joined the expedition for a year collecting along much of the same trail Lewis and Clark traveled (none of those plants, however, had yet been described). When rumors began to circulate of a war between the United States and Great Britain, Nuttall headed back to England with his specimens. While in London, Nuttall met Pursh and shared some of his own specimens for Pursh's *Flora*. Nuttall returned to the United States as soon as the war ended and set off to collect in the mountains of the Southeast. He began compiling his *Flora of North American Plants* in 1817 and a year later two small volumes were published. Following the publication of those volumes Nutall headed to the Mid- and Southwest for an extended collecting trip. By 1825, Nuttall was back in the East and had accepted a position as professor and curator at Harvard. His wanderlust was back by 1834, however, when he accompanied an expedition to the Columbia River and then stayed on collecting throughout the West. In 1836 he returned to the East again, this time to work at the Academy of Natural Sciences in Philadelphia writing up the hundreds of new species he had found. Many of these he would pass on to Torrey and Gray who were by this time embarking on their own *Flora of North America*. The death of Nuttall's uncle in England required his return to take possession of family property. He returned to the States in 1847–48 to complete his work on *The North American Sylva* (1842, 1846, 1849). This three-volume set was lavishly illustrated and accounted for all the trees in North America with special attention paid the Pacific coast. The work is still considered a classic today. Nuttall was the first naturalist to champion the use of a natural system of classification in the United States. Above all, he was a field botanist and after his example future botanists would strive for a combination of field and museum work.

Thomas Nuttall, "Preface" (excerpt), *The Genera of North American Plants and a Catalogue of the Species, to the Year 1817* (1818)

A desire to advance the science of Botany by any additional remarks and facts which might be in my possession, connected with an endeavour to instruct the ignorant, in this engaging science, are the motives which have induced the author to the prosecution of a laborious but gratifying task.

How much he has drawn from every popular source of information and thus advanced the merit of this little publication by the labours of others almost every page can testify.

The tacit evidence of Botanists to the accuracy of the prevailing definitions of genera and species, afford, as it were, an almost inviolable sanction to the labours of their authors, and appear to stamp with temerity every attempt at subversion. The limits of genera, however, since the times of Linnaeus, reverting in a measure to their former simplicity, have now been greatly reduced, and more particularly so, since Botany, assuming a philosophical character, lays claim to a classification by natural affinities. In this interesting and now prevailing view of the subject, a reduction of heterogenous materials to their natural types, has led the way to the construction of genera better according with the plan of nature.

One of the strongest, and perhaps most important objections urged against these improvements is the confusion which they are innocently the means of introducing into Botanical nomenclature, and indeed it must be acknowledged that the concussion of revolution whether in science or politics, even to fulfil the most important object, but little accords with our natural desire of harmony. And yet the same love of revolution might also have been urged with equal force against the great Linnaeus, who in the zenith of his fame, but seldom spared the labours of his predecessors or contemporaries when they stood in the way of his darling system.

But we are at length inclined to believe, that the last and most perfect of systems, perfect because the uncontaminated gift of Nature, is about to be conferred upon and confirmed by the Botanical world. The great plan of natural affinities, sublime and extensive, eludes the arrogance of solitary individuals, and requires the concert of every Botanist and the exploration of every country towards its completion. Can we deny the perception of a prevailing affinity throughout the vegetable kingdom, and carp at the anomalous character of a few individuals? But even here the science begins to triumph, when we perceive that the anomalies diminish by the accession of objects.

Thomas Nuttall, "Pileated Woodpecker," *Manual of the Ornithology of the United States and Canada* (1832)

CHAR.   General color greenish black; side stripe of white from the bill down the sides of the neck; chin, throat, and part of wings white or pale yellow. Male with scarlet crown, crest, and cheek patch. Females with crest partly black and no scarlet on cheek. Length about 18 inches.

Nest.   In a deep forest or the seclusion of a swampy grove; excavated in high trees, and lined only with fine chips.

Eggs.   4–6; snow white and glossy; 1.25 × 1.00.

This large and common Woodpecker, considerably resembling the preceding species, is not unfrequent in well-timbered forests from Mexico and Oregon to the remote regions of Canada, as far as the 63d degree of north latitude; and in all the intermediate region he resides, breeds, and passes most of the year, retiring in a desultory manner only into the Southern States for a few months in the most inclement season from the North and West. In Pennsylvania, however, he is seen as a resident more or less throughout the whole year; and Mr. Hutchins met with him in the interior of Hudson Bay, near Albany River in the month of January. It is, however, sufficiently singular, and shows perhaps the wild timidity of this northern chief of his tribe, that though an inhabitant towards the savage and desolate sources of the Mississippi, he is unknown at this time in all the maritime parts of the populous and long-settled State of Massachusetts. In the western parts of the State of New York he is sufficiently common in the uncleared forests, which have been the perpetual residence of his remotest ancestry. From the tall trees which cast their giant arms over all the uncleared river lands, may often be heard his loud, echoing, and incessant cackle as he flies restlessly from tree to tree, presaging the approach of rainy weather. These notes resemble *ekerek rek rek rek rek rek rek* uttered in a loud cadence with gradually rises and falls. The marks of his industry are also abundantly visible on the decaying trees, which he probes and chisels with great dexterity, stripping off wide flakes of loosened bark to come at the burrowing insects which chiefly compose his food. In whatever engaged, hast and wildness seem to govern all his motions, and by dodging and flying from place to place as soon as observed, he continues to escape every appearance of danger. Even in the event of a fatal wound he still struggles with unconquerable resolution to maintain his grasp on the trunk to which he trusts for safety to the very instant of death. When caught by a disabling wound, he still holds his ground against a tree, and strikes with bitterness the suspicious hand which attempts to grasp him, and, resolute for his native liberty, rarely submits to live in confinement. Without much foundation, he is charged at times with tasting maize. I have observed one occasionally making a hearty repast of holly and smilax berries.

   This species is being driven back by "civilization," and is now found only in the deeper forests. Mr. William Brewster reports that a few pairs still linger in the northern part of Worcester County, Massachusetts.

CONSTANTINE SAMUEL RAFINESQUE
(1783–1840)

Naturalist.
Born in Constantinople, Turkey. Died in Philadelphia, Pennsylvania.

LARGELY self-educated, Rafinesque spent his youth in Marseilles, France, but immigrated to the Unites States at 19 to become an apprentice in a Philadelphia mercantile house. He was already a budding naturalist, spending much of his free time collecting plants and animals from Pennsylvania to Virginia and meeting other naturalists. He left in 1805 to work as secretary to the U.S. consul in Sicily but also made a side living in Italy trading in medicinal plants. He began to seriously collect and study fish at this time as well. He returned to the U.S. in 1815 and soon joined the newly formed Lyceum of Natural History. He set off in 1818 on a collecting trip down the Ohio River and began the first comprehensive survey of the river's fish. The result was his *Ichthyologia Ohiensis* (1820). It was during this trip that he visited with Audubon. In his near manic drive to name things, Rafinesque named a bat in Audubon's home that had not yet been described and then named a fish from one of Audubon's drawings that was a complete fabrication on Audubon's part. It was on returning from the Ohio River collecting trip that Rafinesque stopped in Kentucky to visit a former employer and walked into a teaching position at Transylvania University. By 1826, however, he had been let go – one story says for having an affair with the president's wife, another for never going to class. Either way, the eccentric Rafinesque returned to Philadelphia with 40 crates packed full of specimens from his time in Kentucky. For the remainder of his life he earned a living in a variety of ways, but continued to describe and write about the specimens he had collected. He was, perhaps, best known for his proclivity in devising scientific names. He devised 6,700 in botany alone, although relatively few are still in use today. He did at times seem to have a more modern insight into theoretical issues: the impermanence of species, for example, or the significance of fossils in dating sedimentary geological strata, but this was coupled with a near monomaniacal compulsion to name, often as not renaming things that had already been named. David Starr Jordon called him the "Daniel Boone of American science." Others would name him far less kindly. One hears the term quixotic applied to Rafinesque, but just as frequently troublesome. He wrote prolifically, including a medical flora that was a moderately popular success and an Indian hoax document, *the Walum Olam hoax*, which would prove to be a modest scandal. He died penniless and largely forgotten in a rented attic room in Philadelphia.

C. S. Rafinesque, "Introduction," *Ichthyologia ohiensis, or natural history of the fishes inhabiting the river Ohio and its tributary streams* (1820)

NOBODY had ever paid any correct attention to the fishes of this beautiful river, nor indeed of the whole immense basin, which empties its water into the Mississippi, and hardly twelve species of them had ever been properly named and described, when in 1818 and 1819, I undertook the labour of collecting, observing, describing, and delineating those of the Ohio. I succeeded the first year in ascertaining nearly eighty species among them, and this year I added about twenty more, making altogether about one hundred species of fish, whereof nine tenths are new and undescribed.

Many of them have compelled me to establish new genera, since they could not properly be united with any former genus; and I could have increased their number, had I been inclined, as will be seen in the course of this ichthyology; but I have in many instances proposed sub-genera and sections instead of new genera. I sent last spring to Mr. Blainville of Paris, a short account of some of them, to be published in his Journal of Natural History, in a Tract named *Prodromus of seventy new genera of Animals and fifty new genera of Plants from North America*, and I now propose to publish a complete account of all the species I have discovered. I am confident that they do not include the whole number existing in the Ohio, much less in the Mississippi; but as they will offer a great proportion of them, and, as the additional species may be gradually described in supplements, I venture to introduce them to the acquaintance of the American and European naturalist; being confident that they will not be deemed an inconsiderable addition to our actual knowledge of the finny tribes. To the inhabitants of the western states, to those who feed daily upon them, their correct and scientific account ought to be peculiarly agreeable. I trust they will value the exertions through which I have been able to accomplish so much in so short a period of time, and I wish I could induce them to lend me their aid, in the succession of my studies of those animals, by communicating new facts, details, and rare species. I may assure them that their kind help shall be gratefully received and acknowledged.

The science of Ichthyology has lately received great additions in the United States. A few of the Atlantic fishes had been formerly enumerated by Catesby, Kalm, Forster, Garden, Linnaeus, Schoepf, Castiglione, Bloch, Bosc, and Lacepede; but Dr. Samuel L. Mitchell has increased out knowledge, with about one hundred new species at once, in his two memoirs on the Fishes of New-York, the first published in 1814, in the Transactions of the Literary and Philosophical Society of New-York, and the second in the American Monthly Magazine in 1817. Mr. Lesueur was the first naturalist who visited Lake Erie and Lake Ontario, where he detected a great number of new species, which he has already begun to publish in the Journal of the Academy of Sciences of Philadelphia, and which he means to introduce in his General History of American Fishes, a work on the plan of Wilson's Ornithology, which he has long had in contemplation. And I have added thereto about forty new species, which I discovered in Lake Champlain, Lake George, the Chesapeake, the Hudson, near New-York, Philadelphia, the Atlantic, &c. and published in my *Precis des Decouvertes*, my Memoirs on Sturgeons, my decades and tracts in the American Monthly Magazine, the American Journal of Science, &c. besides three new fishes of the Ohio, published in the Journal of the Academy of Philadelphia.

Many other fishes on the United States have been partially described by Bartram, Carver, Lewis and Clarke and other travelers. It is reasonable to suppose that several others have escaped their notice, and my discoveries in the Ohio prove this assertion. I calculate that we know at present about five hundred species of North American fishes, while ten years ago we hardly knew one hundred and twenty. Among that number about one half are fresh water fishes, and one fourth at least belong to the waters of the western states; but, although there are fifty other species imperfectly known, I should not wander far from reality if I should conjecture that, after all, we merely know one third of the real numbers, when we consider that the whole of the Mexican Provinces is a blank in Ichthyology, as well as California, the North West Coast, the Northern Lakes, and all the immense basin of the Missouri and Mississippi, except the eastern branch of the Ohio: all those regions having never been explored by any real naturalists. From those who are actually surveying the river Missouri much may be expected; but I venture to foretell that many of the fishes of the Ohio will be found common to the greatest part of the streams communicating with it, and therefore throughout the Mississippi and Missouri, whence the ichthyology of the Ohio, will be a pretty accurate specimen of the swimming tribes of all the western waters; while in Mexico, the North West Coast, and in the basin of the St. Lawrence or even in the Floridian waters, a total difference of inhabitants may be detected; since I have already ascertained that out of one hundred species of Ohio fishes, there are hardly two similar to those of the Atlantic streams.

I have in contemplation to visit many other western streams and lakes, where I have no doubt to reap many plentiful harvests of other new animals; meantime communications on the fishes of every western stream are solicited from those, who may be able and willing to furnish them.

It is probable that some of the fishes of the Mississippi are anadromic or come annually from the gulf of Mexico to spawn in that stream and its lower branches; but all the fishes of the Ohio remain permanently in it, or at utmost travel down the Mississippi during the winter, although the greatest proportion dwell during that season in the deep spots of the Ohio. This is proved by their early appearance at the same time in all the parts of the river and even as high as Pittsburgh. This happens even with the Sturgeons and Herrings of the Ohio, which are in other countries periodical fishes, travelling annually from the sea to the rivers in the spring, and from the rivers to the sea in the fall.

Fishes are very abundant in the Ohio, and are taken sometimes by thousands with the seines: some of them are salted; but not so many as in the great lakes. In Pittsburgh,

Cincinnati, Louisville, &c. fish always meets a good market, and sells often higher than meat; but at a distance from those towns you may buy the best fish at the rate of one or two cents the pound. It affords excellent food, and, if not equal to the best sea fish, it comes very near it, being much above the common river fish of Europe: the most delicate fishes are the Salmon-perch, the Bubbler, the Buffaloe-fish, the Sturgeons, the Catfishes, &c. It is not unusual to meet such fishes of the weight of thirty to one hundred pounds, and some monstrous ones are occasionally caught, of double that weight. The most usual manners of catching fish in the Ohio are, with seines or harpoons at night and in shallow water, with boats carrying a light, or with the hooks and lines, and even with baskets.

I am sorry to be compelled to delay the publication of my figures of all the fishes now described: these delineations shall appear at another period.

C. S. Rafinesque, "Notices of Materia Medica, or new medical properties of some American Plants," *Western Minerva or, American Annals of Knowledge and Literature* (1820)

1. THE *Erythronium albidum* of Nuttall is a common plant near Lexington, where it is called Lambs-tongue. The root or bulb is used against the Scrophula [sic], being stewed with milk or cream, and applied to the scrophulous [sic] sores, which it will cure. The yellow flowered species or *Erythronium luteum*, has probably the same property. – Communicated by Mr. Crockett, a medical student.

2. The Bear-grass is not a grass; but a fine rare plant peculiar to the western states, the *Helonias angustifolia* of Michaux, which I have ascertained to be different from *Helonias* and called *Cyanotris pratensis*. Its bulb or root is employed in Kentucky and near Lexington for the cure of the Inflamed Breast: it is mashed and applied to the part as a poultice – Communicated by Mr. Crockett.

3. The *Helenium autumnale* is called Sneezeweed in Kentucky, owing to its strong sternutory [sic] properties. If the flowers are dried and snuffed, they will occasion a strong fit of sneezing.

4. The root and leaves of the *Evonymus atropurpureus*, which are called Arrow-wood or Wahoon in Kentucky, are used with efficiency in the Influenza, Cough, Colds &c. in the shape of tea or decoction.

5. Dr. Samuel Brown having procured and shown me the plant which is said to occasion in Kentucky the Milk Fever, I have ascertained that it is the *Euphorbia peploides* (*E. peplus* of Pursh, not Lenneus) which is not uncommon on the cliffs and rocky situations in Kentucky. When eaten by cows through chance, it gives them a fever, and their milk becomes poisonous, producing the milk fever in those who drink it.

6. Many plants are called Gentian in Kentucky and used as succedanea of the foreign Gentian; they are *Triosteum major* and *Tr. minor*, *Sabatia angularis*, *Gentiana amarelloides*, and several other species of this last genus.
7. The root of the large Plantain, *Plantago major*, has lately been recommended in Europe as a good febrifuge.
8. The Water Horehound, or *Lycopus virginicas*, has lately been discovered to be an excellent remedy in Hemoptysis or spitting of blood. It is used in decoction or tea-like.

C. S. R.

JOHN JAMES AUDUBON (1785–1851)

Ornithologist and artist.
Born in Santo Domino, Haiti. Died in New York City, New York.

AUDUBON was the son of a French naval officer and adventurer, Captain Jean Audubon, and his mistress, Jeanne Rabine. His mother died early on, and by the age of three Audubon was in western France being raised by his father and stepmother. At 18 Audubon was sent to America to escape conscription in Napoleon's army and to manage Mill Grove, a small farm his father had purchased as an investment. Audubon fell in love with the countryside and began his study and drawing of American birds. During this time Audubon conducted the first bird banding in America. At the age of 20 he returned to France, traveled with his family for a year, and secured his father's approval of his marriage to Lucy Bakewell, whose father owned a farm near to Mill Grove. The young couple did not stay long at Mill Grove but moved to Kentucky where Audubon opened a store. It was during this time that he received his visit from Alexander Wilson. Audubon found the wonders of Kentucky so compelling that he often chose to roam the woods sketching over working in his store. By 1819 he faced bankruptcy. This only seems to make him more determined to pursue nature and art, and so he headed to New Orleans to teach drawing to the daughter of a wealthy plantation owner and roam the woods in search of birds. His method included shooting the birds with fine shot and fixing them with wires to achieve a natural position. He then set his birds in rich, natural habitats. The total effect was a vivid contrast to the stiff representations that preceded him. By 1824 he had a portfolio he deemed worthy of shopping around and went to Philadelphia for that purpose but was soundly rebuffed by George Ord and others. The backwoodsman dress and personae he had adopted probably did not help his reception in Philadelphia. Audubon persisted, however, and in 1826 left for London. His reception there was vastly different and immediate. The British loved the personae – it was after all the height of the Continental Romantic period – but even more than that they loved the art. He found willing backers and was able to engage the services of the gifted engraver Robert Havell, Jr. The entire project would take a decade to complete but the first volume was ready by 1827. In the end, *The Birds of America* would contain 435 colored plates of 1,065 individual birds, issued in 4 volumes (1827–38). Each plate appeared on a double-elephant folio sheet (2ft × 3ft). He followed *The Birds of America* with a companion collection of life history essays, *Ornithological Biographies*. Audubon returned to the U.S. in 1829, traveling back to England several more times (often with Lucy and their sons) to oversee the process, and when home devoted his time to illustrating more birds as well as documenting mammals (the latter would become *The Viviparous Quadrupeds of North America*). The scientists at the Academy of Natural Sciences who had rebuffed him in 1824 had seen the error of their ways by 1831 and he was elected an "honorary member." He was also elected a fellow of London's Royal Society, only the second American after Franklin. A highly successful, less expensive, 7-volume octavo edition of *The Birds of America* was published in 1842 in the United States.

John James Audubon, "Passenger Pigeon,"
Ornithological Biography (1831)

THE Passenger Pigeon, or, as it is usually named in America, the Wild Pigeon, moves with extreme rapidity, propelling itself by quickly repeated flaps of the wings, which it brings more or less near to the body, according to the degree of velocity which is required. Like the Domestic Pigeon, it often flies, during the love season, in a circling manner, supporting itself with both wings angularly elevated, in which position it keeps them until it is about to alight. Now and then, during these circular flights, the tips of the primary quills of each wing are made to strike against each other, producing a smart rap, which may be heard at a distance of thirty or forty yards. Before alighting, the Wild Pigeon, like the Carolina Parrot and a few other species of birds, breaks the force of its flight by repeated flappings, as if apprehensive of receiving injury from coming too suddenly into contact with the branch or the spot of ground on which it intends to settle.

I have commenced my description of this species with the above account of its flight, because the most important facts connected with its habits relate to it migrations. These are entirely owing to the necessity of procuring food, and are not performed with the view of escaping the severity of a northern latitude, or of seeking a southern one for the purpose of breeding. They consequently do not take place at any fixed period or season of the year. Indeed, it sometimes happens that a continuance of a sufficient supply of food in one district will keep these birds absent from another for years. I know, at least, to a certainty, that in Kentucky they remained for several years constantly, and were nowhere else to be found. They all suddenly disappeared one season when the mast was exhausted, and did not return for a long period. Similar facts have been observed in other States.

Their great power of flight enables them to survey and pass over an astonishing extend of country in a very short time. This is proved by facts well known in America. Thus, Pigeons have been killed in the neighbourhood of New York, with their crops full of rice, which they must have collected in the fields of Georgia and Carolina, these districts being the nearest in which they could possibly have procured a supply of that kind of good. As their power of digestion is so great that they will decompose food entirely in twelve hours, they must in this case have travelled between three hundred and four hundred miles in six hours, which shews their speed to be at an average about one mile in a minute. A velocity such as this would enable one of these birds, were it so inclined, to visit the European continent in less than three days.

This great power of flight is seconded by as great a power of vision, which enables them, as they travel at the swift rate, to inspect the country below, discover their food with facility, and thus attain the object for which their journey has been undertaken. This I have also proved to be the case, but having observed them, when passing over a sterile part of the country, or one scantily furnished with food suited to them, keep high in the air, flying with an extended front, so as to enable them to survey hundreds of acres at once. On the contrary, when the land is richly covered with food, or the trees abundantly hung with mast, they fly low, in order to discover the part most plentifully supplied.

Their body is of an elongated oval form, steered by a long well-plumed tail, and propelled by well-set wings, the muscles of which are very large and powerful for the size of the bird. When an individual is seen gliding through the woods and close to the observer, it passes like a thought, and on trying to see it again, the eye searches in vain; the bird is gone.

The multitudes of Wild Pigeons in our woods are astonishing. Indeed, after having viewed them so often, and under so many circumstances, I even now feel inclined to pause, and assure myself that what I am going to relate is fact. Yet I have seen it all, and that too in the company of persons who, like myself, were struck with amazement.

In the autumn of 1813, I left my house at Henderson, on the banks of the Ohio, on my way to Louisville. In passing over the Barrens a few miles beyond Hardensburgh, I observed the pigeons flying from north-east to south-west, in greater numbers than I thought I had ever seen them before, and feeling an inclination to count the flocks that might pass within the reach of my eye in one hour, I dismounted, seated myself on an eminence, and began to mark with my pencil, making a dot for every flock that passed. In a short time finding the task which I had undertaken impracticable, as the birds poured in in countless multitudes, I rose, and counting the dots then put down, found that 163 had been made in twenty-one minutes. I travelled on, and still met more the farther I proceeded. The air was literally filled with Pigeons; the light of noon-day was obscured as by an eclipse; the dung fell in spots, not unlike melting flakes of snow; and the continued buzz of wings had a tendency to lull my senses to repose.

Whilst waiting for dinner at YOUNG's inn, at the confluence of Salt-River with the Ohio, I saw, at my leisure, immense legions still going by, with a front reaching far beyond the Ohio on the west, and the beech-wood forests directly on the east of me. Not a single bird alighted; for not a nut or acorn was that year to be seen in the neighbourhood. They consequently flew so high, that different trials to reach them with a capital rifle proved ineffectual; not did the reports disturb them in the least. I cannot describe to you the extreme beauty of their aerial evolutions, when a Hawk chanced to press upon the rear of a flock. At once, like a torrent, and with a noise like thunder, they rushed into a compact mass, pressing upon each other towards the centre. In these almost solid masses, they darted forward in undulating and angular lines, descended and swept close over the earth with inconceivable velocity, mounted perpendicularly so as to resemble a vast column, and, when high, were seen wheeling and twisting within their continued lines, which then resembled the coils of a gigantic serpent.

Before sunset I reached Louisville, distant from Hardensburgh fifty-five miles. The Pigeons were still passing in undiminished numbers, and continued to do so for three days in succession. The people were all in arms. The banks of the Ohio were crowded with men and boys, incessantly shooting at the pilgrims, which there flew lower as they passed the river.   Multitudes were thus destroyed. For a week or more, the population fed on no other flesh than that of Pigeons, and talked of nothing but Pigeons. The atmosphere, during the time, was strongly impregnated with the peculiar odour which emanates from the species.

 It is extremely interesting to see flock after flock performing exactly the same evolutions which had been traced as it were in the air by a preceding flock. Thus, should a Hawk have charged on a group at a certain spot, the angles, curves, and undulations that have been described by the birds, in their efforts to escape from the dreaded talons of the plunderer, are undeviatingly followed by the next group that comes up. Should the bystander happen to witness one of these affrays, and, struck with the rapidity and elegance of the motions exhibited, feel desirous of seeing them repeated, his wishes will be gratified if he only remains in the place until the next group comes up.

It may not, perhaps, be out of place to attempt an estimate of the number of Pigeons contained in one of those mighty flocks and of the quantity of food daily consumed by its members. The inquiry will tend to shew the astonishing bounty of the great Author of Nature in providing for the wants of his creatures. Let us take a column of one mile in breadth, which is far below the average size, and suppose it passing over us without interruption for three hours, at the rate mentioned above of one mile in the minute. This will give us a parallelogram of 180 miles by 1, covering 180 square miles. Allowing two pigeons to the square yard, we have One billion, one hundred and fifteen millions, one hundred and thirty-six thousand pigeons in one flock. As every pigeon daily consumes fully half a pint of food, the quantity necessary for supplying this vast multitude must be eight millions seven hundred and twelve thousand bushels per day.

As soon as the Pigeons discover a sufficiency of food to entice them to alight, they fly round in circles, reviewing the country below. During their evolutions, on such occasions, the dense mass which they form exhibits a beautiful appearance, as it changes its direction, now displaying a glistening sheet of azure, when the backs of the birds come simultaneously into view, and anon, suddenly presenting a mass of rich deep purple. They then pass lower, over the woods, and for a moment are lost among the foliage, but again emerge, and are seen gliding aloft. They now alight, but the next moment, as if suddenly alarmed, they take to wing, producing by the flappings of their wings a noise like the roar of distant thunder, and sweep through the forests to see if danger is near. Hunger, however, soon brings them to the ground. When alighted, they are seen industriously throwing up the withered leaves in quest of the fallen mast. The rear ranks are continually rising, passing over the main-body, and alighting in front, in such rapid succession, that the whole flock seems still on the wing. The quantity of ground thus swept is astonishing, and so completely has it been cleared, that the gleaner who might follow in their rear would find his labour completely lost. Whilst feeding, their avidity is at times so great that in attempting to swallow a large acorn or nut, they are seen gasping for a long while, as if in the agonies of suffocation.

On such occasions, when the woods are filled with these Pigeons, they are killed in immense numbers, although no apparent diminution ensues. About the middle of the day, after their repast is finished, they settle on the trees, to enjoy rest, and digest their food. On the ground they walk with ease, as well as on the branches, frequently jerking their beautiful tail, and moving the neck backwards and forwards in the most graceful manner. As the sun begins to sink beneath the horizon, they depart *en masse* for the roosting-place, which not unfrequently is hundreds of miles distant, as has been ascertained by persons who have kept an account of their arrivals and departures.

Let us now, kind reader, inspect their place of nightly rendezvous. One of these curious roosting-places, on the banks of the Green River in Kentucky, I repeatedly visited. It was, as is always the case, in a portion of the forest where the trees were of great magnitude, and where there was little underwood. I rode through it upwards of forty miles, and, crossing it in different parts, found its average breadth to be rather more than three miles. My first view of it was about a fortnight subsequently to the period when they had made choice of it, and I arrived there nearly two hours before sunset. Few Pigeons were then to be seen, but a great number of persons, with horses and wagons, guns and ammunition, had already established encampments on the borders. Two farmers from the vicinity of Russelville, distant more than a hundred miles, had driven upwards of three hundred hogs to be fattened on the pigeons which were to be slaughtered. Here and there, the people employed in plucking and salting what had already been procured, were seen sitting in the midst of large piles of these birds. The dung lay several inches deep, covering the whole extent of the roosting-place, like a bed of snow. Many trees two feet in diameter, I observed, were broken off at no great distance from the ground; and the branches of many of the largest and tallest had given way, as if the forest had been swept by a tornado. Every thing proved to me that the number of birds resorting to this part of the forest must be immense beyond conception. As the period of their arrival approached, their foes anxiously prepared to receive them. Some were furnished with iron-pots containing sulfur, others with torches of pine-knots, many with poles, and the rest with guns. The sun was lost to our view, yet not a Pigeon had arrived. Everything was ready, and all eyes were gazing on the clear sky, which appeared in glimpses amidst the tall trees. Suddenly there burst forth a general cry of "Here they come!" The noise which they made, though yet distant, reminded me of a hard gale at sea, passing through the rigging of a close-reefed vessel. As the birds arrived and passed over me, I felt a current of air that surprised me. Thousands were soon knocked down by the pole-men. The birds continued to pour in. The fires were lighted, and a magnificent, as well as wonderful and almost terrifying, sight presented itself. The Pigeons, arriving by thousands, alighted everywhere, one above another, until solid masses as large as hogsheads were formed on the branches all round. Here and there the perches gave way under the weight with a crash, and falling to the ground, destroyed hundreds of the birds beneath, forcing down the dense groups with which every stick was loaded. It was a scene of uproar and confusion. I found it quite useless to speak, or even to shout to those persons who were nearest to me. Even the reports of the guns were seldom heard, and I was made aware of the firing only by seeing the shooters reloading.

No one dared venture within the line of devastation. The hogs had been penned up in due time, the picking up of the dead and wounded being left for the next morning's employment. The Pigeons were constantly coming, and it was past midnight before I perceived a decrease in the number of those that arrived. The uproar continued the whole night; and as I was anxious to know to what distance the sound reached, I sent off a man, accustomed to perambulate the forest, who, returning two hours afterwards, informed me he had heard it distinctly when three miles distant from the spot. Towards the approach of the day, the noise in some measure subsided, long before objects were distinguishable, the Pigeons began to move off in a direction quite different from that in which they had arrived the evening before, and at sunrise all that were able to fly had disappeared. The howlings of the wolves now reached out ears, and the foxes, lynxes, cougars, bears, raccoons, oppossons, and pole-cats were seen sneaking off, whilst eagles and hawks of different species, accompanied by a crowd of vultures, came to supplant them, and enjoy their share of the spoil.

It was then that the authors of all this devastation began their entry amongst the dead, the dying, and the mangled. The pigeons were picked up and piled in heaps, until each had as many as he could possibly dispose of, when the hogs were let loose to feed on the remainder.

Persons unacquainted with these birds might naturally conclude that such dreadful havoc would soon put an end to the species. But I have satisfied myself, by long observation, that nothing but the gradual diminution of our forests can accomplish their decrease, as they not unfrequently quadruple their numbers yearly, and always at least doubt it. In 1805 I saw schooners loaded in bulk with Pigeons caught up the Hudson River, coming in to the wharf at New York, when the birds sold for a cent a piece. I knew a man in Pennsylvania, who caught and killed upwards of 500 dozens in a clap-net in one day, sweeping sometimes twenty dozens or more at a single haul. In the month of March 1830, they were so abundant in the markets of New York, that piles of them met the eye in every direction. I have seen the Negroes at the United States' Salines or Saltworks of Shawanee Town, wearied with killing Pigeons, as they alighted to drink the water issue from the leading pipes, for weeks at a time; and yet in 1826, in

Louisiana, I saw congregated flocks of these birds as numerous as ever I had seen them before, during a residence of nearly thirty years in the United States.

The breeding of the Wild Pigeons, and the places chosen for that purpose, are points of great interest. The time is not much influenced by season, and the place selected is where food is most plentiful and most attainable, and always at a convenient distance from water. Forest-trees of great height are those in which the Pigeons form their nests. Thither the countless myriads resort, and prepare to fulfil one of the great laws of nature. At this period the note of the Pigeon is a soft *coo coo coo coo*, much shorter than that of the domestic species. The common notes resemble the monosyllables *kee-kee-kee-kee*, the first being the loudest, the others gradually diminishing in power. The male assumes a pompous demeanour, and follows the female whether on the ground or on the branches, with spread tail and drooping wings, which it rubs against the part over which it is moving. The body is elevated, the throat swells, the eyes sparkle. He continues his notes, and now and then rises on the wing, and flies a few yards to approach the fugitive and timorous female. Like the domestic Pigeon and other species, they caress each other by billing, in which action, the bill of the one is introduced transversely into that of the other, and both parties alternately disgorge the contents of their crop by repeated efforts. These preliminary affairs are soon settled, and the Pigeons commence their nests in general peace and harmony. They are composed of a few dry twigs, crossing each other, and are supported by forks of the branches. On the same tree from fifty to a hundred nests may frequently be seen: – I might say a much greater number, were I not anxious, kind reader, that however wonderful my account of the Wild Pigeon is, you may not feel disposed to refer it to the marvellous. The eggs are two in number, of a broadly elliptical form, and pure white. During incubation, the male supplies the female with food. Indeed, the tenderness and affection displayed by these birds towards their mates, are in the highest degree striking. It is a remarkable fact, that each brood generally consists of a male and a female.

Here again, the tyrant of the creation, man, interferes, disturbing the harmony of this peaceful scene. As the young birds grow up, their enemies, armed with axes, reach the spot, to seize and destroy all they can. The trees are felled, and made to fall in such a way that the cutting of one causes the overthrow of another, or shakes the neighbouring trees so much, that the young Pigeons, or *squabs*, as they are named, are violently hurried to the ground. In this manner also, immense quantities are destroyed.

The young are fed by the parents in the manner described above; in other words, the old bird introduces its bill into the mouth of the young one in a transverse manner, or with the back of each mandible opposite the separations of the mandibles of the young bird, and disgorges the contents of its crop. As soon as the young birds are able to shift for themselves, they leave their parents, and continue separate until they attain maturity. By the end of six months they are capable of reproducing their species.

The flesh of the Wild Pigeon is of a dark colour, but affords tolerable eating. That of young birds from the nest is much esteemed. The skin is covered with small white filmy scales. The feathers fall off at the least touch, as has been remarked to be the case in the Carolina Turtle, I have only to add, that this species, like others of the same genus, immerses its head up to the eyes while drinking.

In March 1830, I bought about 350 of these birds in the market of New York, at four cents a piece. Most of these I carried alive to England, and distributed amongst several noblemen, presenting some at the same time to the Zoological Society.

JOSEPH HENRY (1797–1878)

Physicist.
Born in Albany, New York. Died in Washington, District of Columbia.

JOSEPH Henry is widely regarded as the leading American scientist between Benjamin Franklin and Josiah Willard Gibbs. Largely self-educated, Henry never attended any college yet eventually became a distinguished professor at Princeton University (then the College of New Jersey) and first director of the Smithsonian Institute. At 13, Henry was apprenticed to a watchmaker and silversmith where he acquired skills that would later serve him well. He was an avid reader and at 16 read *Popular Lectures on Experimental Philosophy, Astronomy, and Chemistry*, which began his fascination with science and especially the science of magnetism. At 21 he enrolled in Albany Academy (where the tuition was free) and by 1823 was so advanced that he was teaching the science courses there. In 1826 the school appointed him Professor of Mathematics and Natural Philosophy. It was during these years, while teaching 7 hours a day, that he conducted some of his major scientific experiments. He built the most powerful electromagnet to date for Yale University and discovered the electromagnetic phenomenon of self-inductance and mutual inductance (the latter also discovered independently by Faraday who published on it first). Henry invented the electromagnet relay on which the electrical telegraph was later based and is credited with the invention of the first electric motor. He was a professor at Princeton University from 1832 until 1846, where he continued research on electromagnetics as well as such diverse subjects as phosphorescence, ballistics, and capillary action. He left Princeton and his research reluctantly in 1846 to head up the new Smithsonian Institute where he helped establish its tradition of excellence and research. His movement to connect weather reporting stations ultimately led to the U.S. Weather Bureau. He was also one of the founders of the National Academy of Science as well as its president from 1867 until the time of his death. In 1893, 15 years after his death, the standard electrical unit of inductive resistance was named the Henry in his honor.

Joseph Henry, "On the Production of Currents and Sparks of Electricity from Magnetism," *American Journal of Science and Arts* (July 1832)

ALTHOUGH the discoveries of Oersted, Arago, Faraday, and others, have placed the intimate connection of electricity and magnetism in a most striking point of view, and although the theory of Ampere has referred all the phenomena of both these departments of science to the same general laws, yet until lately one thing remained to be proved by experiment, in order more fully to establish their identity; namely, the possibility of producing electrical effects from magnetism. It is well known that surprising magnetic results can readily be obtained from electricity, and at first sight it might be supposed that electrical effects could with equal facility be produced from magnetism; but such has not been found to be the case, for although the experiment has often been attempted it has nearly as often failed.

It early occurred to me, that if galvanic magnets, on my plan, were substituted for ordinary magnets, in researches of this kind, more success might be expected. Besides their great power, these magnets possess other properties, which render them important instruments in the hands of the experimenter; their polarity can be instantaneously reversed, and their magnetism suddenly destroyed or called into full action, according as the occasion may require. With this view, I commenced, last August, the construction of a much larger galvanic magnet than, to my knowledge, had before been attempted, and also made preparations for a series of experiments with it on a large scale, in reference to the production of electricity from magnetism. I was, however, at the time, accidentally interrupted in the prosecution of these experiments and have not been able since to resume them, until within the last few weeks, and then on a much smaller scale than was at first intended. In the mean time, it has been announced in the 117th number of the Library of Useful Knowledge, that the result so much sought after has at length been found by Mr. Faraday of the Royal Institution. It states that he has established the general fact, that when a piece of metal is moved in any direction, in front of a magnetic pole, electrical currents are developed in the metal, which pass in a direction at right angles to its own motion, and also that the application of this principle affords a complete and satisfactory explanation of the phenomena of magnetic rotation. No detail is given of the experiments, and it is somewhat surprising that results so interesting, and which certainly form a new era in the history of electricity and magnetism, should not have been more fully described before this time in some of the English publications; the only mention I have found of them is the following short account from the Annals of Philosophy for April, under the head of Proceedings of the Royal Institution.

"Feb. 17. – Mr. Faraday gave an account of the first two parts of his researches in electricity; namely, Volta-electric induction and magneto-electric induction. If two wires, A and B, bc placed side by side, but not in contact, and a Voltaic current be passed through A, there is instantly a current produced by induction in B, in the opposite direction. Although the principal current in A be continued, still the secondary current in B is not found to accompany it, for it ceases after the first moment, but when the principal current is stopped then there is a second current produced in B, in the opposite direction to that of the first produced by the inductive action, or in the same direction as that of the principal current.

"If a wire, connected at both extremities with a galvanometer, be coiled in the form of a helix around a magnet, no current of electricity takes place in it. This is an experiment which has been made by various persons hundreds of times, in the hope of evolving electricity from magnetism, and as in other cases in which the wishes of the experimenter and the facts are opposed to each other, has given rise to very conflicting conclusions. But if the magnet be withdrawn from or introduced into such a helix, a current of electricity is produced whilst the magnet is in motion, and is rendered evident by the deflection of the galvanometer. If a single wire be passed by a magnetic pole, a current of electricity is induced through it which can be rendered sensible."

Before having any knowledge of the method given in the above account, I had succeeded in producing electrical effects in the following manner, which differs from that employed by Mr. Faraday, and which appears to me to develop some new and interesting facts. A piece of copper wire, about thirty feet long and covered with elastic varnish, was closely coiled around the middle of the soft iron armature of the galvanic magnet, described in Vol. XIX of the American Journal of Science, and which, when excited, will readily sustain between six hundred and seven hundred pounds. The wire was wound upon itself so as to occupy only about one inch of the length of the armature which is seven inches in all. The armature, thus furnished with the wire, was placed in its proper position across the ends of the galvanic magnet, and there fastened so that no motion could take place. The two projecting ends of the helix were dipped into two cups of mercury, and there connected with a distant galvanometer by means of two copper wires, each about forty feet long. This arrangement being completed, I stationed myself near the galvanometer and directed an assistant at a given word to immerse suddenly, in a vessel of dilute acid, the galvanic battery attached to the magnet. At the instant of immersion, the north end of the needle was deflected 30° to the west, indicating a current of electricity from the helix surrounding the armature. The effect, however, appeared only as a single impulse, for the needle, after a few oscillations, resumed its former undisturbed position in the magnetic meridian, although the galvanic action of the battery, and consequently the magnetic power was still continued. I was, however, much surprised to see the needle suddenly deflected from a state of rest to about 20° to the east, or in a contrary direction when the battery was withdrawn from the acid, and again deflected to the west when it was re-immersed. This operation was repeated many times in succession, and uniformly with the same result, the armature, the whole time, remaining immoveably attached to the poles of the magnet, no motion being required to produce the effect, as it appeared to take place only in consequence of the instantaneous development of the magnetic action in one, and the sudden cessation of it in the other.

This experiment illustrates most strikingly the reciprocal action of the two principles of electricity and magnetism, if indeed it does not establish their absolute identity. In the first place, magnetism is developed in the soft iron of the galvanic magnet by the action of the currents of electricity from the battery, and secondly the armature, rendered magnetic by contact with the poles of the magnet, induces in its turn, currents of electricity in the helix which surrounds it; we have thus as it were electricity converted into magnetism and this magnetism again into electricity.

Another fact was observed which is somewhat interesting in as much as it serves, in some respects, to generalize the phenomena. After the battery had been withdrawn from the acid, and the needle of the galvanometer suffered to come to a state of rest after the resulting deflection, it was again deflected in the same direction by partially detaching the armature from the poles of the magnet to which it continued to adhere from the action of the residual magnetism, and in this way, a series of deflections, all in the same direction, was produced by merely slipping off the armature, by degrees, until the contact was entirely broken. The following extract from the register of the experiments exhibits the relative deflections observed in one experiment of this kind.

At the instant of immersion of the battery, deflec.　40° west.
　　　”　　　　”　　　”　　　　”　　　”　　　”　　18　east.
Armature partially detached,　　　　　　　　　”　　　7　east.
Armature entirely detached,　　　　　　　　　　”　　12　east.

The effect was reversed in another experiment, in which the needle was turned to the west in a series of deflections by dipping the battery but a small distance into the acid at first and afterwards immersing it by degrees.

From the foregoing facts, it appears that a current of electricity is produced, for an instant, in a helix of copper wire surrounding a piece of soft iron whenever magnetism is induced in the iron; and a current in an opposite direction when the magnetic action ceases; also that an instantaneous current in one or the other direction accompanies every change in the magnetic intensity of the iron.

Since reading the account before given of Mr. Faraday's method of producing electrical currents I have attempted to combine the effects of motion and induction; for this purpose a rod of soft iron ten inches long and one inch and a quarter in diameter, was attached to a common turning lathe, and surrounded with four helices of copper wire in such a manner that it could be suddenly and powerfully magnetized, while in rapid motion, by transmitting galvanic currents through three of the helices; the fourth being connected with the distant galvanometer was intended to transmit the current of induced electricity: all the helices were stationary while the iron rod revolved on its axis within them. From a number of trials in succession, first with the road in one direction then in the opposite, and next in a state of rest, it was concluded that no perceptible effect was produced on the intensity of the magneto-electric current by a rotatory motion of the iron combined with its sudden magnetization.

The same apparatus however furnished the means of measuring separately the relative power of motion and induction in producing electrical currents. The iron rod was first magnetized by currents through the helices attached to the battery and while in this state one of its ends was quickly introduced into the helix connected with the galvanometer; the deflection of the needle, in this case, was seven degrees. The end of the road was next introduced into the same helix while in its natural state and then suddenly magnetized; the deflection, in this instance amounted to thirty degrees, showing a great superiority in the method of induction.

The next attempt was to increase the magneto-electric effect while the magnetic power remained the same, and in this I was more successful. Two iron rods six inches long and one inch in diameter, were each surrounded by two helices and then placed perpendicularly on the face of the armature, and between it and the poles of the magnet so that each rod formed as it were a prolongation of the poles, and to these the armature adhered when the magnet was excited. With this arrangement, a current from one helix produced a deflection of thirty seven degrees; from two helices both on the same rod fifty two degrees, and from three fifty nice degrees: but when four helices were used, the deflection was only fifty five degrees, and when to these were added the helix of smaller wire around the armature, the deflection was no more than thirty degrees. This result may perhaps have been somewhat affected by the want of proper insulation in the several spires of the helices, it however establishes the fact that an increase in the electric current is produced by using at least two or three helices instead of one. The same principle was applied to another arrangement which seems to afford the maximum of electric development from a given magnetic power; in place of the two pieces of iron and the armature used in the last experiments, the poles of the magnet were connected by a single rod or iron, bent into the form of a horse-shoe, and its extremities filed perfectly flat so as to come in perfect contact with the faces of the poles: around the middle of the arch of this horse-shoe, two strands of copper wire were tightly coiled one over the other. A current from one of these helices deflected the needle one hundred degrees, and when both were used the needle was deflected with such force as to make a complete circuit. But the most surprising effect was produced when instead of passing the current through the long wires to the galvanometer, the opposite ends of the helices were held nearly in contact with each other, and the magnet suddenly excited; in this case a small but vivid spark was seen to pass between the ends of the wires and this effect was repeated as often as the state of intensity of the magnet was changed.

In these experiments the connection of the battery with the wires from the magnet was not formed by soldering, but by two cups of mercury which permitted the galvanic action on the magnet to be instantaneously suspended and the polarity to be changed and rechanged without removing the battery from the acid; a succession of vivid sparks was obtained by rapidly interrupting and forming the communication by means of one of these cups; but the greatest effect was produced when the magnetism was entirely destroyed and instantaneously reproduced by a change of polarity.

It appears from the May No. of the Annals of Philosophy, that I have been anticipated in this experiment of drawing sparks from the magnet by Mr. James D. Forbes of Edinburgh, who obtained a spark[1] on the 30th of March; my experiments

---

[1] From a natural magnet.

being made during the last two weeks of June. A simple notification of his result is given, without any account of the experiment, which is reserved for a communication to the Royal Society of Edinburgh; my result is therefore entirely independent of his and was undoubtedly obtained by a different process.

I have made several other experiments in relation to the same subject, but which more important duties will not permit me to verify in time for this paper. I may however mention one fact which I have not seen noticed in any work and which appears to me to belong to the same class of phenomena as those before described: it is this; when a small battery is moderately excited by diluted acid and its poles, which must be terminated by cups of mercury, are connected by a copper wire not more than a foot in length, no spark is perceived when the connection is either formed or broken: but if a wire thirty or forty feet long be used, instead of the short wire though no spark will be perceptible when the connection is made, yet when it is broken by drawing one end of the wire from its cup of mercury a vivid spark is produced. If the action of the battery be very intense, a spark will be given by the short wire; in this case it is only necessary to wait a few minutes until the action partially subsides and until no more sparks are given from the short wire; if the long wire be now substituted a spark will again be obtained. The effect appears somewhat increased by coiling the wire into a helix; it seems also to depend in some measure on the length and thickness of the wire; I can account for these phenomena only by supposing the long wire to become charged with electricity which by its reaction on itself projects a spark when the connection is broken.

OLIVER WENDELL HOLMES (1809–1894)

Physician, man of letters.
Born and died in Cambridge, Massachusetts.

Holmes initially read law at Harvard but then switched to medicine. He studied at Harvard and Paris and received his medical degree in 1836. He practiced medicine for ten years, during which time he wrote the medical essay provided here in which he first identified the contagiousness of puerpera fever and the vector of infection: the physician. Holmes joined the faculty at Harvard in 1847 as professor of anatomy and physiology. He was Dean of the Harvard Medical School until 1882. Somehow, in addition to all this, he managed to become one of the most celebrated American writers, poets, essayists, and novelists of the nineteenth century. His son, Oliver Wendell Holmes, Jr., would prove equally accomplished as a war hero, Supreme Court judge and author of groundbreaking legal discourse which argued that law should evolve along with the society it serves.

Oliver Wendell Holmes, "The Contagiousness of Puerperal Fever" (excerpts), *New England Quarterly Journal of Medicine* (1843)

From **PART I**

In collecting, enforcing and adding to the evidence accumulated upon this most serious subject, I would not be understood to imply that there exists a doubt in the mind of any well-informed member of the medical profession as to the fact that puerperal fever is sometimes communicated from one person to another, both directly and indirectly. In the present state of our knowledge upon this point I should consider such doubts merely as a proof that the sceptic had either not examined the evidence, or, having examined it, refused to accept its plain and unavoidable consequences. I should be sorry to think, with Dr. Rigby, that it was a case of "oblique vision"; I should be unwilling to force home the *argumentum ad hominem* of Dr. Blundell, but I would not consent to make a *question* of a momentous fact which is no longer to be considered as a subject for trivial discussions, but to be acted upon with silent promptitude. It signifies nothing that wise and experienced practitioners have sometimes doubted the reality of the danger in question; no man has the right to doubt it any longer. No negative facts, no opposing opinions, be they what they may, or whose they may, can form any answer to the series of cases now within the reach of all who choose to explore the records of medical science.

If there are some who conceive that any important end would be answered by recording such opinions, or by collecting the history of all the cases they could find in which no evidence of the influence of contagion existed, I believe they are in error. Suppose a few writers of authority can be found to profess a disbelief in contagion, – and they are very few compared with those who think differently, – is it quite clear that they formed their opinions on a view of all the facts, or is it not apparent that they relied mostly on their own solitary experience? Still further, of those whose names are quoted, is it not true that scarcely a single one could, by any possibility, have known the half or the tenth of the facts bearing on the subject which have reached such a frightful amount within the last few years? Again, as to the utility of negative facts, as we may briefly call them, – instances, namely, in which exposure has not been followed by disease, – although, like other truths, they may be worth knowing, I do not see that they are like to shed any important light upon the subject before us. Every such instance requires a good deal of circumstantial explanation before it can be accepted. It is not enough that a practitioner should have had a single case of puerperal fever not followed by others. It must be known whether he attended others while this case was in progress, whether he went directly from one chamber to others, whether he took any, and what, precautions. It is important to know that several women were exposed to infection derived from the patient, so that allowance may be made for want of predisposition. Now if of negative facts so sifted there could be accumulated a hundred for every one plain instance of communication here recorded, I trust it need not be said that we are bound to guard and watch over the hundredth tenant of our fold, though the ninety and nine may be sure of escaping the wolf at its entrance. If any one is disposed, then, to take a hundred instances of lives, endangered or sacrificed out of those I have mentioned, and make it reasonably clear that within a similar time and compass *ten thousand* escaped the same exposure, I shall thank him for his industry, but I must be permitted to hold to my own practical conclusions, and beg him to adopt or at least to examine them also. Children that walk in calico before open fires are not always burned to death; the instances to the contrary may be worth recording; but by no means if they are to be used as arguments against woollen frocks and high fenders.

I am not sure that this paper will escape another remark which it might be wished were founded in justice. It may be said that the facts are too generally known and acknowledged to require any formal argument or exposition, that there is nothing new in the positions advanced, and no need of laying additional statements before the profession. But on turning to two works, one almost universally, and the other extensively, appealed to as authority in this country, I see ample reason to overlook this objection. In the last edition of Dewees's Treatise on the "Diseases of Females" it is expressly said, "In this country, under no circumstance that puerperal fever has appeared hitherto, does it afford the slightest ground for the belief that it is contagious." In the "Philadelphia Practice of Midwifery" not one word can be found in the chapter devoted to this disease which would lead the reader to suspect that the idea of contagion had ever been entertained. It seems proper, therefore, to remind those who are in the habit of referring to the works for guidance that there may possibly be some sources of danger they have slighted or omitted, quite as important as a trifling irregularity of diet, or a confined state of the bowels, and that whatever confidence a physician may have in his own mode of treatment, his services are of questionable value whenever he carries the bane as well as the antidote about his person.

The practical point to be illustrated is the following: *The disease known as Puerperal Fever is so far contagious as to be frequently carried from patient to patient by physicians and nurses.*

Let me begin by throwing out certain incidental questions, which, without being absolutely essential, would render the subject more complicated, and by making such concessions and assumptions as may be fairly supposed to be without the pale of discussion.

1. It is granted that all the forms of what is called puerperal fever may not be, and probably are not, equally contagious or infectious. I do not enter into the distinctions which have been drawn by authors, because the facts do not appear to me sufficient to establish any absolute line of demarcation between such forms as may be propagated by contagion and those which are never so propagated. This general result I shall only support by the authority of Dr. Ramsbotham, who gives, as the result of his experience, that the same symptoms belong to what he calls the infectious and the sporadic forms of the disease, and the opinion of Armstrong in his original Essay. If others can show any such distinction, I leave it to them to do it. But there are cases enough that show the prevalence of the disease among the patients of a single practitioner when it was in no degree epidemic, in the proper sense of the term. I may refer to those of Mr. Roberton and of Dr. Peirson, hereafter to be cited, as examples.

2. I shall not enter into any dispute about the particular *mode* of infection, whether it be by the atmosphere the physician carries about him into the sick-chamber, or by the direct application of the virus to the absorbing surfaces with which his hand comes in contact. Many facts and opinions are in favour of each of these modes of transmission. But it is obvious that, in the majority of cases, it must be impossible to decide by which of these channels the disease is conveyed, from the nature of the intercourse between the physician and the patient.

3. It is not pretended that the contagion of puerperal fever must always be followed by the disease. It is true of all contagious diseases that they frequently spare those who appear to be fully submitted to their influence. Even the vaccine virus, fresh from the subject, fails every day to produce its legitimate effect, though every precaution is taken to insure its action. This is still more remarkably the case with scarlet fever and some other diseases.

4. It is granted that the disease may be produced and variously modified by many causes besides contagion, and more especially by epidemic and endemic influences.

But this is not peculiar to the disease in question. There is no doubt that smallpox is propagated to a great extent by contagion, yet it goes through the same records of periodical increase and diminution which have been remarked in puerperal fever. If the question is asked how we are to reconcile the great variations in the mortality of puerperal fever in different seasons and places with the supposition of contagion, I will answer it by another question from Mr. Farr's letter to the Registrar-General. He makes the statement that "*five* die weekly of smallpox in the metropolis when the disease is not epidemic," and adds, "The problem for solution is,—Why do the five deaths become 10, 15, 20, 31, 58, 88, weekly, and then progressively fall through the same measured steps?"

5. I take it for granted that if it can be shown that great numbers of lives have been and are sacrificed to ignorance or blindness on this point, no other error of which physicians or nurses may be occasionally suspected will be alleged in palliation of this; but that whenever and wherever they can be shown to carry disease and death instead of health and safety, the common instincts of humanity will silence every attempt to explain away their responsibility.

From PART III

It is true that some of the historians of the disease, especially Hulme, Hull, and Leake, in England; Tonellé, Dugès, and Baudelocque, in France, profess not to have found puerperal fever contagious. At the most they give us mere negative facts, worthless against an extent of evidence which now overlaps the widest range of doubt, and doubles upon itself in the redundancy of superfluous demonstration. Examined in detail, this and much of the show of testimony brought up to stare the daylight of conviction out of countenance, proves to be in a great measure unmeaning and inapplicable, as might be easily shown were it necessary. Nor do I feel the necessity of enforcing the conclusion which arises spontaneously from the facts which have been enumerated by formally citing the opinions of those grave authorities who have for the last half-century been sounding the unwelcome truth it has cost so many lives to establish.

"It is to the British practitioner," says Dr. Rigby, "that we are indebted for strongly insisting upon this important and dangerous character of puerperal fever.[1]

The names of Gordon, John Clarke, Denman, Burns, Young,[2] Hamilton,[3] Haighton,[4] Good,[5] Waller,[6] Blundell, Gooch, Ramsbotham, Douglas,[7] Lee, Ingleby, Locock,[8]

[1] *British and Foreign Med. Rev.* for January, 1842.
[2] *Encyc. Britannica*, xiii, 467, art., "Medicine."
[3] *Outlines of Midwifery*, p. 109.
[4] *Oral Lectures*, etc.
[5] *Study of Medicine*, ii, 195.
[6] *Medical and Physical Journal*, July 1830.
[7] *Dublin Hospital Reports* for 1822.
[8] *Library of Practical Medicine*, i, 373.

Abercrombie,[9] Alison,[10] Travers,[11] Rigby, and Watson[12] many of whose writings I have already referred to, may have some influence with those who prefer the weight of authorities to the simple deductions of their own reason from the facts laid before them. A few Continental writers have adopted similar conclusions.[13] It gives me pleasure to remember that, while the doctrine has been unceremoniously discredited in one of the leading journals,[14] and made very light of by teachers in two of the principal medical schools of this country, Dr. Channing has for many years inculcated, and enforced by examples, the danger to be apprehended and the precautions to be taken in the disease under consideration.

I have no wish to express any harsh feeling with regard to the painful subject which has come before us. If there are any so far excited by the story of these dreadful events that they ask for some word of indignant remonstrance to show that science does not turn the hearts of its followers into ice or stone, let me remind them that such words have been uttered by those who speak with an authority I could not claim.[15] It is as a lesson rather than as a reproach that I call up the memory of these irreparable errors and wrongs. No tongue can tell the heart-breaking calamity they have caused; they have closed the eyes just opened upon a new world of love and happiness; they have bowed the strength of manhood into the dust; they have cast the helplessness of infancy into the stranger's arms, or bequeathed it, with less cruelty, the death of its dying parent. There is no tone deep enough for regret, and no voice loud enough for warning. The woman about to become a mother or with her new-born infant upon her bosom, should be the object of trembling care and sympathy wherever she bears her tender burden or stretches her aching limbs. The very outcast of the streets has pity upon her sister in degradation when the seal of promised maternity is impressed upon her. The remorseless vengeance of the law, brought down upon its victim by a machinery as sure as destiny, is arrested in its fall at a word which reveals her transient claim for mercy. The solemn prayer of the liturgy singles out her sorrows from the multiplied trials of life, to plead for her in the hour of peril. God forbid that any member of the profession to which she trusts her life, doubly precious at that eventful period, should hazard it negligently, unadvisedly, or selfishly!

There may be some among those whom I address who are disposed to ask the question, What course are we to follow in relation to this matter? The facts are before them, and the answer must be left to their own judgment and conscience. If any

[9] *Researches on Diseases of the Stomach*, etc., p. 181.
[10] *Library of Practical Medicine*, i, 96.
[11] *Further Researches on Constitutional Irritation*, p. 128.
[12] *London Medical Gazette*, February 1842.
[13] See *British and Foreign Medical Review*, vol. iii, p. 525, and vol. iv, p. 517. Also *Ed. Med. And Surg. Journal* for July 1824, and *American Journal of Med. Sciences* for January 1841.
[14] *Phil. Med. Journal*, vol. xii, p. 364.
[15] Dr. Blundell and Dr. Rigby in the works already cited.

should care to know my own conclusions, they are the following; and in taking the liberty to state them very freely and broadly, I would ask the inquirer to examine them as freely in the light of the evidence which has been laid before him.

1. A physician holding himself in readiness to attend cases of midwifery should never take any active part in the post-mortem examination of cases of puerperal fever.
2. If a physician is present at such autopsies, he should use thorough ablution, change every article of dress, and allow twenty-four hours or more to elapse before attending to any case of midwifery. It may be well to extend the same caution to cases of simple peritonitis.
3. Similar precautions should be taken after the autopsy or surgical treatment of cases of erysipelas, if the physician is obliged to unite such offices with his obstetrical duties, which is in the highest degree inexpedient.
4. On the occurrence of a single case of puerperal fever in his practice, the physician is bound to consider the next female he attends in labor, unless some weeks at least have elapsed, as in danger of being infected by him, and it is his duty to take every precaution to diminish her risk of disease and death.
5. If within a short period two cases of puerperal fever happen close to each other, in the practice of the same physician, the disease not existing or prevailing in the neighborhood, he would do wisely to relinquish his obstetrical practice for at least one month, and endeavor to free himself by every available means from any noxious influence he may carry about with him.
6. The occurrence of three or more closely connected cases, in the practice of one individual, no others existing in the neighborhood, and no other sufficient cause being alleged for the coincidence, is *prima facie* evidence that he is the vehicle of contagion.
7. It is the duty of the physician to take every precaution that the disease shall not be introduced by nurses or other assistants, by making proper inquiries concerning them, and giving timely warning of every suspected source of danger.
8. Whatever indulgence may be granted to those who have heretofore been the ignorant causes of so much misery, the time has come when the existence of a *private pestilence* in the sphere of a single physician should be looked upon, not as a misfortune, but a crime; and in the knowledge of such occurrences the duties of the practitioner to his profession should give way to his paramount obligations to society.

ELIAS LOOMIS (1811–1889)

Mathematician and astronomer.
Born in Willington, Connecticut. Died in New Haven, Connecticut.

Loomis graduated from Yale in 1830 and taught in various venues until Yale invited him back in 1833 to teach elementary instruction in Latin, mathematics, and natural philosophy. He also began scientific research on the earth's magnetic field, the altitude of shooting stars, astronomical measurements, among other considerations. In 1836 he went to Western Reserve College in Ohio as their mathematics and natural history professor. The salary was minimal but the trustees did appropriate $4,000 for Loomis to travel to Europe to purchase instruments, books, and meet with other scientists. While at Western Reserve, he supervised the construction of an observatory (only the third college observatory in the U.S.), and in 1842 he produced the first weather map, tracking a storm that occurred in 1836 across the East coast. When the college's funds began to dry up around 1844, Loomis moved to New York University (then the University of the City of New York). There he published widely on electricity and telegraphy. He returned to Yale in 1860 and focused on meteorology, publishing on auroras, sunspots, magnetic storms, and cyclones. He was a prolific writer of textbooks, selling a total of around 600,000 copies by the end of his career. He was elected to the National Academy of Sciences in 1873, and between 1874 and 1889 published twenty-three papers on meteorology. He could be best distinguished from earlier meteorologist like Redfield and Espy by his application of both experimental and observational components in his science.

Elias Loomis, "On the Two Storms Which Were Experienced throughout the United States, in the Month of February, 1842" (excerpt), *Transactions of the American Philosophical Society* (1845)

Generalization

I. Direction of the Wind

The question which has, within a few years, excited most interest in Meteorology, is, whether storms are rotary or centripetal, and it has been supposed that either diagram, No. 1 or No. 2, must represent the motion of the wind.

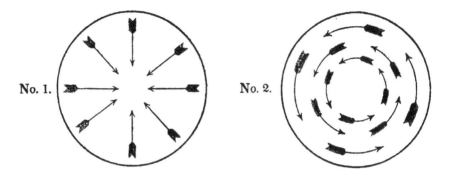

The accompanying charts will show that neither of these diagrams faithfully represents the storms here investigated, and it is doubtful whether either of them ever accurately represents the motion of the wind over any large portion of the earth's surface. The storm of December 15, 1839, has been quoted as a strong case of the revolving kind, and if we take only a semicircle of two hundred miles radius on the north-west side of the centre, the correspondence is very good (see Mr. Redfield's diagram, *Philosophical Transactions* Volume viii., page 81) though, even here, most of the arrows are inclined to the circumferences drawn, and, as Mr. Redfield has remarked, the direction is generally *inward* towards the centre of the storm. But if we take the other half of the semicircle, which Mr. Redfield has omitted, the circles are not completed. There have been several cases, of limited extent, in which the winds might be represented by figure No. 1; but I have seen no account of a storm, of a hundred or more miles in diameter, in which the winds could be faithfully represented by either of the above diagrams. A combination of the two is, however, frequently seen. No. 3 is intended to represent the motions of the wind February 16, 1842, at sunset, near the centre of the storm. Substantially the same diagram is applicable for the morning of the seventeenth, so far as appears from the observation, and tolerably well for the morning of the sixteenth, although there is here more irregularity, the storm being of larger dimensions and less violent. The lower half of this diagram represents, very faithfully, the observations for the morning of December 21, 1836; and it may be inferred that if observations could be procured

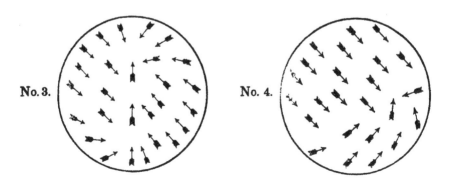

from more northern stations, they would show the other half of the diagram. The same figure represents, pretty accurately, the storm of January 26, 1839, also the great British storm of August 17, 1840, and that of January 6, 1839, except that in the south-west quarter of the latter, the winds were, perhaps, more south-westerly.

Diagram No. 4, represents the storm of March 17, 1838, the north-east wind covering more than half the region of rain and snow. The storm of December 15, 1839, was quite similar to this last, though the north-east wind covered a smaller portion of the storm. In some storms there is a predominance of north-east winds, in others of the south-east; in some north-west, and others south-west winds; but in all we find certain common characteristics, namely, an inward motion, with a tendency to circulate against the sun. These are the ordinary features of those storms which are accompanied by a gust and sudden oscillation of the barometer. The storm of February 3, 1842, can hardly be comprehended under this class. It was so intensive, the motion of the barometer so slow, and the winds moderate on the second and third, that it seems more properly to belong to a distinct class; although on the fourth and fifth, when the winds sprung up with some force, we find the characteristics mentioned above.

II. Progress of the Storm

A second question of some importance is, in what direction and with velocity does the storm advance? It is not difficult to perceive that the general progress of storms is eastward, but, when we undertake to assign the direction with minute precision, we feel the necessity of a precise definition of the terms "storm," and the "centre of the storm." I define the centre of a storm to be the point of greatest barometric depression. The storm of February 16 travelled, then, from sunrise of the sixteenth to sunrise of the seventeenth, five hundred and sixty statute miles in a direction of about north 53° 18' east, although to sunset of the sixteenth, its direction had been about north 74° 46' east. It seems probable that its path was not a straight line, although it is possible that its centre, on the morning of the seventeenth, was farther south than I have here assigned, there being a deficiency of observations on the southern quarter. Velocity twenty-three miles per hour. The storm of February 2 and 3, remained nearly stationary. On the fourth and fifth its progress can be pretty accurately traced. In travelled north 62° 17' east, eight hundred and sixty miles in twenty-four hours, or thirty-six miles per hour. The direction of the storm of December 20, 1836, cannot be precisely assigned. It was clearly eastward, but having only observations from the southern half of the storm, it is impossible to assign, accurately, the place of greatest barometric depression.

Conclusion

The following, then, is my view of the origin of such storms as I have been investigating. This generalization will probably include the greater part of winter storms, but will require some modification when applied to summer storms.

Imagine a time perfectly clear, when the wind is from the west, or a few degrees south of west, with the barometer and thermometer at their mean heights. This may be regarded as the normal state of the atmosphere, and the whole body of air from its upper limit to the surface of the earth is moving on harmoniously in one direction. How is rain produced in such at atmosphere? The first requisite appears to be a change of direction of the lower stratum of air. This appears, in winter, to be frequently the effect of a preceding storm. The prevalent westerly current being temporarily checked in it progress by a violent storm, soon acquires force sufficient to break down all opposition. It supplants the rarefied air of the storm, and not only restores the barometer to it mean height, but the momentum of the excited mass carries it considerably above the mean. This excess of pressure causes a reverse current a little to the westward of a violent storm; and hence we sometimes have a long series of violent storms succeeded by an unusually high barometer affords ground for expecting a second storm within one or two days. But this explanation will not apply to all cases, for then, if the barometer should ever settle down to its mean height all over the globe, we never could have another storm. The case here supposed is not likely ever to happen, but even if it should, we cannot admit the consequences attributed to it. Admit such a case to occur, and the sun's heat would be competent to generate a new storm. Different portions of the earth's surface absorb the sun's rays in unequal degrees, and afford unequal quantities of moisture for evaporation. The result is, that we find bodies of air in close proximity of unequal density, arising from unequal temperature or humidity. Either case would be sufficient to cause a deflection of the lower stratum of air from its normal direction. Suppose, then, we have the mass of the atmosphere pursuing its wonted course from west to east, while a stratum of a mile or so in height, next to the earth's surface, blows in some different direction. If this direction be from the south to the north, then this current must be cooled in its progress by change of latitude. This effect may be aided by the inequalities of the earth's surface, and by friction upon the upper stratum of colder air. At the surface of the earth, where the temperature is probably five or ten degrees above the dew point, no remarkable effect may follow. But at a certain elevation, the air is always saturated with vapour. A very slight reduction of temperature causes cloud, and its density and extent will be proportioned to the energy of the causes in operation. If the wind should blow from the north, it might happen that no cloud would be formed; but if the direction should be easterly, being partly opposed to the normal current, some portion of this mass would almost necessarily be elevated from the earth's surface, and being cooled, its vapour be condensed. The first stage of the process, then, is an abnormal current at the earth's surface; the second is the production of cloud. At this stage the sky is covered with a veil which checks radiation; the thermometer rises above the mean from this cause, and also from the heat liberated in the condensation. This only adds to the energy of the first abnormal current. More cloud is thus formed, and presently the particles of water having acquired sufficient size, fall rapidly to the earth.

The wind being southerly, the thermometer rises. A portion of the atmosphere being thus unusually heated and loaded with vapour, while the upper limit of the atmosphere remains nearly invariable, the barometer necessarily falls. Thus, these causes might continue to operate a long time, acquiring energy by their action. A limit, however, is soon attained. The rarefaction thus produced creates a tendency in the surrounding colder and heavier air, to rush in and occupy its place. Moreover, if the wind be at all easterly, as is usually the case, it partially obstructs the progress of the normal current. This temporary retardation only gives it accumulated energy, and it is soon reinstated with unwonted violence. When the rarefaction is considerable, this rush of air upon the last half of a storm is not generally in the precise direction of the upper current, but more northerly; this air being the denser. Our southerly wind is thus supplanted by a violent north-wester. We have thus a great rarefaction and elevation of the temperature under a south or south-east wind with rain, extending over a large territory. This may be called the third phase of the storm, although it differs from the second only in intensity. There is now a general rush of heavier air to fill this void. This rush is chiefly from the north; but an independent cause, that which imparts direction to the upper current, would give us a west wind. Under these two forces the resulting current is chiefly north-west, but every where upon the borders, the tendency will be inward. The air thus flowing inward towards a central area, forces upwards the warmer air which rises in the middle, and being cooled by elevation, discharges a greater quantity of rain. The currents moving centrally from every point of the compass, interfere with each other, and pursue their routes spirally inward. We have thus a species of rotation, which, in the centre of the storm may have a destructive violence, as at Mayfield, February 4, 1842. This is the fourth phase, and is the case of a violent storm fully organized. This west or north-west wind carries the storm off from a fixed locality, and it is thus transferred successively to points farther and farther east. But this action cannot continue indefinitely. There is a cause in operation which will soon terminate its violence. This westerly wind travels more rapidly than the easterly. The rarefaction at the centre of the storm is a cause which acts equally upon both winds; but the one is opposed to the upper current, and the other nearly coincides with it: hence the one is accelerated, and the other retarded. This result is, that at successive points, farther and farther east, the same storm (after the north-west wind has begun to blow with great violence) has a less duration; the thermometer rises to a less height, the barometer has a smaller oscillation, and thus at a point far eastward, the oscillation becomes nearly extinct, and the only peculiarity observed in the wind is a stronger westerly current succeeding a calm. This is the fifth and final phase of the storm.

★ ★ ★

IT appears to me, that if the course of investigation adopted with respect to the two storms of February, 1842, were systematically pursued, we should soon have some settled principles in meteorology. If we could be furnished with two meteorological charts of the United States, daily, for one year, charts showing the state of the barometer, thermometer, winds, sky, &c., for every part of the country, it would settle for ever the laws of storms. No false theory could stand against such an array of testimony. Such a set of maps would be worth more than all which has been hitherto done in meteorology. Moreover, the subject would be well nigh exhausted. But one year's observation would be needed; the storms of one year are probably but a repetition of those of the preceding. Instead, then, of the guerilla warfare which has been maintained for centuries with indifferent success, although at the expense of great self-devotion on the part of individual chiefs, is it not time to embark in a general meteorological crusade? A well arranged system of observations spread over the country, would accomplish more in one year, than observations at a few insulated posts, however accurate and complete, continued to the end of time. The United States are favourably situated for such an enterprise. Observations spread over a smaller territory would be inadequate, as they would not show the extent of any large storm. If we take a survey of the entire globe, we shall search in vain for more than one equal area, which could be occupied by the same number of trusty observers. In Europe there is opportunity for a like organization, but with this incumbrance, that it must needs embrace several nations of different languages and governments. The United States, then, afford decidedly the most hopeful field for such an enterprise. Shall we hesitate to embark in it; or shall we grope timidly along, as in former years? There are but few questions of science which can be prosecuted in this country to the same advantage as in Europe. Here is one where the advantage is in our favour. Would it not be wise to devote our main strength to the reduction of this fortress? We need observers spread over the entire country at distances from each other not less than fifty miles. This would require five or six hundred observers for the United States. About half this number of registers are now kept, in one shape or another, and the number, by suitable efforts, might probably be doubled. Supervision is needed to introduce uniformity throughout, and to render some of the registers more complete. Is not such an enterprise worthy of the American Philosophical Society? The general government have, for more than twenty years, done something, and have lately manifested a disposition to do more for this object. If private zeal could be more generally enlisted, the ward might soon be ended, and men would cease to ridicule the idea of our being able to predict an approaching storm.

Part Two: 1846–1876

Warriors

WARRIORS

It is telling that two of the most useful background sources for this period, George H. Daniels' 1967 essay, "The Process of Professionalization in American Science," and Robert V. Bruce's Pulitzer Prize winning book, *The Launching of Modern American Science: 1846–1876*, should both have titles with words of becoming, for this was indeed a dynamic period. As Daniels puts it, "Most of the controversies within science can be understood in terms of tensions inherent in the transition from one mode of scientific activity to another" (p. 151); and Bruce notes more specifically, "By 1846 the influence of technology was proving to be a match for that of abundant land, and by 1876 the way of collective, organized enterprise clearly dominated American life. The tension between these two ways to wealth were mirrored in American science during the nineteenth century, and not by coincidence" (p. 5). The scientists of this period seemed aware that they were both the raw material and the potential victors (in bounty and fame) of these dynamic years, which may account in part for the urgency, competition, and tension that marked it. If there is one word that connects all the various themes of this period, it is tension, and so it is around these tensions that I focus this introduction to Part Two.

Lazzaroni / U.S. Coast Survey / Dudley Observatory / National Academy of Sciences

If Louis Agassiz could be considered the face of American science in the antebellum period and Joseph Henry the brains, then Alexander Dallas Bache would be its body. For a span of almost 25 years, Bache was an acknowledged leader, advocate, and benefactor of the scientific community in America, and it is less his science than his motivation of science that marks him as an outstanding force. Hugh Richard Slotten's book, *Patronage, Practice, and the Culture of American Science: Alexander Dallas Bache and the U.S. Coast Survey*, is an excellent place to start for an in-depth understanding of Bache's role in antebellum science and the variety of tensions he had to negotiate during his tenure. (Capt. Albert E. Theberge has also written a detailed history of this period at the Survey, available at NOAA's website.) Since all of the items listed in the subheading above are best understood in reference to Bache and his circle of influence, I will try to touch upon the basics enough to peak an interest for further study.

Alexander Dallas Bache was the grandson of Benjamin Franklin. He graduated from the United States Military Academy at West Point before becoming professor of natural philosophy and chemistry at the University of Pennsylvania. As might be expected, he was also an active member of the Franklin Institute for the Promotion of Mechanic Arts and of the American Philosophical Society in Philadelphia. In 1843 he was appointed superintendent of the United States Coast Survey on the death of the survey's first superintendent, Ferdinand Hassler. Hassler had seen the survey through a fairly rough beginning and had from the start established firm, modern scientific standards for the organization, but he had also struck many as elitist and difficult. In contrast to Hassler, Bache, like his grandfather, had a talent for being at once a member of the elite, capable of using all his political, social, and scientific connections to great advantage, while still paying proper, populist attention to the democratic realities of the unique American landscape.

Bache and Joseph Henry had been friends as well as colleagues for a long time (even travelling together in Europe in the late 1830s), and Bache was instrumental in bringing Henry to the Smithsonian in 1846. From the beginning, Henry and Bache knew they wanted to affect the course of science in America. With Agassiz ensconced at Harvard beginning in 1847 and the newly formed (or reformed) American Association for the Advancement of Science or AAAS as meeting ground, an informal group of like-minded scientists began to coalesce around Bache, Henry, and Agassiz that included, among others, Benjamin Peirce, Benjamin A. Gould, Oliver Wolcott Gibbs, and James Dwight Dana. This group came to be known as the Lazzaroni, that is, scientific beggars – a playful literary reference to a group of Italian beggars associated with the Hospital of St. Lazarus in Naples. Much of the growth of science in the years preceding the Civil War can be traced, overtly and covertly, to the actions of this circle. Joseph Henry's oft-quoted rant on his fears for science was shared by all associated with the Lazzaroni, "We are overrun in his country with charlatanism. Our newspapers are filled with puffs of Quackery and every man who can burn phosphorous in oxygen and exhibit a few experiments to a class of young ladies is called a man of science" (Miller, p. 7). In this quote the principle goal of the Lazzaroni circle is made clear: a standard in judging what is and is not good science and a clearer understanding of who should judge. In addition, the group considered that original research was the surest way to win scientific respect overseas. This being said, the Lazzaroni group then proceeded to exclude, or ignore, many of the more forward-thinking scientists of the time: Asa Gray, Agassiz's colleague and one-time supporter at Harvard; brothers William Barton Rogers, who would go on to found MIT, and Henry Darwin Rogers, the first American deemed enlightened enough to be hired across the pond in Glasgow; the polymath Joseph Leidy, professor at the University of Pennsylvania; Henry Draper, perhaps one of the earliest experimentalists in America; and Matthew Fontaine Maury, Bache's adversarial counterpart at the Naval Observatory. Still, the Lazzaroni held sway, drafting the constitution of the AAAS, defining the curriculum and the professional orientation of Harvard's Lawrence Scientific School, filling key academic and federal jobs with scientists of their choosing, and bringing into being the legislation that created the National Academy of Sciences.

In their zeal, however, the Lazzaroni made two major missteps that eventually lead to the end of their claim to moral high ground and power. The first was their mishandling of the Dudley Observatory. The observatory was the centerpiece of a plan for a national university in Albany, New York. The whole concept was greeted with excitement and fanfare, but in the end became a classic case of the tensions possible between societal expectations (specifically from those supplying the money) and scientific expectations (specifically from those supplying the expertise). The original driving force behind the university was two of Albany's leading figures, Dr. James Armsby (a physician) and Thomas Olcott (a banker). Before too long the university idea had foundered, but the plan for the observatory was secured by the contribution of $10,000 from a wealthy widow, Blandina Dudley, in memory of her husband. The observatory trustees, lead by Armsby and Olcott, sought the advice of Benjamin Peirce on the type of telescope to be purchased and on a possible head for the observatory when astronomer and fellow trustee O. M. Mitchell could not take up the post. Peirce suggest Benjamin A. Gould, an excellent astronomer and mathematician who had studied and worked in Germany under Gauss for years and was also working for the Coast Survey. Gould took up the post in 1855 and the trustees secured Peirce, Henry, Bache, and Gould as advisory Scientific Council to help manage the observatory's affairs. Some tensions between the money men and the science men would be expected, but Gould's personality was exceptionally abrasive and condescending toward the trustees, and the scientific council perhaps backed him too long thinking that only scientists had the proper expertise to reprimand other scientists. But expertise was not the source of tension; social and political savvy were and in those fields Gould was not as well schooled as he might have supposed. In the end, Gould locked the trustees out of the building as a rather extreme show of power, and both sides resorted to defending themselves in the press (with Gould getting personal and vindictive). When editors of these same papers began to write commentaries in defense of the trustees and to question the honesty and judgment of the scientific council, Henry feared all his efforts to improve the image and place of science might be in jeopardy. He retreated from opposing the trustees first and Peirce soon followed. By January 1859, the trustees had legally prevailed and Gould was evicted from the observatory. He returned to his job at the Coast Survey where he performed exceptionally within a society he better understood and which better understood him.

The second and some could argue last mistake the Lazzaroni made was committed during one of their triumphs: the founding of the National Academy of Sciences. Since the 1840s, Bache had talked about an American institution modeled on the French Academy. During the 1850s the "democratic unruliness of the AAAS" (Bruce, p. 301) convinced the Lazzaroni of the need for a more elevated scientific authority. The Lazzaroni worked for years attempting some sort of ultimate arbitrating entity for science at the national level, but it was the Civil War and the federal need for information, guidance, and innovations in war that finally brought about the NAS.

Section 3 of the 1863 Act of Incorporation states why the government too was ready for such an alliance: "the Academy shall, whenever called upon by any department of the Government, investigate, examine, experiment, and report upon any subject of science or art, the actual expense of such investigations, examinations, experiments, and reports to be paid from appropriations which may be made for the purpose, but the Academy shall receive no compensation whatever for any services to the Government of the United States." The U.S. government, like any frugal acquirer, had struck a bargain with science at cost and science, like any image-conscious upstart, had the recognition it desired with the minimalist patronage that would ensure maximum freedom. This was a monumental moment in nationhood and in American science. It came about partly because an obstructive South seceded, leaving a more activist congress in its wake, and partly because the Lazzaroni, who had been discussing and pushing the need for the Academy for years, were prepared to act. In this the Lazzaroni did both nation and science a great service, but then they overreached.

Joseph Henry had made it known that he had doubts about the Academy and feared it would be perceived as too undemocratic and elitist. Knowing his reluctance, Bache, Agassiz, Peirce, Gould, and Rear Admiral Charles H. Davis met without Henry's knowledge, drafted the act, and handpicked the 50 scientists who would make up the Academy. If calls of undemocratic elitism were a possibility before, they had now become a certainty. In fact, a great many of the most distinguished scientists named for inclusion had serious misgivings about the process. Joseph Henry, the most acclaimed American scientist of the day, had been sidestepped by his own Lazzaroni members, and if he had decided to boycott the Academy, most notable scientists outside the Lazzaroni would have followed his lead. He was not, however, that sort of man. Henry decided to work from the inside to build reconciliation. He started by righting a slight made against his own assistant at the Smithsonian, Spencer Baird, who had been kept out of the original 50. Henry joined forces with Asa Gray and James D. Dana to elect Spencer Baird to the Academy over the strong opposition of Louis Agassiz and the Lazzaroni still loyal to him. Baird was elected and the Lazzaroni's hold on American science began to evaporate. Although Bache was elected the first NAS president in 1863, in 1864 he suffered a debilitating stroke. Out of a sense of duty to his old friend and to the science community they had together so carefully fostered, Henry guided the organization through the years until Bache's death in 1867 and then served as its president until his own death in 1878.

The Darwin Debates

At the same time that the Lazzaroni were trying to recover from the controversy of the Dudley Observatory, another – far more important – controversy for science was about to break: Darwin's *On the Origin of Species*. Asa Gray had been corresponding with Darwin since 1855 and as early as 1857 had seen an abstract of Darwin's notions on speciation. In the fall of 1859, concurrent with publication of *On the Origin of*

Species, Darwin wrote 11 personal notes to prominent scientists giving them advance notice and attempting a modest promotion of his ideas. Two of these missiles were sent to America, a two-page letter to his friend Gray and a three-sentence note to Louis Agassiz. Darwin knew well the controversy he was about to unleash and who the combatants in America would be.

By January 1860, Agassiz was already on the attack. He began his opposition at a meeting of the American Academy of Arts and Sciences by restating his long-held view that modern species and fossil species had no relationship to one another. His catastrophic theory stated that species were periodically wiped out and then recreated by God. William Barton Rogers – the same Rogers who opposed the Lazzaroni and would figure prominently in the promotion of Baird to the NAS in 1864 – was ready to take Agassiz on. Agassiz and Rogers would debate (not formal debates as much as prepared discussions) *On the Origin of Species* four times at the Boston meetings between February and April. No record of their exact words exists but an account of the meetings and a general summary of their arguments are recorded in the society's *proceedings*, and various contemporaries would write of the debates in letters and journals. (W. M. Smallwood's essay "The Agassiz-Rogers Debate on Evolution" provides an excellent summary of various accounts with background.) The consensus was that Rogers won all four debates. Agassiz was more of a lecturer than a debater, and he had come up against a scientist in Rogers who was not only his equal in intellect but not particularly taken with Agassiz's style or substance. Rogers rattled Agassiz's confidence and forced him into several sloppy arguments. On geological questions in particular, Rogers had first hand and up-to-date knowledge, Agassiz did not. On the whole, the details of this debate remained within the community of scientists, where it reinforced growing doubts about Agassiz's reputation and scientific views. Smallwood searched the Boston papers, January–April 1860, and found no reference to the debate between Rogers and Agassiz. The debate between Gray and Agassiz, however, would be argued specifically in print and for public consumption.

Gray's first review of *On the Origin of Species* appeared in the March 1860 issue of the *American Journal of Science* and was positive but brief. Considering the debate between Rogers and Agassiz that was going on at the time, Gray must have known he would eventually need to defend Darwin more comprehensively. Agassiz's provided his view of Darwin's work in the same journal in July, labeling the book as mistaken, untrue, unscientific, and "mischievous in its tendency." Gray would respond in a three-part series in the *Atlantic Monthly* (the opening salvo of which is provided here) and in his collection *Darwiniana* (1876), which would appear three years after Agassiz's death. In between, he was far more interested in botanizing than proselytizing, although he kept up his correspondence with Darwin. Agassiz, however, loved a good one-sided argument and would lecture and publish widely on his particular take on special creation for another decade, but almost exclusively for public consumption and in non-scientific journals.

Civil War (1861–65)

The real impact of the Civil War upon American science is still a subject of some debate. I have already discussed how it figured in the enactment of the National Academy of Sciences legislation. In his study, Robert Bruce gives it relatively small measure, finding "Of the two leading civilian institutions in scientific Washington, the Civil War harnessed one, the Coast Survey, and hobbled the other, the Smithsonian Institution" (p. 297). Hugh R. Slotten, in his study of Bache, views the same events as an overall advantage to science. What all can agree on, however, is that it effectively shut down the public debate on Darwin. Sidney Ratner for his "Evolution and the Rise of the Scientific Spirit in America" searched through the leading scientific, religious, and popular magazines between 1861 and 1865 and found only four articles on evolution (p. 108). Perhaps when compared to the war's sheer loss of life and the unrealistic early expectations, scientific theory didn't seem to matter. Perhaps American confidence in self, nation, and God were already too shaken to take on yet another readjustment of our place in the universe. Perhaps the war so wore away at the imagination that it was hard to chance thoughts that might lead to unknown places. More than a few American scientists in the Northeast write of their science as an escape from the chaotic horrors of war back to small wonders and orderly thought. Agassiz told his students that the best way they could serve the war effort was to stay at their desks and devote themselves to learning. For Asa Gray, the war brought wider possible applications for Darwin's ideas, leading him to note privately that "natural selection crushes out weak nations" and "Blessed are the *strong*, for they shall inherit the earth" (Pauly, pp. 41–42). Such thoughts must have crossed many a mind during the four-year national horror. The North, taken by surprise at the beginning, would find its footing by 1863 – helped in part by the U.S. Coast Survey's knowledge of land and shore. Much of science in the North, however, did not appear to miss a beat, even though money was tight until the economic rebound of 1863. In the South, however, many of the scientific institutions were damaged or destroyed, and even those that survived too often retreated into a reactionary stance.

Leaving Agassiz / *The American Naturalist* / Neo-Lamarckians

Around the middle of the 1860s, several of Agassiz's museum assistants quit. Addison E. Verrill went on to work for the Smithsonian, while four others – Alpheus Hyatt, Alpheus S. Packard, Jr., Frederick W. Putnam, and Edward S. Morse – left to join the new Peabody Academy of Science at Salem, Massachusetts. (Hyatt and Packard had left Agassiz's lab earlier to serve in the Union Army but then returned.) These were not men who undervalued Agassiz or what he could teach them. Rather, they were exemplars of Agassiz's teaching and research talents, but they were also one more example that the great man could not evolve. The reasons for the revolt were cumulative but centered around their disagreement with Agassiz's harsh take on

Darwinism, his insistence on the fixity of species, and his inability to recommend them for tenure appointments or facilitated their career aspirations. The final straw was probably when Agassiz posted his "Regulations for the Museum of Comparative Zoology," which prohibited assistants from pursuing their own research interests on museum time.

The four men who went to work at the Peabody Academy started *The American Naturalist* in 1866, with the first issue appearing the following year. While the journal always published a wide range of scientists, the founders certainly used the journal to publish their own research and promote their own point of view. This view from the beginning was an amalgam of ideas swirling around the problem of speciation. In 1866 Hyatt began to work out his own ideas on evolution. Hyatt acknowledged Agassiz's ontogeny recapitulates phylogeny idea but then suggested that through the addition and deletion of phylogenetic stages a Lamarckian-like progressive evolution – specifically the inheritance of acquired traits – was possible. (This was in contrast to the Darwinian process of natural selection and in defiance of Agassiz's species fixity). E. D. Cope had come to similar conclusions at about this time and joined forces with Hyatt in the promotion of what became known as neo-Lamarckianism. The neo-Lamarckian take on evolution held considerable sway until the end of the century and greatly influenced the next generation of scientists.

The Taconic Controversy

Bruce summed this controversy up perfectly when he commented, "The strata in question, like the controversy itself, are extraordinarily disturbed and confusing" (p. 247). In addition, the confusion was twofold, as the Taconic controversy in America mirrored the Cambrian controversy in Britain. Both grew out of inadequate geological theories to clarify classification and age relationship issues for particularly troublesome ranges of strata. This inability was understandable considering that the theory and mechanism for explaining the complexity of these ranges was not really developed until the advent of plate tectonics in the 1960s. The controversy, starting in the early 1840s and lasting until the end of the century, played out in the pages of geological reports and scientific articles (especially the *American Journal of Science* under Dana's editorship), but initially arose from a conflict between James Hall and Ebenezer Emmons, both working for the New York Geological Survey. Eventually nearly every geologist in American would chime in (mostly against Emmons), and several revamping's of the original controversy would play out after new input from Canadian geologists (pro-Hall) and from French geologist Joachim Barrande (pro-Emmons). There would even be a libel suit over negative statements made by Agassiz and Hall concerning which geological chart should rightly be used in maps and textbooks. In 1849 the New York legislature, tired of the bickering, cut appropriations and expelled both Hall and Emmons from their State House quarters. Cecil J. Schneer's 1978 essay on the controversy is perhaps the most thorough, detailed explanation of the principals and principles involved. However, a nineteenth century contemporary, N. H. Winchell, writing in 1886 when none of the issues were yet resolved, offers a fairly simplified explanation in his "The Taconic Controversy in a Nutshell":

> The same mistake was made by Emmons at first as by his opponents. None of them imagined they had to deal with two different and unconformable formations. The strata were all either Taconic or Hudson River. Emmons approached them from one side, the primordial, and his opponents from the opposite direction. Each had evidence to support his claim; and, viewed from his own stand-point, each was right. (p. 34)

And, more to the point, each was wrong.

"The Bones War"

The biographical details and scientific accomplishments of E. D. Cope and O. C. Marsh are provided in the biographies fronting their anthologized essays, so I will focus here on their infamous "Bones War." To summarize, the rivalry between Cope and Marsh began and ended with authority over dinosaur fossils; it was fed by ego, personal animosity, and independent financing. In the pursuit of this title, both men sent small armies of bone collectors throughout the West, spent a small fortune in the process, and ended up opening their personal careers and their field of science to some derision. Admittedly they did discover and describe a great many new fossils, but they did so at a considerable cost not only to themselves, but possibly to untold fossils through hurried and excessive collection techniques and simple spite.

The rivalry did not really begin until 1868. Cope had been working with Joseph Leidy, off and on, in the marl pits of New Jersey since 1858, when the first nearly complete dinosaur skeleton was found. In 1866 Cope, now a curator at the Academy of Natural Science in Philadelphia, discovered the second nearly complete dinosaur skeleton. Marsh, newly at Yale, came to visit Cope and his work in 1868. The visit went well; the two men hunted for bones and discussed their interests. After Marsh left, however, Cope discovered that Marsh had surreptitiously returned to see the marl pit supervisors to offer them money for any future finds. After that, as both men turned their attention to fossil collecting in the West, the competition became prodigious in its proportions. Each side sent out spies and agents to attempt to out buy, steal, or maneuver the other side. During the worst of it, Cope reportedly had a train full of Marsh's fossils diverted from New Haven to Philadelphia, and Marsh blew up a fossil site rather than have Cope find it. In addition to warring in the field, each warred in print about the other's scientific standards and authority. Cope was constantly accusing Marsh of stealing, and when Leidy noticed that one of Cope's dinosaur reconstructions had the head on the wrong end; Marsh did his best to make it a *cause célèbre*. (In 1981 Yale's Peabody Museum discovered that Marsh too had made a wrong-headed reconstruction.)

In the 1880s when John Wesley Powell was made director of the newly formed U.S. Geological Survey, Marsh was his choice for the survey's vertebrate palaeontologist. Soon thereafter Cope's government funding for Western survey dried up. Then in 1891 a series of articles in the New York Herald (running over a week mid-January) brought the feuding to national attention. Powell, Cope, and Marsh were all cited in the articles. Marsh seemed to get the better of Cope by revealing Cope's mistaken reconstruction of a dinosaur skeleton – with head where tail should be – some 20 years earlier. A year later, however, Marsh too was forced out of his federal support. Poorer but hardly wiser, their animosity lasted until Cope's death in 1897, or slightly thereafter. Supposedly, in a final attempt to pull ahead, as it were, Cope had his skull preserved for eventual comparison with Marsh's own. For once, Marsh resisted the bait, kept his head, and the final measure of both men was never taken.

1847   **Joseph Leidy, "On the Fossil Horses of America," Proceedings of the Academy of Natural Sciences of Philadelphia.**

       Maria Mitchell discovers comet – Comet Mitchell 1847 VI.

       David J. Peck becomes the first African American to receive an MD from an American college.

       John William Draper, "Examination of the Radiations of Red-Hot Bodies. The Production of Light by Heat," *American Journal of Science and Arts.*

1848   George Phillips Bond discovers Hyperion, the eighth moon of Saturn.

       Asa Gray publishes *Manual of the Botany of the Northern United States* (Boston). Five editions during Gray's lifetime.

       Maria Mitchell elected to the American Academy of Arts and Science.

1849   Arnold Henri Guyot, *Earth and Man, or Lectures on Comparative Physical Geography in Its Relation to the History of Mankind* (Boston).

       U.S. Department of the Interior established.

1850   William Cranch Bond and George Phillips Bond discover the C-ring of Saturn.

1853   **Joseph Leidy, "A Flora and Fauna Within Living Animals,"** *Smithsonian Contributions to Knowledge* **(Washington, D.C.).**

1856   **James D. Dana, "On the Origin of the Geographical Distribution of Crustacea,"** *Annals and Magazine of Natural History.*

1857   **Louis Agassiz, "Section IX: Range of Geographical Distribution," Essay on Classification (Boston).**

1859   Charles Darwin, *Origin of Species* (London). Initial print of 1,000 copies sells out in one day.

       Missouri Botanical Garden established in St. Louis by Henry Shaw.

       Harvard's Museum of Comparative Zoology opened under the direction of Agassiz.

1860   Debate over Charles Darwin's Origin of Species between William Barton Rogers and Louis Agassiz at four consecutive meetings of the Boston Society of Natural History.

       Louis Agassiz, "On the Origin of Species," *American Journal of Science.*

       Asa Gray, "Darwin on the Origin of Species," *Atlantic Monthly.*

1861   Civil War begins.

       Massachusetts Institute of Technology chartered by the state.

1863   Legislature establishing the National Academy of Sciences enacted.

       James D. Dana, "On Cephalization," *New Englander.*

1864   George Perkins Marsh, *Man and Nature; or, Physical Geography as Modified by Human Action* (New York).

       Agassiz's students, Hyatt, Morse, Packard, Putnam, Scudder and Verrill leave Harvard.

1865   Civil War ends.

1866   Edward D. Cope and Alpheus Hyatt independently theorize on biological recapitulation in relation to evolution of species; beginning of Neo-Lamarckians.

1867   Benjamin Peirce superintendent of the U.S. Coast Survey.

       Daniel Kirkwood, "On the Theory of Meteors," *Proceedings of the American Association for the Advancement of Science.*

1867–72 U.S. Army's Geological Exploration of the Fortieth Parallel.

1869   John Wesley Powell explores canyons of the Green and Colorado rivers.

1870   **Benjamin Peirce, "Linear Associative Algebras," read before the National Academy of Sciences. C. S. Peirce would have it printed in the** *American Journal of Mathematics* **in 1881.**

       Chauncey Wright, "Limits of Natural Selection," *North American Review.*

1871   **Chauncey Wright, "The Genesis of Species,"** *North American Review.*

Charles Darwin, *The Descent of Man* (London).

E. D. Cope, "The Method of Creation of Organic Form," *Proceedings of the American Philosophical Society.*

1872   **Asa Gray, "Sequoia and Its History,"** *American Naturalist.*

1873   **James D. Dana "On some Results of the Earth's Contraction from cooling, including a discussion of the Origin of Mountains, and the nature of the Earth's Interior,"** *American Journal of Science.*

1874   **O. C. Marsh, "Fossil Horses in America,"** *American Naturalist.*

1875   **O. C. Marsh, "Odontornithes, or Birds with Teeth,"** *American Naturalist.*

1876   Thomas Henry Huxley visits United States and lectures on evolution.

Edward A. Bouchet becomes the first African American to receive a Ph.D. in physics.

Alpheus Spring Packard, Jr. *Life Histories of Animals, Including Man, or Outlines of Comparative Embryology* (New York).

David Starr Jordan, *Manual of Vertebrates of the Northern United States* (Chicago).

American Chemical Society begins.

Thomas Alva Edison sets up laboratory in Menlo Park to research and develop inventions, precursor of industrial research labs.

Johns Hopkins University opens.

JOSEPH LEIDY (1823–1891)

Comparative anatomist, zoologist, botanist, parasitologist, palaeontologist, illustrator. Born in Philadelphia, Pennsylvania. Died in Philadelphia, Pennsylvania.

In 1844, at the age of 21, Leidy received his medical degree from the University of Pennsylvania. Except for a time during the Civil War when he served as surgeon at Satterlee General Hospital in Philadelphia, Leidy never really practiced medicine. Instead, he became prosector to the professor of anatomy at the University of Pennsylvania and focused his attention on the study of anatomy, physiology, and natural history, on publishing original research and descriptive work, and on teaching. At 22 Leidy was elected into both the Boston Society of Natural History and the Academy of Natural Sciences of Philadelphia. It was at the Academy, working with their ever-growing fossil collection, that he added the study of palaeontology to his list of skill sets. He published his first paleontological paper, "On the Fossil Horse," in 1847. It was an auspicious beginning as this paper would later be used by both Huxley and O. C. Marsh as support for the theory of evolution. Leidy would go on to write over 200 papers on palaeontology and over 800 papers in total, including many natural history descriptions of new specimens obtained by various federal surveys or sent to him by other naturalists. He traveled through Europe in 1848, visiting museums and hospitals, and began to lecture on microscopic anatomy upon his return. His medical school mentor, Paul B. Goddard, first introduced him to the new achromatic compound microscope, and Leidy made adept use of it in his research and descriptive papers. For instance, in addition to the "Flora and Fauna" paper anthologized here, consider his 1879 protozoology monograph "Fresh-Water Rhizopods of North America" and his 1881 study of endosymbionts in termites "The Parasites of the Termites". In 1853, he was appointed professor of anatomy at the University of Pennsylvania. He would remain in that position until he became director of the school's newly established department of biology in 1884. Leidy was a scientist uncomfortable with controversy. The Cope-Marsh bones war that initially grew out of his own fossil studies disgusted him and was part of the reason he withdrew from active acquisition of fossils, although he continued to be an inspiration to both his ex-student Cope and the Yale professor, Marsh. He never took a strong stand during the evolution debates, but he did heartily welcome Darwin's *Origin of Species*, commenting in a letter to Darwin, "I feel as though I had hitherto groped about in darkness and that all of a sudden a meteor flashed upon the skies." His work, however, much of which would prove probative for the theory of evolution, spoke more eloquently than many a long-winded defense. The observations he made in his introduction to "Flora and Fauna within Living Animals" concerning the origin of life and spontaneous generation were years in advance of Pasteur and Darwin. A polymath of the first order, Leidy tended to offer insights and let them go rather than follow them through to theoretical conclusions, perhaps for fear of controversy, perhaps because that was simply how his mind worked. Leonard Warren's 1998 biography of Leidy gives an excellently rounded view of the quietly brilliant Leidy in *Joseph Leidy: The Last Man Who Knew Everything.* He is considered the Father of American Vertebrate Palaeontology, the Founder of American Parasitology, and one of the finest medical illustrators of his time.

Joseph Leidy, "On the Fossil Horses of America," *Proceedings of the Academy of Natural Sciences of Philadelphia* (1847)

THE fact of the existence of fossil remains of the horse in America has been generally received with a good deal of incredulity, arising, perhaps, from the mere fact being stated of their having been found, often without even mentioning the associate fossils, and in all cases, previous to Mr. Owen,[1] without describing the specimen. At present their existence being fully confirmed, it is probably as much a wonder to naturalists as was the first sight of the horses of the Spaniards to the aboriginal inhabitants of the country, for it is very remarkable that the genus Equus should have so entirely passed away from the vast pastures of the western world, in after ages to be replace by a foreign species to which the country has proved so well adapted; and it is impossible, in the present state of our knowledge, to conceive what could have been the circumstances which have been so universally destructive to the genus upon one continent, and so partial in its influence upon the other.

The remains are by no means unfrequent, and according to William Cooper, the author of a paper entitled "Notices of Big-Bone Lick," in Featherstonhaugh's "Journal of Geology,"[2] the first printed notice of them occurs in Mitchell's "Catalogue of Organic Remains,"[3] upon referring to which, I find mentioned pp. 7, 8, that a cervical vertebra and teeth of the horse were found associated with the *Mastodon*, &c., in a tract extending from the base of the Neversink Hills to Bordentown, New Jersey. The author of "Notices, &c., also mentions the remains of the horse being found at Big-Bone Lick, but speaks doubtfully as to the authenticity of such remains having been found in a fossil state in this country, and says, p. 208, "I saw nothing in support of it myself, nor have I met with any person who could answer for such a fact from his own careful observation."

[1] Zoology of the Voyage of the Beagle, part 1. By R. Owen, Esq. London, 1840.
[2] Philada., 1831, vol. 1, p. 208.
[3] New York, 1826.

Dr. Harlan[4] mentions the sparing existence of fossil remains of the horse, which, from the heading of his chapter, he has referred to the same species as the existing Equus Caballus.

The most satisfactory account, however, with which I am acquainted, is give by Mr. Owen in the Zoology of the Voyage of the Beagle, Part 1, Fossil Mammalia, p. 109, and is derived from two teeth obtained by Mr. Darwin in South America. One of them, a superior molar, was far decomposed, and Mr. Owen observes, "every point of comparison that could be established, proved it to differ from the tooth of the common Equus caballus only in a slight inferiority of size." The other a superior left molar, was found with the *Mastodon*, &c., in the province of Entre Rios, and is figured (Pl. xxxii. figs. 13 and 14) in the work. One of the figures represents an antero-lateral view of the tooth, and is rather smaller in size, and is much more curved than in the corresponding tooth of the recent E. caballus. The other figure represents the crown of the tooth and indicates the diameters to be somewhat less. From what Mr. Owen remarks in the "British Fossil Mammalia,"[5] this is a species which he proved to be distinct from all European fossil and existing species, and from the greater degree of curvature of the upper molars[6] he has designated it under the name of Equus curvidens. In the cabinet of the Academy there are a number of specimens of American fossil horse teeth, which I refer to two distinct and well marked species.

The first of these I consider as identical with the Equus curvidens, of which there are ten specimens of permanent molars, one a superior posterior molar of the left side, and five inferior molars of the right side, and four of the left side. These are all obtained from that celebrated fossil bone deposit, Big-Bone Lick, Kentucky, where they were associated with the *Megalonyx, Mastodon*, &c., and are a part of a donation to the Academy by Mr. J. P. Wetherill. The external cementum is almost entirely removed, and the color, which is brown in the inferior molars, a bluish black in the superior molars, corresponds with that of their fossil associates. They are very little inferior in size, both in length and diameter, to the corresponding teeth of the recent E. caballus. The lateral diameter of the inferior molars hardly varies at all, the difference existing in the transverse diameter, which gives to the teeth a rather more compressed appearance. The superior posterior molar tooth in all species of Equus is much curved, so that but little difference is observable in this respect in the fossil specimen. The bodies of the inferior molars are considerably more curved laterally than is usually in the corresponding teeth of the recent horse, which fact, however, was not to be expected from the greater degree of curvature in the superior molars.

The enamel folds generally are more delicate, but I do no find sufficient peculiarity in their course to render them characteristic. On comparing the crowns of these fossil molar teeth, with the recent species, I find a remarkable degree of resemblance to exist, and in fact, greater differences may be found in this respect, in different individuals of

---

[4] Med. And Phys. Researches, Philada., 1835, p. 267.

[5] London, 1846, p. 398.

[6] Odontography. By R. Owen, Esq., London, 1840–45, vol. 1, p. 575.

the existing species. The posterior part of the enamel folding of the posterior tooth is rather narrower, and has a deeper groove upon the outer side than I have seen in the recent tooth. The superior molars lead to more positive results than the inferior, yet it is necessary to be very careful, for if we do perceive more differences in these particular teeth in different species, than exists in the inferior teeth, so also do we find a greater variation among them in different individuals of the same species. This variation in the same species is very striking in the case of the posterior tooth of the recent horse, as may be seen by comparing any number of specimens. In this particular tooth in the recent horse, there is always a disposition to the others. Sometimes it appears as if the disposition existed, but for want of room in the process of development of the tooth, the ordinary posterior, isolated enamel fold becomes united by an isthmus to the peripheral fold. In the fossil tooth no disposition of the kind has existed, so that it has more the appearance of the other molars, and indicates a less amount of room for development, and consequently a smaller jaw.

From the foregoing description it will be perceived that I have fixed upon no absolute characters for determining this species with any degree of accuracy, and that this is not possible, I may state upon the authority of Cuvier, who acknowledged his incompetency to find characters, "assez fixes," to pronounce upon any species of horse, examined by him, from an isolated bone,[7] and it is therefore only from their being fossil American teeth coinciding with the E. curvidens as described by Mr. Owens, more than with any other species, so far as I am capable of judging, which has made me refer them to that species.

The second species is founded upon twelve specimens of teeth which have been deposited in the cabinet of the Academy by our enterprising fellow-member, Dr. M. W. Dickeson, and is one only of the many important results of his palaeontological researches in the southwestern part of the United States. Ten of these interesting relics, consisting of five superior and five inferior molars, Dr. Dickeson states,[8] were obtained, together with remains of the *Megalonyx*, *Ursus*, the *os hominis innominatum fossile*, &c., in the vicinity of Natchez, Mississippi, from a stratum of tenacious blue clay underlying a diluvial deposit. The remaining two, both right superior posterior molars, are rolled or water-worn, and were found, as Dr. D. informs me, upon one of the Natchez Islands, in the Mississippi River. All the specimens have the exterior cementum entirely removed, with the exception of one inferior molar of the right side, in which it still exists upon the external face, and much of the inferior cementum, and part even of the dentine, is also destroyed, so that the enamel folds everywhere stand out in strong relief.

These teeth are larger than those of any species heretofore known, recent or fossil, and must have belonged to a horse, which, in point of magnitude, was a fit contemporary for the *Mastodon*, *Elephas*, &c., and worthy of the large continent which produced it, and I have therefore named it *Equus Americanus*.

[7] Cuvier, Ossemens Fossiles, 4 Ed. T. 3, p. 216.
[8] Proc. Acad. Nat. Sci., vol 3, p. 106.

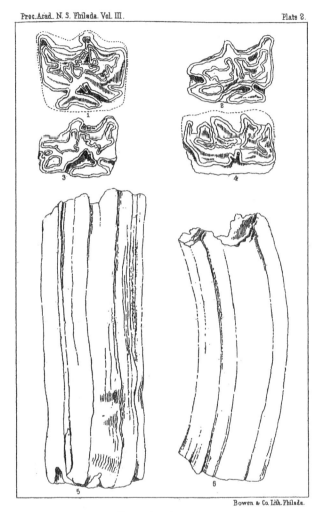

Bowen & Co. Lith. Philada.

Equus Americanus.

References to the Figures

Figs. 1 and 6, and 4 and 5, were taken from the same specimens.

Fig. 1   Crown of a superior middle molar of the left side; antero-posterior measurement 1.2 in., transverse 1.1 in.

Fig. 2.   Crown of the superior posterior molar of the right side; antero-posterior measurement 1.3 in., transverse 1.9 in.

Fig. 3.   Do. from another specimen.

Fig. 4.   Crown of an inferior middle molar of the right side; antero-posterior measurement 1.24 in., trasverse. 7 in.

Fig. 5.   Internal view of an inferior middle molar of the left side; greatest length 4.15 in; depth of its curve .15 in.

Fig. 6.   Antero-lateral view of a superior middle molar of the left side; greatest length 3.9 in.' depth of curve .3 in.

Two of the inferior molar teeth measure 4.3 inches in length, with a lateral diameter of 1.25 of an inch, and a transverse diameter of .7 of an inch. Two also, of the superior molars, measure 3.9 inches in length externally with a lateral diameter of 1.2 of an inch, and a transverse diameter of 1.1 of an inch. The inferior molars are curved from without inwards, instead of laterally, as is usual. The superior molars are curved to a degree intermediate to that of the *Equus caballus* and *Equus curvidens*.

The enamel folds are one-fourth thicker than in the recent horse and the isolated enamel folds of the superior molars are much more plicated, resembling in this respect the *Equus plicidens*, Owen. In one of the two superior posterior molars, there is an additional or third isolated enamel fold, which is oval, and two or three times larger than in the recent horse, and in the other there is a fourth, small, round, isolated enamel fold. Both of these teeth indicate a greater amount of room for development, and consequently a larger jaw. Other and considerable differences will be noticed upon comparing the figures 2 and 3, representing the crowns of these teeth, especially at the posterior part, which might lead to the supposition that they belonged to distinct species, but from the general characters of the two specimens, added to reasons before stated, relative to the amount of variation existing in the corresponding tooth of the recent horse, I cannot but think they both belong to Equus Americanus.

There is in the cabinet but one remaining tooth, to which I shall refer. This is an inferior middle molar of the left side, in an excellent state of preservation, and is a beautiful specimen; the whole of the exterior cementum being preserved without a fissure, apparently through the agency of oxide of iron and siliceous matter, which have rendered it as hard as the dentine itself. It was found with the bones of the *Mastodon*, *Megatherium*, *Harlanus*, &c., in making the excavation for the Brunswick Canal, near Darien, Georgia, and was kindly presented to the Academy by Mr. J. H. Couper.

It is straight, and although not longer than the corresponding tooth of the recent horse, with a very little increase in the diameters, character enough cannot be found in it to consider it distinct from the Equus Americanus.

Joseph Leidy, "A Flora and Fauna Within Living Animals" (excerpt), *Smithsonian Contributions to Knowledge* (1853)

THE recent excellent works by Dujardin,[1] Diesing,[2] and Robin,[3] upon animals and vegetable parasites of living animals, render another systematic record of the labors in this field almost superfluous; and the object of the present memoir is simply to give

---

[1] Histoire Naturelle des Helminthes, Paris, 1845.
[2] Systema Helminthum, Vindobonae, 1850.
[3] Des Vegetaux qui croissent sur l'homme et sur les animaux vivants, Paris, 1847.

the result of a series of observations, commenced several years ago, upon associated entozoan and entophyta, constituting a flora and fauna within animals.

The existence of entozoan, or of animals living within other species has, from the most remote times, attracted attention, on account of the peculiarity of their position, the unpleasant ideas associated with them, the sufferings they frequently induce, and the difficulty of explaining their mode or origin.

The entozoan have always constituted the strongest support of the doctrine of equivocal or spontaneous generation, one which has found distinguished disciples even to the present time; but since the days when barnacles were supposed to originate from the foam of the ocean, and ducks and geese to be developed from barnacles, this belief has been so weakened by the accumulation of facts, undenied and undeniable by the supporters of the doctrine, that it bids fair soon to be little more than an echo of the past.

The existence of vegetable parasites within animals, or of entophyta, from their minuteness, remained unknown, until the microscope of Leeuwenhoek detected the algoid filaments of the human mouth; but it is only within a comparatively recent period that any large number of distinct parasitic plants has been discovered.

The very great majority of modern observations indicate that entozoa and entophyta are produced from germs derived from parents, and having a cyclical development.

A great difficulty in determining the course of this development, particularly with entozoan, is, that their various stages of existence are passed under totally different circumstances; sometimes within one organ and then another of the same animal; sometimes in several animals; and, at other times, even quite external to and independent of the animals they infest. If, however, an entozoon preserved the same form throughout its migrations, the difficulty just mentioned would be easily over come; but such is not the case; for the alteration of form is frequently, and probably always, so great that two successive conditions cannot be recognized as the same.[4]

---

[4] Thus, almost everybody is familiar with the *Gordius*, or hair-worm, vulgarly supposed to be a transformed horse-hair. The animal is rather common in brooks and creeks in the latter part of summer and in autumn, occurring from a few inches to a foot in length. Its color passes through all shades of brown to black, and it is perfectly hairlike in its form, except that in the male the tail end is bifurcated, in the female trifurcated (American species). No one has yet been able to trace the animal it its origin! The female deposits in the water, in which it is found, millions of eggs, connected together in long cords. In the course of three weeks, the embryos escape from the eggs, of a totally different form and construction from the parents. Their body is only the 1–450th of an inch long, and consists of two portions; the posterior cylindrical, slightly dilated and rounded at the free extremity, where it is furnished with two short spines; and the anterior broader, cylindrical, and annulated, having the mouth furnished with two circlets of protractile tentaculae and a club-shaped proboscis. No one has yet been able to determine what becomes of the embryo in its normal cyclical course. Those which I have observed, always died a few days after escaping from the egg.

The grasshoppers in the meadows below the city of Philadelphia are very much infested with a species of *Gordius*, probably the same as the former, but in a different stage of development. More than

When, however, entozoan have been traced to their highest condition of development, they have always been found to possess well-characterized organs of reproduction, and the females contain such multitudes of eggs as to render it no longer surprising to find intestinal worms so frequently in vast quantities. The entophyta, when fully studied, have been satisfactorily traced to sporules.

Under the circumstances above mentioned, it is very unreasonable even to suppose the necessity of spontaneous generation for animals, which, in such very numerous instances have been proved to possess as great capabilities of reproduction as those whose cyclical development is more evident; and it remains for the supporters of the doctrine to present one single direct observation, before even its probability can be asserted.

To learn fully the nature, origin, and most favorable conditions of entozoic and entophytic life, we must commence our investigations with a clear view of the character and conditions of life in general.

An attentive study of geology proves that there was a time when no living bodies existed upon the earth.

The oblately spheroidal form of the earth, and the physical constitution of its periphery, indicate that it was once in a molten state.

A progressively increasing temperature in descending into the interior of the earth beyond the solar influence, with the phenomena of volcanoes, earthquakes, hot springs, etc., are strong evidences that the central mass of this planet yet preserves its early igneous condition.

---

half the grasshoppers in the locality mentioned contain them; but those in drier places, as in the fields west and north of Philadelphia, are quite rarely infested, for I have frequently opened large numbers without finding one worm.

The number of *Gordii* in each insect varies from one to five, their length from three inches to a foot; they occupy a position in the visceral cavity, where they lie coiled among the viscera, and often extend from the end of the abdomen forward through the thorax even into the head; their bulk and weight are frequently greater than all the soft parts, including the muscles, of their living habitation. Nevertheless, with this relatively immense mass of parasites, the insects jump about almost as freely as those not infested.

The worms are milk-white in color, and undivided at the extremities. The females are distended with ova, but I have never observed them extruded.

When the bodies of grasshoppers, containing these entozoan, are broken and lain upon moist earth, the worms gradually creep out and pass below its surface. Some specimens which crawled out of the bodies of grasshoppers, and penetrated into earth contained in a bowl, last August, have undergone no change, and are alive at the present time (November, 1852).

In the natural condition, when the grasshoppers die, the worms creep from the body and enter the earth; for, suspecting the fact, I spent an hour looking over a meadow for dead grasshoppers, and, having discovered five, beneath two of them, several inches below the surface, I found the *Gordii* which had escaped from the corpses.

Some of the worms put in water lived for about four weeks, and then died from the growth of *Achlya prolifera*. What is their cyclical development?

The facts presented in this note serve well to show the difficulties in ascertaining the developmental history of entozoan.

The period which elapsed was incalculably great before the earth-crust upon its liquid nucleus had sufficiently cooled by the radiation of its heat for living beings to become capable of existing upon its surface. Not until the temperature had been reduced below the boiling point of water (212° F.), could life have originated, for water in its liquid condition is necessary to the simplest phenomenon of life. It is even highly probable that no living thing appeared upon the earth's surface until its temperature had fallen below 165°. This ordinarily is the highest point at which albumen coagulates,[5] a substance in the liquid form, probably existent in all living beings, and essential to the performance of the simplest vital phenomenon.

Living beings, characterized by a peculiar structure and series of phenomena, appeared upon earth at a definite though very remote period.

Composed of the same ultimate elements which constitute the earth, they originated in the pre-existing materials of their structure.

Living beings originate in a formless liquid matter. The first step in organization is the appearance of a solid particle. An aggregation of organic particles constitutes the spherical, vesicular, nucleolated, nucleated body, the organic cell, the type of the physical structure or organization of living beings.

The phenomena which characterize the living being are: 1. Origin, or birth; 2. nutrition and assimilation; 3. exuration;[6] 4. development and growth; 5. reproduction; 6. death. These, in the aggregate, constitute life.

The origin or birth of a living being is the appearance of its first particle, whether directly from inorganic nature or from a parent. There is a birth to every organic cell.

Nutrition and assimilation are associated in all living actions, being coeval with the birth of a living being, and ceasing only upon its death.

During life, particles of the living structure become effete, and are removed by consumption or exuration, through the agency of the oxygen of the atmosphere. This process has been confounded with that of respiration, a function of especial organs, the lungs, branchiae, and trachaea, which exist in higher animals only; it is really secondary to the more important process of life, exuration. Exuration occurs in plants as well as in animals; in germination of the seed and in inflorescence it is very evident. In the growing plant, exuration is usually masked by the peculiar character and activity of the process of nutrition; but, at night, when the nutrition of the plant is at rest, the exuration becomes marked in the evolution of carbonic acid.

Development and growth are definite in each species of living being.

Reproduction perpetuates the structure as well as the species of the living being.

---

[5] Vegetable albumen coagulates from 140° to 160° F.; animal albumen, from 145° to 165°. – Turner's *Chemistry*, American edition, pp. 740, 744.

Albumen in the liquid state "on being heated to 140° begins to give indications of coagulating: if the solution is very dilute, the temperature may be raised to 165° with the occurrence of this change: and when present in very small quantity, the albumen may not separate till the fluid boils, or even until the ebullition has been prolonged for a short time." – Simon's *Chemistry of Man*, Am. Ed. 1846, p. 24.

[6] *Exuro*, I consume.

Death commences with life in the destruction of effete molecules of structure; it is the cessation of all life-phenomena in the individual, or is the last phenomenon of life.

To life, requires certain indispensable conditions never absent from life; always preceding it. These consist of the specific components of the living body together with the constant presence of water, air, and a definite range of temperature.

The constituent matter of living beings necessarily precedes the phenomena of life.

Without water there can be not movement to indicate life.

Air is necessary to exuration. No living being is found out of its influence. The minutest radicle of a plant never penetrates into the earth beyond the access of air.

The range of temperature necessary to life is between 35° F.[7] and 135° F.[8]

Life cannot exist independently of any one of the above-mentioned conditions. In very many instances, the removal of certain of the indispensable conditions of life-action may take place without the destruction of the power of living when these conditions are restored. Thus many plants and animals, seeds and eggs, may be dried; yet, upon supplying moisture to them, with all the other conditions, they will again present the characteristic phenomena of life.

The indispensable conditions of life are susceptible of a great variety of modification, within a definite range, without its destruction.

Accompanying a variation of the essential conditions of life, is presented the immense number of specific and individual peculiarities of living beings. The variation consists in the difference of the relative quantities of the indispensable conditions required and supplied, in addition to a modifying influence of light, probably of electricity, and possibly of some other but yet unknown agency.

The absence or presence of light is a highly important modifying condition to those indispensable to life. With the solar light, we find the green plant, which constitutes the basis of life with most terrestrial animals. Without it, the green plant and its dependent animals could not exist, but another race, now represented by certain cryptogamia, and the animal denizens of dark caverns, might inhabit the earth.

A species of plant or animal may be defined to be an immutable organic form, whose characteristic distinctions may always be recognized by a study of its history.

Any species may present individual forms not characteristic; for all, in the progress of development and course of life, are liable to modification within definite limits, which cannot be transcended without cessation of action. The original proposition is,

---

[7] The so-called red-snow, *Protococcus nivalis*, Agardh., an algous plant of polar and alpine regions, grows and reproduces only upon thawing snow, though it may be found beneath virgin snow and in a temperature far below zero; nevertheless, in such circumstances it has ceased all activity, and may remain so for a long period. The plant is remarkably indestructible. I have a specimen contained in melted snow-water, yet alive and of a red color, December, 1852, which was brought by the enterprising traveler, Dr. E. K. Kane, U.S.N., from Cape Beverly, latitude 76.10, during the Grinnell expedition for 1850–51, in search of Sir John Franklin.

[8] Certain algae grow in thermal springs, of the temperature of 117° F.

however, not affected, for no one has ever been able to demonstrate the transmutation of one species into another.

The most ordinary and extensive modifications of species from the characteristic type are presented by arrests of development. Hence, the necessity of studying the history or cyclical course of a species in order to be capable of always recognizing it.

A modification of condition beyond the range of specific life-action, must, necessarily, result in the extinction of the species.

The study of the earth's crust teaches us that very many species of plants and animals became extinct at successive periods, while other races originated to occupy their places. This probably was the result, in many cases, of a change in exterior conditions incompatible with the life of certain species, and favorable to the primitive production of others. But such a change does not always satisfactorily explain the extinction of species.[9]

Probably every species has a definite course to run in consequence of a general law; an origin, an increase, a point of culmination, a decline, and an extinction. Within this course there may occur, under the influence of ordinary circumstances, cycles of temporary increase and diminution, until, finally, the entire machine of life of the species runs down.

The historical period of man is too short to ascertain with certainty whether such a view be correct, but it appears to be favored by analogy. The power of reproduction is limited in each individual. Plants may be reproduced to an incalculable extent by cuttings, but ultimately the power to reproduce in this manner becomes exhausted. The perennial plant puts forth phyton after phyton, but the seed is necessary to its perpetuation. Numerous lower animals are reproduced to a vast extent by segmentation or allied processes, but ultimately a recurrence to sexual admixture becomes necessary for the preservation of the species.[10] Sexual admixture, limited to a few families of a

---

[9] Thus, there are numerous instances of species of animals which have become extinct, and their place supplied by others so closely allied, that it is difficult to comprehend how the exterior conditions for their existence should be so different; as in the case of the *Equus primigenius*, *E. Americanus*, &c., which have given place to the *E. caballus*, the *Bos primigenius*, whose place is supplied by the *Bos Taurus*, the *Bison latifrons* by the *Bison Americanus*, &c.

[10] Instances in favor of this view are numerous; among others, I have met with a striking example in the case of a worm, to which I have given the same of *Stylaria fossularis* (Proc. Acad. Nat. Sci. V. 287). This worm is found abundantly in ditches in the neighborhood of Philadelphia, during warm weather, and is constantly observed to be undergoing division. Individuals, a third of an inch long, are usually found to consist of two divisions, and occasionally of three, in various stages of progress towards separation. The divisions are composed of about twenty-two annulations, each possessed of a pair of fasciculi of five podal spines and two bristles. The head consists of a large lobe, with a long digit-like appendage, and presents an eye upon each side of a large mouth. The latter opens into a capacious pharynx, which afterwards contracts into a cylindrical oesophagus, continuous with a well-developed intestine, but within the animal no trace of a generative apparatus can be perceived.

In the course of a season, a single individual may reproduce some millions simply by segmentation; but as cold weather approaches, we find the animal to lose this power – not resulting from the influence

species, soon ends in their extinction. Finally, the complex living being, from birth to death, produces an immensity of living organisms, the organic cells; but the egg and seed are necessary to insure the species against extinction.

Living beings did not exist upon earth prior to their indispensable conditions of action, but wherever these have been brought into operation concomitantly, the former originated; and for such an immensity of time and vastness in quantity, have they existed, that most of the superficial rocks of the earth's crust are composed of their remains.

The stratum of life has been always subjected to the destructive agency of earthquakes, volcanoes, and torrents; but it is wonderful how soon, under the play of the life-conditions, the new surface again teems with living beings. Here and there, upon the wide area of the earth, an igneous rock peeps out as if to observe the monopoly of life, but even this, in the progress of time, has its steep sides hidden by lichens and its summit enveloped in verdue.

Of the life, present everywhere with its indispensable conditions, and coeval in its origin with them, what was the immediate cause? It could not have existed upon earth prior to its essential conditions; and is it, therefore, the result of these?

There appear to be but trifling steps from the oscillating particle of inorganic matter, to a *Bacterium*; from this to a *Vibrio*, thence to a *Monas*, and so gradually up to the highest orders of life! The most ancient rocks containing remains of living beings, indicate the contemporaneous existence of the more complex as well as the simplest of organic forms; but, nevertheless, life may have been ushered upon earth, through oceans of the lowest types, long previously to the deposit of the oldest palaeozoic rocks as know to us!!

The primitive species of living beings which appeared upon earth, and those which have been successively and periodically produced, must have been the result of pre-existing natural conditions, or the former alone originated in this manner, and the latter were the result of their transmutation under the influence of varying exterior conditions, or all species in all times originated directly through supranatural agency.

The last mode, of course, can only be an inference, in absence of all other facts; and if living beings did not originate in this way, it follows they are the result of natural conditions.

Be this as it may, the most prolonged and closest observations, and the most carefully conducted experiments have not led to the proof of a single instance of spontaneous or equivocal generation even of one of the simplest of all living beings; but, on the

---

of the cold, but from exhaustion of the power; because, even if the worms be placed in a warm situation, as in the window of a warm room, where the sun may shine upon the vessel containing them, they are observed to cease division. The loss of power of this mode of reproduction, is, however, compensated for by another succession of developments.

The worms grown to an inch in length and are composed of sixty annulations, each being provided with double the previous number of podal spines. Within the body, an androgynous generative apparatus becomes developed; within the ovaries are developed ova, and within the testis, spermatozoa. Two individuals copulate, eggs contained in bottle-shaped cases are extruded, and ultimately the parent dies.

After some weeks, in a warm situation, the ova are hatched, the young escape and move freely about, and soon commence to reproduce their numbers by division.

contrary, they all lead farther and farther from or entirely disprove it, and thus involve the whole subject in obscurity.

Schulze[11] performed an experiment to test the possibility of equivocal generation under the play of the indispensable conditions of life, free from access to any preexisting vegetable or animal germs.

A glass vessel half filled with a mixture of various dead vegetable and animal substances in water, was heated to 212° F., so as to destroy any living bodies which might exist within. To the vessel was then adapted a pair of Liebig's bulbs, one of which contained sulphuric acid, the other a solution of potassa, and through these only could the exterior air have access to its interior. The apparatus was then placed in a window, where it received the full influence of light and the necessary temperature for the production of life. The air within the vessel was daily renewed from May until August, by blowing through the sulphuric acid, from which it could suffer no change, except to be deprived of moisture and organic particles. During all that time, not even the simplest animal or vegetable forms were produced, while in an open vessel containing the same mixture, in the same situation, there were observed on the following day numerous Vibrios and Monades, and to these were soon added larger animalculae.

This interesting experiment of Schulze, I repeated with three different vessels; in one of which was a mixture of ditch-mud and confervae with water; in a second, decaying wood with water; and, in the third, a clod of earth with growing grass, earthworms, and water. Exactly under the same circumstances, they were supplied with fresh air from time to time, from July of 1850 to December of 1851, and in the end the results were the same as in the experiment of Schulze.

These experiments, however, may not be conclusive; other conditions may be required; what might have been the influence of a long-continued current of electricity, under the same circumstances, upon the mixtures?

The experiments of Crosse and Weekes, in the production of the so-called *Acarus Crossii*, were performed under the least favorable circumstances to the origination of life, and, although neither an Acarus nor a Homunculus has been created under the inspection of man, yet ridicule or prejudice should not prevent us from making every observation and experiment which can bear on the subject, in order that we may say positively whether living beings do or do not originate from the inorganic world under natural and still-existing conditions.[12]

---

[11] Notice of the Result of an Experimental Observation made regarding Equivocal Generation. By F. Schulze, Berlin. The *Edinburgh New Philosophical Journal*, vol. xxiii. 1837, p. 165.

[12] The experiments of Crosse and Weekes appear to me exceedingly absurd; for, in the first case, how were the carbon and nitrogen of the animal body to be derived by the play of a voltaic current upon a solution of silicate of potassa? If they previously existed in the water, was it not quite as probably that the ova of Acari were there also? Again, when the solution of ferrocyanide of potassium was made the womb of life by the electrical current, why could not the embryology of the new being be observed? An Acarus is a highly complex animal, presenting a well-developed tegumentary, muscular, and nervous system, and a digestive, respiratory, and generative apparatus. The gap between the inorganic world and the Acarus is greater than that between the latter and man!

To the present, we are totally unacquainted with the mode of the primitive origin of living beings!

Each species once in existence, very generally, and probably universally, requires two distinct elements, denominated sexual, for its perpetuation.

The power of reproduction by segmentation, budding, or the production of numerous successions of asexual fertile generations is, probably, in all cases limited; the species necessarily reverting to sexual admixture for its perpetuation.

The statements of Horkel[13] and Schleinden,[14] and their followers, that true sexes do not exist in the phanerogamia have been most amply refuted by the more careful observations of Amici,[15] Hugo von Mohl,[16] Carl Müller,[17] Hoffmeister,[18] Gasparini,[19] Tulasne,[20] and others.

Sexual elements have also been detected in most of the crytogamia, and, in a little time, will probably be discovered in all. They have been observed in Ferns,[21] Mosses,[22] Algae,[23] but not yet among Fungi.

Having thus taken a cursory glance at the laws of life in general, we have next to consider those especially operating in the production of parasitic life.

Within living beings, *i.e.* within their cavities or the parenchyma of the organs, of course all the indispensable conditions of life exist, and consequently we cannot wonder at their being infested with other living beings adapted to their parasitic position. Nevertheless, although the conditions of life are necessarily ever present in living beings, yet these frequently do not contain parasites. There are many circumstances besides those essential to life in general, which influence of these circumstances is the convenience or ease of access, or of entrance to the living body infested.

Within the living, closed, organic cell parasites very rarely if ever exist, because it is liquid matter only which can endosmose through cell-membrane, and, therefore, solid germs cannot enter,[24] and hence the unfrequency of true entozoan in vegetables. Entozoa

---

[13] Monatsb.d. Berline. Adad. 1836.

[14] Archiv F. Naturges. 1837; Nova Acta C. L. C. Acad. Nat. Cur. Vol. xix. 1839, p. 27; Grundzuge d. Wissenschaftliche Botanik.

[15] Giornale botanico Italiano, An. 2; Annales des Sciences Naturelles, Botanique, 1847, t. vii. P. 193.

[16] Botanische Zeitung, 1847; An. Sc. Nat., Bot. t. ix. 1848.

[17] Ibid.

[18] Ibid. and t. xi. 1849.

[19] Ibid.

[20] Ibid. t. xii. 1849.

[21] Tulasne, ibofid. t. xi. P. 5; ibid. p. 114.

[22] Schimper: Recherches Anatom. Et Morpholog. Sur les Mousses. Strasbourg, 1848. This author says, page 55: *"Mais je tiens le fait que jamais une mousse ne parvient a la fructification quand elle se trouve hors de l'influence des organs que je considere comme des organes males."*

[23] Thwaites, in *Annals and Magazine of Natural History*, 1847, vol. xx. p. 9; ibid. p. 343; ibid. 1848, 2d ser. vol. i. p. 161.

[24] In some experiments upon the endosmosis of solid matter through organic cell-membrane, I found that particles of carmine, diffused in water, which I estimated to measure about the 52,000th of an inch, would no more penetrate the cell-membrane than the largest masses.

may and do penetrate through living tissues, but it is entirely by the mechanical process of boring.

The intestinal canal of animals is most frequently infested by entoparasites on account of the ease with which their germs enter with the food.

Aquatic animals are more troubled by entozoan than those which are terrestrial, because the water affords a better medium of access than the air.

Terrestrial animals, on the other hand, are more infested by ectoparasites because their covering of hair, wool, and feathers is more favorable to their protection and reproduction. A low degree of organic activity and slowly digestible food favor the development of entoparasites, and hence they are more frequently in the relatively sluggish herbivora than in the carnivora. Comparatively indigestible food, and such as contains but a small proportion of nutritive matter, from its long retention in the alimentary canal, favors the development of entozoic and entophytic germs more than that in which the contrary conditions prevail.

Animals subsisting upon the endosmosed juices of the tissues of other animals, and of plants, are rarely infested by parasites, as in the case of the hemipterous insects, aphides, etc., because such food is necessarily free from parasites or their germs. Entozoa themselves, on this account, are not infested.

On the other hand, if the liquid food be open to the air, parasitic germs may be readily introduced into the animal, as in the case of the common house-fly, which often contains myriads of a species of *Bodo*.

Food swallowed in large morsels favors the introduction of attached parasites; hence these are frequently found in reptiles, and even in birds, which are, among the vertebrata, of the highest organic activity.

Animals of feeble organic activity using solid food, which is very slowly digested, and contains little nutriment, are rarely free from parasites. This is the case with the coleopterous insect *Passalus*, and the myriapod *Julus*.

Cooking food is of advantage in destroying the germs of parasites; and hence man, notwithstanding his liability to the latter, is less infested than most other mammalia. Did instinct originally lead him to cook his food, to avoid the introduction of parasites?

Entozoa are more abundant than entophyta, because the power of voluntary movement favors them in their transmigrations, and renders them less liable to expulsion from the intestinal canal.

Influence of Parasites in the Production of Disease. – In many animals entozoan and entophyta are almost never absent, and probably when in their natural habitation, and few in number, or not of excessive size, are harmless, as observed by Dujardin in the introduction of his excellent work on Intestinal Worms: "Les helminthes se développent dans un site qui leur convient, sans nuire plus que les lichens sur l'écorce d'un affaiblissement provenant d'une tout autre cause, d'une mauvaise alimentation, du séjour dans un lieu froid et humide, etc.: sans cela, les helminthes naissent et meurent dans le corps de leurs hôtes, et peuvent paraître et disparaître alternativement sans inconvénients."

Many important diseases have been supposed to originate from parasitic animals and vegetables. The former are not the true entozoan, for these are too large, and may be detected by the naked eye, but they are considered to be animalculae so small that they cannot be discovered even with the highest powers of the microscope. But, independent of the face that the existence of such entities is a mere suspicion, none of the well-known animalculae are poisonous. At various times, I have purposely swallowed large draughts of water containing myriads of *Monas, Vibrio, Euglenia, Volvox, Leucophrys, Paramecium, Vorticella,* etc. without ever having perceived any subsequent effect.

The production of certain diseases, however, through the agency of entophyta, is no longer a subject of doubt; as in the case of Muscardine in the Silk-worm, the Mycoderm of Porrigo favosa in Man, etc.; but that malarial and epidemic fevers have their origin in cryptogamic vegetables or spores requires yet a single proof.[26] If such were the case, these minute vegetables or spores, conveyed through the air, and introduced into the body in respiration, could be detected. The minutest of all known livings beings is the *Vibrio lineola* of Müller, measuring only the 36,000th of an inch, and the smallest known vegetable spore is very much larger than this, whilst particles of inorganic matter can be distinguished the 200,000th of an inch in size.

I have frequently examined the rains and dews of localities in which intermittents were epidemic upon the Schuylkill and Susquehanna Rivers, but without being able to detect animalculae, spores, or even any solid particles whatever. I have examined the air itself for such bodies, by passing a current through clear water. This was done by means of a bottle, with two tubes passing through a cork stopper; one tube dipping into the water, the other reaching not quite to its surface. By sucking upon the latter tube, a current of air passed through the former, and was deprived in its course of any solid particles. Ordinarily, when the atmosphere was still, early in the morning, or in the evening, neither spores nor animalcules could be detected. When piles of decaying sticks or dry leaves were stirred up, or the dust was blown about by the wind, a host of most incongruous objects could be obtained from the air; none, however, which could be supposed capable of producing disease.

To assert, under these circumstances, that there are spores and animalculae capable of giving rise to epidemics, but not discernible by any means at our command, is absurd, as it is only saying in other words that such spore and animalculae are liquid and dissolved in the air, or in a condition of chemical solution. That the air may be poisoned by matters incapable of detection by the chemist is proved by the emanations from such plants as the *Rhus vernix, Hippomane mancinella,* etc.

---

[25] The inhabitants of the United States appear to be less infested with entozoan than those of any other part of the world. This probably arises to a great extent from the more nutritious character of their food; even the poorest laborer being daily supplied with abundance of wholesome flesh, producing a tendency to high organic activity, which is unfavorable to parasitic development.

[26] See an ingenious little work by my distinguished friend Dr. J. K. Mitchell, "On the Cryptogamous Origin of Malarious and Epidemic Fevers."

JOHN WILLIAM DRAPER (1811–1882)

Chemist, photographer, historian.
Born in St. Helens, Merseyside, England. Died in Hastings, New York.

EDUCATED at home with tutors before attending the Woodhouse Grove School, Draper read chemistry at the University of London until the death of his father. He then immigrated with his family to the United States and resumed his studies at the medical school of the University of Pennsylvania. In 1839 he became professor of chemistry at New York University and helped found its school of medicine. He served as professor of chemistry and physiology at the medical school from 1840–50, as its president from 1850–73, and then professor of chemistry again until 1881. Early on Draper's research focused on radiant energy, and his study of the effect of light upon chemicals led him to photography. He made improvements on Daguerre's process that lead to portrait photography, and a daguerreotype he took of his sister in 1840 is to this day the oldest surviving photographic portrait. In the field of photochemistry, he developed the proposition (1842) that only light rays that are absorbed can produce chemical change. It came to be known as the Grotthuss-Draper law when his name was teamed with that of a lesser known Baltic chemist Theodor Grotthuss. He was the first to take astrophotographs and the first to take a photograph of the moon which actually showed lunar features. He was a pioneer in microphotography and was actively engaged with Samuel F. B. Morse in the production of the electromagnetic telegraph. Sometime in the 1850s he began to turn his attention to history, writing *The History of the Intellectual Development of Europe* (1862), the basic thesis of which was that history is the result of physical realities–climate, soil, natural resources–interacting with human activity. He viewed history as a science and society as an organism striving within the larger ecological realities of the natural world. While his knowledge of history was not as rigorous or as deep as his knowledge of science, all his history books sold well because of the vividness of his style and his frank analysis of history. His other history works were *A History of the American Civil War* (1867–70, 3 vols) and the widely read and controversial *History of the Conflict between Religion and Science* (1874). Interestingly, a paper entitled "On the Intellectual Development of Europe, considered with Reference to the Views of Mr. Darwin and others, that the Progression of Organisms is determined by law," which Draper read at the famous 1860 meeting of the British Association, was the opening act to Huxley's bearding of Bishop Wilberforce. A writer for *Macmillan's Magazine* recalling the historical debate mentioned that he found the presentation of the American Dr. Draper "somewhat dry." Surely any opening act would have seemed "dry" in comparison to what followed. In 1875 Draper received the Rumford medal from the American academy of science and arts for his research in "Radiant Energy." He published his *Scientific Memoirs* in

1878. His son Henry Draper, following in his father's footsteps, would make important contributions in spectroscopy but sadly only survived his father by a year.

John William Draper, "Examination of the Radiations of Red-Hot Bodies. The Production of Light by Heat" (excerpt), *American Journal of Science and Arts* (1847)

ALTHOUGH the phenomenon of the production of light by all solid bodies, when their temperature is raised to a certain degree, is one of the most familiar, no person so far as I know has hitherto attempted a critical investigation of it. The difficulties environing the inquiry are so great that even among the most eminent philosophers a diversity of opinion has prevailed respecting some of the leading facts. Thus Sir Isaac Newton fixed the temperature at which bodies become self-luminous as 635°; Sir Humphrey Davy at 812°; Mr. Wedgwood at 947°; and Mr. Daniel at 980°. As respects the nature of the light emitted, there are similar contradictions. In some philosophical works of considerable repute it is stated that when a solid begins to shine it first emits red and then white rays; in others it is asserted that a mixture of blue and red light is the first that appears.

I have succeeded in escaping or overcoming many of the difficulties of this problem, and have arrived at satisfactory solutions of the main points; and as the experiments now to be described lead to some striking and perhaps unexpected analogies between light and heat, they commend themselves to our attention, as having a bearing on the question of the identity of those principles. It is known that heretofore I have been led to believe in the existence of cardinal distinctions not only between these, but also other imponderable agents, and I may therefore state that when this investigation was first undertaken it was in the expectation that it would lead to results very different from those which have actually arisen.

The following are the points on which I propose to treat:

1. To determine the point of incandescence of platinum, and to prove that different bodies become incandescent at the same temperature.
2. To determine the color of the rays emitted by self-luminous bodies at different temperatures. This is done by the only reliable method – analysis by the prism.
   From these experiments it will appear that as the temperature rises the light increases in refrangibility; and making due allowance for the physiological imperfection of the eye, the true order of the colors in red, orange, yellow, green, blue, indigo, violet.
3. To determine the relation between the brilliancy of the light emitted by a shining body and its temperature.

★ ★ ★

I next pass to the third branch of this investigation – to examine the relation between the temperatures of self-luminous bodies and the intensity of the light they emit, premising it with the following considerations:

The close analogy which has been traced between the phenomena of light and radiant heat lends countenance to the supposition that the law which regulates the escape of heat from a body will also determine its rate of emission of light. Sir Isaac Newton supposed that while the temperature of a body rose in an arithmetical progression, the amount of heat escaping from it increased in a geometrical progression. The error of this was subsequently shown by Martin, Erxleben, and Delaroche, and finally Dulong and Petit gave the true law: "When a hot body cools in vacuo, surrounded by a medium the temperature of which is constant, the velocity of cooling for excess of temperature in arithmetical progression increases as the terms of a geometrical progression, diminished by a constant quantity." The introduction of this constant depends on the operation of the theory of the exchanges of heat; for a body when cooling under the circumstances have supposed is simultaneously receiving back a constant amount of heat from the medium of constant temperature.

While Newton's law represents the rate of cooling of bodies, and therefore the quantities of heat they emit when the range of temperature is limited, and the law of Dulong and Petit holds to a wider extent, there are in the present inquiry certain circumstances to be taken into account not contemplated by those philosophers. Dulong and Petit, throughout their memoir, regard radiant heat as a homogeneous agent, and look upon the theory of exchanges, which is indeed their starting point and guide, as a very simple affair. But the progress of this department of knowledge since their time has shown that precisely the same modifications found in the colors of light occur also for heat; a fact conveniently designated by the phrase "ideal coloration of heat," and further, that the wave-length of the heat emitted depends upon the temperature of the radiating source. It is one thing to investigate the phenomena of the exchanges of heat-rays of the same color, and another when the colors are different. A complete theory of the exchanges of heat must include this principle, and, of course, so too must a law of cooling applicable to any temperature.

There is another fact to some extent considered by Dulong and Petit, but not of such weight in their investigations, where the range of temperature was small, as in these where it rises as high as nearly 3000° Fahr. This is the difference of specific heat of the same body at different temperatures. At the high temperatures herein employed, there cannot be a doubt that the capacity of platinum for heat is far greater than that at a low point. This, therefore, must affect its rate of calorific emission, and probably that for light also.

From these and similar considerations we should be led to expect that as the temperature of an incandescent solid rises, the intensity of the light emitted increases very rapidly.

I pass not to the experimental proofs which substantiate the foregoing reasoning.

The apparatus employed as the source of the light and measure of the temperature was the same as in the preceding experiments – a strip of platinum brought to a known temperature by the passage of a voltaic current of the proper force, and connected with an index which measured its expansion.

The principle upon which the intensities of the light were determined was that originally described by Bouguer, and subsequently used by Masson. After many experiments, I found that it is the most accurate method known.

Any one who will endeavor to determine the intensities of light by Rumford's method of contrasting shadows, or by that of equally illuminated surfaces, will find, when every precaution has been used, that the results of repeated experiments do not accord. There is, moreover, the great defect that when the lights differ in color, it is impossible to obtain reliable results except by resorting to such contrivances as that described in the *Philosophical Magazine*, August, 1844.

Bouguer's principle is far more exact; and where the lights differ in color, that difference actually tends to make the result more correct. As it is not generally known, I will indicate the nature of it briefly:

Let there be placed at a certain distance from a sheet of white paper a candle, so arranged as to throw the shadow of an opaque body, such as a rod of metal, on the sheet. If a second candle be placed also in front of the paper and nearer than the former, there is a certain distance at which its light completely obliterates all traces of the shadow. This distance is readily found, for the disappearance of the shadow can be determined with considerable exactness. When the lights are equal, Bouguer ascertained that the relative distances were as 1:8, and therefore inferred correctly that in the case of his eye the effect of a given light was imperceptible when it was in presence of another sixty-four times as intense. The precise number differs according to the sensibility of different eyes, but for the same organ it is constant.

Upon a paper screen I threw the shadow of a rod of copper, which intercepted the rays of the incandescent platinum; then taking an Argand lamp, surrounded by a cylindrical metal shade, through an aperture in which the light passed, and the flame of which I had found by previous trial would continue for an hour of almost the same intensity, I approached it to the paper sheet, until the shadow cast by the copper disappeared. The distance at which this took place was then measured, and the temperature of the platinum determined.

The temperature of the platinum was now raised, the shadow became more intense, and it was necessary to bring the Argand lamp nearer before it was effaced. When this took place, the distance of the lamp was again measured, and the temperature of the platinum again determined.

In Fig. 5, *a b* is the strip of ignited platinum. It casts a shadow, *h*, of the metal rod *e* on the white screen *f g*; *c* is a metallic cylinder containing an Argand lamp, the light of which issues through an aperture, *d*, and extinguishes the shadow on the screen.

In this manner I obtained several series of results, one of which is given in the following table. They exhibited a more perfect accordance among each other than I had anticipated.

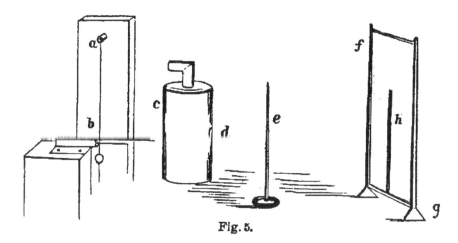

Fig. 5.

The intensity of the light of the platinum is of course inversely proportional to the square of the distance of the Argand lamp at the moment of the extinction of the shadow.

Table of the Intensity of Light emitted by Platinum at Different Temperatures.

Temperature of the platinum.	Distance of Argand lamp.		Mean.	Intensity of light.
	Experiment I.	Experiment II.		
980°	. . . .	. . . .	. . . .	0.00
1900	54.00	54.00	54.00	0.34
2015	39.00	41.00	40.00	0.62
2130	24.00	24.00	24.00	1.73
2245	18.00	19.00	18.50	2.92
2360	14.50	15.50	15.00	4.40
2475	11.50	12.00	11.75	7.24
2590	9.00	9.00	9.00	12.34

In this table the first column gives the temperatures under examination in Fahrenheit degrees; the second and third the distances of the Argand lamp from the screen in English inches, in two different sets of experiments; the fourth the mean of the two; and the fifth the corresponding intensity of the light.

The results thus obtained proved that the increase in the intensity of the light of the ignited platinum, though slow at first, became very rapid as the temperature rose. At 2590° the brilliancy of the light was more than thirty-six times as great as it was at 1900°.

Thus, therefore, the theoretical anticipation founded on the analogy of light and heat was completely verified, the emission of light by a self-luminous solid as its temperature rises being in greater proportion than would correspond to mere difference of temperature.

To place this in a more striking point of view, I made some corresponding experiments in relation to the heat emitted. No one thus far had published results for high temperatures, or had endeavored to establish through an extensive scale the principle of Delaroche, that "the quality of heat which a hot body gives off in a given time, by way of radiation to a cold body situated at a distance, increases, other things being equal, in a progression more rapid than the excess of the temperature of the first above that of the second."

As the object thus proposed was mainly to illustrate the remarkable analogy between light and heat, the experiments now to be related were arranged so as to resemble the foregoing; that is to say, as in determining the intensities of light emitted by a shining body at different temperatures, I had received the rays upon a screen placed at an invariable distance, and then determined their value by photometric methods, so in this case I received the rays of heat upon a screen placed at an invariable distance, and measured their intensity by thermometric methods. In this instance the screen employed was, in fact, the blackened surface of a thermo-electric pile. It was arranged at a distance of about one inch from the strip of ignited platinum, a distance sufficient to keep it from any disturbance from the stream of hot air arising from the metal; care was also taken that the multiplier itself was placed so far from the rest of the apparatus that its astatic needles could not be affected by the voltaic current igniting the platinum, or the electro-magnetic action of the wires or rheostat used to modify the degrees of heat.

In Fig. 6, *a b* is the ignited platinum strip, *c* the thermo-electric pile, *d d* the multiplier.

The experiments were conducted as follows: The needles of the thermo-multiplier standing at the zero of their scale, the voltaic current was passed through the platinum,

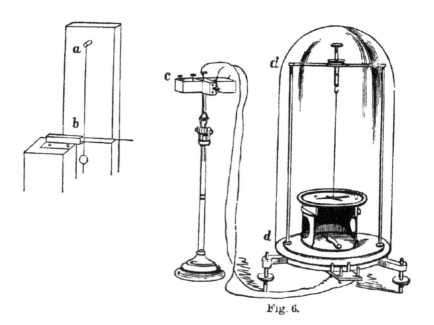

Fig. 6.

which immediately rose to the corresponding temperature, and radiated its heat to the face of the pile. The instant this current passed, the needles of the multiplier moved, and kept steadily advancing on the scale. At the close of one minute the deviation of the needle and the temperature of the platinum were simultaneously noted, and then the voltaic current was stopped.

Sufficient time was now given for the needles of the multiplier to come back to zero. This time varied in the different cases, according to the intensity of the heat to which the pile has been exposed; in no instance, however, did it exceed six minutes, and in most cases was much less. A little consideration will show that the usual artifice employed to drive the needles back to zero by warming the opposite face of the pile was not admissible in these experiments.

The needles having regained their zero, the platinum was brought again to a given temperature, and the experiment conducted as before. The following table exhibits a series of these results.

In this table the first column given the temperatures of the platinum in Fahrenheit degrees; the second and third, two series of experiments expressing the arc passed over by the needle at the close of a radiation lasting one minute, each number being the mean of several successive trials, and the fourth the mean of the two. It therefore gives the radiant effect of the incandescent platinum on the thermo-multiplier for the different temperatures.

Of course it is understood that I here take the angular deviations of the needle as expressing the force of the thermo-electric current, or, in other words, as being proportional to the temperatures. This hypothesis, it is known, is admissible.

Table of the Intensity of Radiant Heat emitted by Platinum at Different Temperatures.

Temperature of the platinum.	Intensity of heat emitted.		Mean.
	Experiment I.	Experiment II.	
980°	.75	1.00	.87
1095	1.00	1.20	1.10
1210	1.40	1.60	1.50
1325	1.60	2.00	1.80
1440	2.20	2.20	2.20
1555	2.75	2.85	2.80
1670	3.65	3.75	3.70
1785	5.00	5.00	5.00
1900	6.70	6.90	6.80
2015	8.60	8.60	8.60
2130	10.00	10.00	10.00
2245	12.50	12.50	12.50
2360	15.50	15.50	15.50

It therefore appears that if the quantity of heat radiated by platinum at 980° be taken as unity, it will have increased at 1440° to 2.5, at 1900° to 7.8, and at 2360° to 17.8, nearly. The rate of increase is, therefore, very rapid. Further, it may be remarked, as illustrative of the same fact, that the quantity of heat radiated by a mass of platinum in passing from 1000° to 1300° is nearly equal to the amount it gives out in passing from common temperatures up to 1000°.

I cannot here express myself with too much emphasis on the remarkable analogy between light and heat which these experiments reveal. The march of the phenomena in all their leading points is the same in both cases. The rapid increase of effect as the temperature rises is common to both.

It is not to be forgotten, however, that in the case of light we necessarily measure its effects by an apparatus which possesses special peculiarities. The eye is insensible to rays not comprehended within certain limits of refrangibility. In these experiments it is requisite to raise the temperature of the platinum almost to 1000° before we can discover the first traces of light. Measures obtained under such circumstances are dependent on the physiological action of the visual organ itself, and hence their analogy with those obtained by the thermometer becomes more striking, because we should scarcely have anticipated that it could be so complete.

Among writers on Optics it has been a desideratum to obtain an artificial light of standard brilliancy. The preceding experiments furnish an easy means of supplying that want, and give us what might be termed a "unit lamp." A surface of platinum of standard dimensions, raised to a standard temperature by a voltaic current, will always emit a constant light. A strip of that metal, one inch long and 1/20 of an inch wide, connected with a lever by which its expansion might be measured, would yield at 2000° a light suitable for most purposes. Moreover, it would be very easy to form from it a photometer by screening portions of the shining surface. An ingenious artist would have very little difficulty, by taking advantage of the movements of the lever, in making a self-acting apparatus in which the platinum should be maintained at a uniform temperature, notwithstanding any change taking place in the voltaic current.

LOUIS AGASSIZ (1807–1873)

Ichthyologist, geologist, naturalist, organismal biologist.
Born in Montier, Switzerland. Died in Boston, Massachusetts.

Jean Louis Rodolphe Agassiz studied at some of the finest universities in Europe: Zurich, Heidelberg, Munich, and Erlangen. By the time he was 23 he had earned both a doctor of philosophy (natural history) and a medical degree. Natural history was his true calling, the medical degree a nod to his family's aspirations. Before he had completed either, however, he was chosen to describe a collection of freshwater fishes from Brazil gathered by the botanist von Martius. Agassiz had completed and published this set of specimens by 1829 and decided that he would become the leading authority on fish in Europe. After completing school, he moved to Paris in order to study comparative fish anatomies at the museums and collections there, and he soon met Humboldt and Cuvier. Cuvier was impressed with the young go-getter and gave Agassiz his own notes for a planned study of fossil fish. When Cuvier died soon after, Agassiz took on the mantle of intellectual heir to Cuvier. For the rest of his life, he would consistently (and stubbornly) defend Cuvier's geological catastrophism and classification of animals. In 1832 he returned to Switzerland to accept the professorship of natural history at Neuchatel. During the 13 years that he taught at the Lyceum of Neuchatel, Agassiz would publish his *History of the Freshwater Fishes of Central Europe* (1839–42) and his monumental *Recherches sur les poissons fossiles* (5 vols; 1833–43) which described over 1700 fossil fish. These two works alone would have made him famous throughout Europe, but then Agassiz took up the study of glaciers. He noticed that in his native Switzerland there were signs of glaciation where no glaciers existed. Previously these had been explained away as remnants of floods or icebergs, but Agassiz synthesized all the known facts on the subject (particularly those of his friend Charpentier), made several trips to the Alpine region, and even built a hut and lived a while upon one of the Aar glaciers to better investigate structure and movement. The result was *Etudes sur les glaciers* (1840), a work that not only fit the facts on the ground–so to speak–but also supported Agassiz's view of catastrophism. In 1842–1846, his *Nomenclator Zoologicus* was published, a huge undertaking of all names used in zoology for genera and groups. Agassiz had achieved the recognition that he had set out to in 1829 and more; unfortunately, this accomplishment did not protect him from a number of personal set backs that culminated about this time. His marriage was in trouble and so was his financial situation, due in part to a failed publishing venture. When his benefactor, the King of Prussia, offered to finance an expedition to America to continue his fish work, Agassiz seized the opportunity and arranged to also give a series of lectures for the Lowell Institute in Boston. From the beginning Agassiz was taken with

American and the new science and life opportunities it offered him, and America was equally taken with what he represented: European enthusiasm over American science. Agassiz had the intellect, skill, and personality to make science seem at once erudite and populist. That first year in America he traveled widely and made a tidy sum from a series of standing-room-only lectures. He soon accepted a professorship in zoology and geology at Harvard and began to settle in for the second phase of his life. Straight off he began a lecture tour and research expedition of the Lake Superior region (*Lake Superior: Its Physical Character, Vegetation and Animals, Compared with those of Other and Similar Regions*) and upon his return learned that his estranged wife had died. Within the next two years he would marry into the Boston Brahmin world of Elizabeth Cabot Cary, send for his three children, and settle once and for all in America. In the following decade he would be awarded the *Prix Cuvier* by the French Academy of Sciences; sell his large private collection of fossils to Harvard and begin the fund-raising and cajoling that would lead to the building of a new Museum of Comparative Zoology at Harvard; pen numerous papers and reviews, both scientific and other; write the first two (and only) volumes of his planned 10 volume *Contributions to the Natural History of the United States*; and prove himself a master at raising money and his own profile. Unfortunately, during this decade Agassiz also severely reviewed Darwin's *Origin of Species* and dismissively viewed the theory of natural selection as a fad that would pass, taking a rather superior, deliberative tone in defense of the "fixity of species" that did not set well with many of his fellow scientists. He also supported theories of the "plural origins of mankind," which played into the hands of racial supremacists and the pro-slavery forces. Agassiz may have felt that the tensions between North and South were a passing fad too – he proved strongly pro-Union once the war began his previous complicity with Southern apologists notwithstanding. Both stances, on fixity of species and plural origins, proved to slowly poison his standing within the scientific community. More and more after the war his writings addressed a general, nonscientific audience. He began to focus more on expeditions and travel writing as science. As his legendary, seemingly inexhaustible, energy began to fail in the early 1870s, he turned his focus on the completion of his Museum of Comparative Zoology. Although he made a few acknowledgments that evolution was not a fad, he could never really give up his belief in the world view of Cuvier and his youthful conquest of Europe.

Louis Agassiz, "Section IX: Range of the Geographical Distribution of Animals," *Essay on Classification* (1857)

THE surface of the earth being partly formed by water and partly by land, and the organization of all living beings standing in close relation to the one or the other of these mediums, it is in the nature of things that no single species, either of animals or plants, should be uniformly distributed over the whole globe. Yet there are some types of the animal as well as of the vegetable kingdom which are equably distributed over the whole surface of the land, and others which are as widely scattered in the sea, while others are limited to some continent or some ocean, to some particular province, to some lake, nay, to some very limited spot of the earth's surface.[1]

As far as the primary divisions of animals are concerned, and the nature of the medium to which they are adapted does not interfere, representatives of the four great branches of the animal kingdom are everywhere found together. Radiata, Mollusks, Articulata, and Vertebrata occur together in every part of the ocean, in the Arctics, as well as under the equator, and near the southern pole as far as man has penetrated; every bay, every inlet, every shoal is haunted by them. So universal is this association, not only at present but in all past geological ages, that I consider it as a sufficient reason to expect that fishes will be found in those few fossiliferous beds of the Silurian System in which thus far they have not yet been found.[2] Upon land we find equally everywhere Vertebrata, Articulata, and Mollusks, and but no Radiata, this whole branch being limited to the waters; but as far as terrestrial animals extend, we find representatives of the other three branches associated, as we find them all four in the sea. Classes have already a more limited range of distribution. Among Radiata, the Polypi, Acalephs, and Echinoderms are not only all aquatic, they are all marine, with a single exception,[3] the genus Hydra, which inhabits fresh waters. Among Mollusks, the Acephala are all aquatic, but partly marine and partly fluviatile, the Gasteropoda partly marine, partly fluviatile and partly terrestrial, while all Cephalopoda are marine. Among Articulata, the Worms are partly marine, partly fluviatile, and partly terrestrial, while many are internal parasites, living in the cavities or in the organs of other animals; the Crustacea are partly marine and partly fluviatile, a few are terrestrial; the Insects are mostly terrestrial or rather aerial, yet some are marine, others fluviatile, and a large number of those, which in their perfect state live in the air, are terrestrial or even aquatic during their earlier stages of growth. Among Vertebrata the Fishes are all aquatic, but partly marine and partly fluviatile; the Reptiles are either aquatic during the early part of their life; the Birds are all aerial, but some more terrestrial and others more aquatic; finally, the Mammalia, though all aerial, live partly in the sea, partly in fresh water, but mostly

---

[1] The human race affords an example of the wide distribution of a terrestrial type; the Herring and the Mackerel families have an equally wide distribution in the sea. The Mammalia of New Holland show how some families may be limited to one continent; the family of *Labyrinthici* of the Indian Ocean, how fishes may be circumscribed in the sea, and that of the Goniodonts of South America in the fresh waters. The Chaca of Lake Baikal is found nowhere else. This is equally true of the Blindfish (*Amblyopsis*) of the Mammoth Cave, and of the *Proteus* of the caverns of Carinthia.

[2] See above, Sect. VII.

[3] I need hardly say in this connection that so-called fresh-water Polyps, Alcyonella, Plumatella, etc., are Bryozoa, and not true Polyps.

upon land. A more special review might show that this localization in connection with the elements in which animals live has a direct reference to peculiarities of structure of such importance, that a close consideration of the habitat of animals within the limits of the classes might in most cases lead to a very natural classification.[4] But this is true only within the limits of the classes, and even here not absolutely, as in some the orders only, or the families only are thus closely related to the elements; there are even natural groups in which this connection is not manifested beyond the limits of the genera, and a few cases in which it is actually confined to the species. Yet in every degree of these connections we find that upon every spot of the globe it extends simultaneously to the representatives of different classes and even of different branches of the animal and vegetable kingdoms; a circumstance which shows that when called into existence in such an association, these various animals and plants were respectively adapted, with all the peculiarities of their kingdom, those of their class, those of their order, those of their genus, and those of their species, to the home assigned to them, and therefore not produced by the nature of the place, or of the element, or any other physical condition.[5] To maintain the contrary would really amount to asserting that wherever a variety of organized beings live together, no matter now great their diversity, the physical agents prevailing there must have in their combined action the power of producing such a diversity of structures as exists in animals, notwithstanding the close connection in which these animals stand to them, or to work out an intimate relation to themselves in beings, the essential characteristics of which have no reference to their nature. In other words, in all these animals and plants there is one side of their organization which has an immediate reference to the elements in which they live, and another which has no such connection, and yet it is precisely this part of the structure of animals and plants which has no direct bearing upon the conditions in which they are placed in nature, which constitutes their essential, their typical character. This proves beyond the possibility of an objection that the elements in which animals and plants live (and under this expression I mean to include all that is commonly called physical agents, physical causes, etc.) cannot in any way be considered as the cause of their existence.

If the naturalists of past centuries have failed to improve their systems of Zoology of introducing considerations derived from the habitat of animals, it is chiefly because they have taken this habitat as the foundation of their primary divisions; but reduced

[4] Agassiz, "The Natural Relations between Animals and the Elements in which They Live," *American Journal of Science*, IX (2d ser., 1850), 369–394.

[5] In the study of the geographical distribution of animals and plants and their relations to the conditions under which they live, too little importance is attached to the circumstance that representations of the most diversified types are everywhere found associated, within limited areas, under identical conditions of existence. These combinations of numerous and most heterogeneous types, under all possible variations of climatic influences, severally circumscribed within the narrowest limits, seems to me to present the most insuperable objection to the supposition that the organized beings, so combined, could in any way have originated spontaneously by the working of any natural law.

to its proper limits, the study of the connection between the structure and the natural home of animals cannot fail to lead to interesting results, among which the growing conviction that these relations are not produced by physical agents, but determined in the plan ordained from the beginning, will not be the least important.

The unequal limitation of groups of a different value upon the surface of the earth produces the most diversified combinations possible, when we consider the mode of association of different families of animals and plants in different parts of the world. These combinations are so regulated that every natural province has a character of its own, as far as its animals and plants are concerned, and such natural associations of organized beings extending over a wider or narrower area are called *Faunae* when the animals alone are considered, and *Florae* when the plants alone are regarded. Their natural limits are far from being yet ascertained satisfactorily everywhere. As the works of Schouw and Schmarda may suffice to give an approximate idea of their extent,[6] I would refer to them for further details and allude here only to the unequal extent of these different faunae and to the necessity of limiting them in different ways, according to the point of view under which they are considered, or rather show that as different groups have a wider or more limited range, in investigating their associations or the faunae, we must distinguish between zoological realms, zoological provinces, zoological counties, zoological fields, as it were; that is, between zoological areas of unequal value over the widest of which range the most extensive types, while in their smaller and smaller divisions we find more and more limited types, sometimes overlapping one another, sometimes placed side by side, sometimes concentric to one another, but always and everywhere impressing a special character upon some part of a wider area, which is thus made to differ from that of any other part within its natural limits.

These various combinations of smaller or wider areas, equally well defined in different types, has given rise to the conflicting views prevailing among naturalists respecting the natural limits of faunae; but with the progress of our knowledge these discrepancies cannot fail to disappear. In some respect every island of the Pacific upon which distinct animals are found may be considered as exhibiting a distinct fauna, yet several groups of these islands have a common character which unites them into more comprehensive faunae, the Sandwich Islands, for instance, compared to the Fejees or to New Zealand. What is true of disconnected islands or of isolated lakes is equally true of connected parts of the mainland and of the ocean.

Since it is well known that many animals are limited to a very narrow range in their geographical distribution, it would be a highly interesting subject of inquiry to ascertain what are the narrowest limits within which animals of different types may be circumscribed, as this would furnish the first basis for a scientific consideration of the conditions under which animals may have been created. The time is passed when

---

[6] I would also refer to a sketch I have published of the Faunae ("Sketch of the Natural Provinces of the Animal World and Their Relation to the Different Types of Man," in J. C. Nott and George R. Gliddon, *Types of Mankind*, Philadelphia, 1854, accompanied with a map and illustrations, pp. lvii–lxxviii).

the mere indication of the continent whence an animal had been obtained could satisfy our curiosity; and the naturalists who, having an opportunity of ascertaining closely the particular circumstances under which the animals they describe are placed in their natural home, are guilty of a gross disregard of the interest of science when they neglect to relate them. Our knowledge of the geographical distribution of animals would be far more extensive and precise than it is now, but for this neglect. Every new fact relating to the geographical distribution of well-known species is as important to science as the discovery of a new species. Could we only know the range of a single animal as accurately as Alphonse de Candolle has lately determined that of many species of plants, we might begin a new era in Zoology. It is greatly to be regretted that in most works containing the scientific results of explorations of distant countries only new species are described, when the mere enumeration of those already known might have added invaluable information respecting their geographical distribution. The carelessness with which some naturalists distinguish species merely because they are found in distant regions, without even attempting to secure specimens for comparison, is a perpetual source of erroneous conclusions in the study of the geographical distribution of organized beings, not less detrimental to the progress of science than the readiness of others to consider as identical animals and plants which may resemble each other closely, without paying the least regard to their distinct origins, and without even pointing out the differences they may perceive between specimens from different parts of the world. The perfect identity of animals and plants living in very remote parts of the globe has often been ascertained, and it is also so well known how closely species may be allied and yet differ in all the essential relations which characterize species, that such loose investigations are no longer justifiable.

This close resemblance of animals and plants in distant parts of the world is the most interesting subject of investigation with reference to the question of the unity of origin of animals and to that of the influence of physical agents upon organized beings in general. It appears to me, that, as facts now point distinctly to an independent origin of individuals of the same species in remote regions, or of closely allied species representing one another in distant parts of the world, one of the strongest arguments in favor of the supposition, that physical agents may have had a controlling influence in changing the character of the organic world, is gone for ever.

The narrowest limits within which certain Vertebrata may be circumscribed is exemplified among Mammalia by some large and remarkable species: the Orang-Outangs upon the Sunda Islands, the Chimpanzee and the Gorilla along the western coast of Africa, several distinct species of Rhinoceros about the Cape of Good Hope, and in Java and Sumatra, the Pinchaque and the common Tapir in South America, and the eastern Tapir in Sumatra, the East Indian and the African Elephant, the Bactrian Camel and the Dromedary, the Llamas, and the different kinds of wild Bulls, wild Goats, and wild sheep, etc.; among birds by the African Ostrich, the two American Rheas, the Emeu (*Dromaeus*) of New Holland, and the Casuary (*Casuarius galeatus*) of the Indian Archipelago, and still more by the different species of doves confined to particular islands in the Pacific Ocean; among Reptiles, by the *Proteus* of the cave of Adelsberg in Carinthia, by the Gopher (*Testudo Polyphemus* Auct.) of our Southern States; among fishes, by the Blind Fish (*Amblyopsis spelaeus*) of the Mammoth Cave. Examples of closely limited Articulata may not be so striking, yet the Blind Crawfish of the Mammoth Cave and the many parasites found only upon or within certain species of animals are very remarkable in this respect. Among Mollusks I would remark the many species of land shells, ascertained by Professor Adams to occur only in Jamaica,[7] among the West India Islands, and the species discovered by the United States Exploring Expedition upon isolated islands of the Pacific and described by Dr. Gould.[8] Even among Radiata many species might be quoted, among Echinoderms as well as among Medusae and Polypi, which are only known from a few localities; but as long as these animals are not collected with the special view of ascertaining their geographical range, the indications of travelers must be received with great caution, and any generalization respecting the extent of their natural area would be premature as long as the counties they inhabit have not been more extensively explored.[9] It is nevertheless true as established by ample evidence that within definite limits all the animals occurring in different natural zoological provinces are specifically distinct. What remains to be ascertained more minutely is the precise range of each species, as well as the most natural limits of the different faunae.

Louis Agassiz, "On the Origin of Species" (excerpt), *American Journal of Science* (1860)

IT seems to me that there is much confusion of ideas in the general statement of the variability of species so often repeated lately. If species do not exist at all, as the supporters of the transmutation theory maintain, how can they vary, and if individuals alone exist, how can the differences which may be observed among them prove the variability of species? The fact seems to me to be that while species are based upon definite relations among individuals which differ in various ways among themselves, each individual, as a distinct being, has a definite course to run from

---

[7] Charles B. Adams, *Contributions to Conchology* (12 nos., New York, 1849–1852). A series of pamphlets, full of original information.

[8] Augustus A. Gould, *Mollusca and Shells*, United States Exploring Expedition, *Reports*, XII (Philadelphia, 1852).

[9] With reference to the Echinoderms and Acalephs, I am able to state that the species of the Atlantic shores of North America, found along the northern states, differ entirely from those of the southern states, and these differ again from those of the Gulf of Mexico.

the time of its first formation to the end of its existence, during which it never loses its identity nor changes its individuality, nor its relations to other individuals belonging to the same species, but preserves the categories of relationship which constitute specific or generic or family affinity, or any other kind or degree of affinity. *To prove that species vary it should be proved that individuals born from common ancestors change the different categories of relationship which they bore primitively to one another.* While all that has thus far been shown is, that there exists a considerable difference among individuals of one and the same species. This may be new to those who have looked upon every individual picked up at random, as affording the means of describing satisfactorily any species; but no naturalist who has studied carefully any of the species now best known, can have failed to perceive that it requires extensive series of specimens accurately to describe a species, and that the more complete such series are, the more precise appear the limits which separate species. Surely the aim of science cannot be to furnish amateur zoologists or collectors, a *recipe* for a ready identification of any chance specimen that may fall into their hands. And the difficulties with which we may meet in attempting to characterize species do not afford the first indication that species do not exist at all, as long as most of them can be distinguished, as such, almost at first sight. I foresee that some convert to the transmutation creed will at once object that the facility with which species may be distinguished is no evidence that they were not derived from other species. It may be so. But as long as no fact is adduced to show that any one well-known species among the many thousands that are buried in the whole series of fossiliferous rocks, is actually the parent of any one of the species now living, such arguments can have no weight; and thus far the supporters of the transmutation theory have failed to produce any such facts.

ASA GRAY (1810–1888)

Botanist.
Born in Sauquoit, New York. Died in Cambridge, Massachusetts.

Gray grew up in a rural community in Oneida County, New York. He took a medical degree from the College of Physicians and Surgeons of the Western District of the State of New York in the town of Fairfield in 1831, but from the beginning he was much more interested in the science side of the course work than the medical side. It was at Fairfield in a chemistry and *materia medica* course that he was first introduced to botany and developed the passion that would lead to his recognition as the most significant American botanist of the nineteenth century. Before he had even finished his medical degree, Gray had contacted John Torrey, who several years prior had completed his *Flora of the Northern and Middle States*. Torrey, who by this time held teaching positions at both the College of New Jersey (later Princeton University) and what would become Columbia University, was impressed with some specimens Gray had prepared and so took the younger botanist under his wing. Gray accepted a position at the Lyceum of Natural History in New York and became Torrey's heir apparent. The two men collaborated for over a decade and the end result was their *Flora of North America*, which broke completely from the Linnean system of classification and used instead the natural system of classification which considered more overall resemblances and attempted pre-phylogenetic (evolutionary) groupings rather than the more limited characteristics of the artificial system. The adoption of this natural system paved the way for the evolutionary considerations that would come to fruition 20 years in the future. Natural distributions were part of this natural analysis, and Gray published two outstanding studies of plant distribution, one in 1856 ("Statistics of the Flora of the Northern United States") and another in 1859 looking at specimens acquired from the U.S. North Pacific Exploring Expedition. The latter study looked at the botany of Japan and its relations to that of North America. Sir J. D. Hooker called it Gray's "*opus magnum*." It was Gray's work on plant distribution that brought him into correspondence with Charles Darwin, a correspondence between scientists and friends. Asa Gray would be one of three scientists to receive an advance copy of Darwin's *Origin of Species* and would prove to be one of his biggest allies in the United States. (See the introduction to this Part II for a discussion of the Gray/ Agassiz evolution dispute.) In 1842 Gray accepted the first Fisher professorship of natural history at Harvard. He would remain there for the next 30 years, training two generations of botanists and quite naturally evolving as the discipline needed; I offer his paper on *Sequoia* to illustrate the point. His *Darwinia: Essays and Reviews pertaining to Darwinism* (1876) and *Manual of the Botany of the Northern United States* (1847) are perhaps his most lasting monuments, along with his personal herbarium of more than 200,000 specimens and his botany library of over 2,000 books which were the beginning of Harvard's present-day botanical holdings. He may have been a poor cousin to Agassiz in raising money, attention, and eyebrows mid-century but by the time of his death, Gray was without doubt the more profoundly respected scientist.

Asa Gray, "Darwin on the Origin of Species," *Atlantic Monthly* (1860)

NOVELTIES are enticing to most people: to us they are simply annoying. We cling to a long-accepted theory, just as we cling to an old suit of clothes. A new theory, like a new pair of breeches ("The Atlantic" still affects the older type of nether garment) is sure to have hard-fitting places; or even when no particular fault can be found with the article, it oppresses with a sense of general discomfort. New notions and new styles worry us, till we get well used to them, which is only by slow degrees.

Wherefore, in Galileo's time, we might have helped to proscribe, or to burn – had he been stubborn enough to warrant cremation – even the great pioneer of inductive research; although, when we had fairly recovered our composure, and had leisurely excogitated the matter, we might have come to conclude that the new doctrine was better than the old one, after all, at least for those who had nothing to unlearn.

Such being our habitual state of mind, it may well be believed that the perusal of the new book "On the Origin of Species by Means of Natural Selection" left an uncomfortable impression, in spite of its plausible and winning ways. We were not wholly unprepared for it, as many of our contemporaries seem to have been. The scientific reading in which we indulge as a relaxation from severer studies had raised dim forebodings. Investigations about the succession of species in time, and their actual geographical distribution over the earth's surface, were leading up from all sides and in various ways to the question of their origin. Now and then we encountered a sentence, like Professor Owen's "axiom of the continuous operation of the ordained becoming of living things," which haunted us like an apparition. For, dim as our conception must needs be as to what such oracular and grandiloquent phrases might really mean, we felt confident that they presaged no good to old beliefs. Foreseeing, yet deprecating, the coming time of trouble, we still hoped, that, with some repairs and make-shifts, the old views might last out our days. *Après nous le déluge*. Still, not to lag behind the rest of the world, we read the book in which the new theory is promulgated. We took it up, like our neighbors, and, as was natural, in a somewhat captious frame of mind.

Well, we found no cause of quarrel with the first chapter. Here the author takes us directly to the barn-yard and the kitchen-garden. Like an honorable rural member of our General Court, who sat silent until, near the close of a long session, a bill requiring all swine at large to wear pokes was introduced, when he claimed the privilege of addressing the house, on the proper ground that he had been "brought up among the pigs, and knew all about them," – so we were brought up among cows and cabbages; and the lowing of cattle, the cackling of hens, and the cooing of pigeons were sounds native and pleasant to our ears. So "Variation under Domestication" dealt with familiar subjects in a natural way, and gently introduced "Variation under Nature," which seemed likely enough. Then follows "Struggle for Existence," – a principle which we experimentally know to be true and cogent, – bringing the comfortable assurance, that man, even upon Leviathan Hobbes's theory of society, is no worse than the rest of creation, since all Nature is at war, one species with another, and the nearer kindred the more internecine – bringing in thousand-fold confirmation and extension of the Malthusian doctrine, that population tends far to outrun means of subsistence throughout the animal and vegetable world, and has to be kept down by sharp preventive checks; so that not more than one of a hundred or a thousand of the individuals whose existence is so wonderfully and so sedulously provided for ever comes to anything, under ordinary circumstances; so the lucky and the strong must prevail, and the weaker and ill-favored must perish; – and then follows, as naturally as one sheep follows another, the chapter on "Natural Selection," Darwin's *cheval de bataille*, which is very much the Napoleonic doctrine, that Providence favors the strongest battalions – that, since many more individuals are born than can possibly survive, those individuals and those variations which possess any advantage, however slight, over the rest, are in the long run sure to survive, to propagate, and to occupy the limited field, to the exclusion or destruction of the weaker brethren. All this we pondered, and could not much object to. In fact, we began to contract a liking for a system which at the outset illustrates the advantages of good breeding, and which makes the most "of every creature's best."

Could we "let by-gones be by-gones," and, beginning now, go on improving and diversifying for the future by natural selection – could we even take up the theory at the introduction of the actually existing species, we should be well content, and so perhaps would most naturalists be. It is by no means difficult to believe that varieties are incipient or possible species, when we see what trouble naturalists, especially botanists, have to distinguish between them – one regarding as a true species what another regards as a variety; when the progress of knowledge increases, rather than diminishes, the number of doubtful instances; and when there is less agreement than ever among naturalists as to what the basis is in Nature upon which our, idea of species reposes, or how the word is practically to be defined. Indeed, when we consider the endless disputes of naturalists and ethnologists over the human races, as to whether they belong to one species or to more, and if to more, whether to three, or five, or fifty, we can hardly help fancying that both may be right – or rather, that the unihumanitarians would have been right several thousand years ago, and the multihumanitarians will be a few thousand years later; while at present the safe thing to say is, that, probably, there is some truth on both sides. "Natural selection," Darwin remarks, "leads to divergence of character; for more living beings can be supported on the same area the more they diverge in structure, habits, and constitution" (a principle which, by the way, is paralleled and illustrated by the diversification of human labor) and also leads to much extinction of intermediate or unimproved forms. Now, though this divergence may "steadily tend to increase," yet this is evidently a slow process in Nature, and liable to much counteraction wherever man does not interpose, and so not likely to work much harm for the future. And if natural selection, with artificial to help it, will produce better animals and better men than the present, and fit them better to "the conditions of existence," why, let it work, say we, to the top of its bent. There is still room enough for improvement. Only let us hope that it always works for good: if not, the divergent lines on Darwin's diagram of transmutation made easy ominously show what small deviations from the straight path may come to in the end.

The prospect of the future, accordingly, is on the whole pleasant and encouraging. It is only the backward glance, the gaze up the long vista of the past, that reveals anything alarming. Here the lines converge as they recede into the geological ages, and point to conclusions which, upon the theory, are inevitable, but by no means welcome. The very first step backwards makes the Negro and the Hottentot our blood-relations; – not that reason or Scripture objects to that, though pride may. The next suggests a closer association of our ancestors of the olden time with "our poor relations" of the quadrumanous family than we like to acknowledge. Fortunately, however – even if we must account for him scientifically – man with his two feet stands upon a foundation of his own. Intermediate links between the *Bimana* and the *Quadrumana* are lacking altogether; so that, put the genealogy of the brutes upon what footing you will, the four-handed races will not serve for our forerunners; – at least, not until some monkey, live or fossil, is producible with great-toes, instead of thumbs, upon his nether extremities; or until some lucky geologist turns up the bones of his ancestor and prototype in France or England, who was so busy "napping the chuckie-stanes" and chipping out flint knives and arrow-beads in the time of the drift, very many ages ago – before the British Channel existed, says Lyell[1] – and until these men of the olden time are shown to have worn their great-toes in a divergent and thumblike fashion. That would be evidence indeed: but until some testimony of the sort is produced, we must needs believe in the separate and special creation of man, however it may have been with the lower animals and with plants.

No doubt, the full development and symmetry of Darwin's hypothesis strongly suggest the evolution of the human no less than the lower animal races out of some simple primordial animal – that all are equally "lineal descendants of some few beings which lived long before the first bed of the Silurian system was deposited."

---

[1] Vide *Proceedings of the British Association for the Advancement of Science*, 1859, and London *Athenaeum*, passim. It appears to be conceded that these "celts" or stone knives are artificial productions, and of the age of the mammoth, the fossil rhinoceros, etc.

But, as the author speaks disrespectfully of spontaneous generation, and accepts a supernatural beginning of life on earth, in some form or forms of being which included potentially all that have since existed and are yet to be, he is thereby not warranted to extend his inferences beyond the evidence or the fair probability. There seems as great likelihood that one special origination should be followed by another upon fitting occasion (such as the introduction of man) as that one form should be transmuted into another upon fitting occasion, as, for instance, in the succession of species which differ from each other only in some details. To compare small things with great in a homely illustration: man alters from time to time his instruments or machines, as new circumstances or conditions may require and his wit suggest. Minor alterations and improvements he adds to the machine he possesses: he adapts a new rig or a new rudder to an old boat: this answers to variation. If boats could engender, the variations would doubtless be propagated, like those of domestic cattle. In course of time the old ones would be worn out or wrecked; the best sorts would be chosen for each particular use, and further improved upon, and so the primordial boat be developed into the scow, the skiff, the sloop, and other species of water-craft, the very diversification, as well as the successive improvements, entailing the disappearance of many intermediate forms, less adapted to any one particular purpose; wherefore these go slowly out of use, and become extinct species: this is *natural selection*. Now let a great and important advance be made, like that of steam-navigation: here, though the engine might be added to the old vessel, yet the wiser and therefore the actual way is to make a new vessel on a modified plan: this may answer to *specific creation*. Anyhow, the one does not necessarily exclude the other. Variation and natural selection may play their part, and so may specific creation also. Why not?

This leads us to ask for the reasons which call for this new theory of transmutation. The beginning of things must needs lie in obscurity, beyond the bounds of proof, though within those of conjecture or of analogical inference. Why not hold fast to the customary view, that all species were directly, instead of indirectly, created after their respective kinds, as we now behold them – and that in a manner which, passing our comprehension, we intuitively refer to the supernatural? Why this continual striving after "the unattained and dim" – these anxious endeavors, especially of late years, by naturalists and philosophers of various schools and different tendencies, to penetrate what one of them calls "the mystery of mysteries," the origin of species? To this, in general, sufficient answer may be found in the activity of the human intellect, "the delirious yet divine desire to know," stimulated as it has been by its own success in unveiling the laws and processes of inorganic Nature – in the fact that the principal triumphs of our age in physical science have consisted in tracing connections where none were known before, in reducing heterogeneous phenomena to a common cause or origin, in a manner quite analogous to that of the reduction of supposed independently originated species to a common ultimate origin – thus, and in various other ways, largely and legitimately extending the domain of secondary causes. Surely the scientific mind of an age which contemplates the solar system as evolved from a common, revolving, fluid mass – which, through experimental research, has come to regard light, heat, electricity, magnetism, chemical affinity, and mechanical power as varieties or derivative and convertible forms of one force, instead of independent species – which has brought the so-called elementary kinds of matter, such as the metals, into kindred groups, and raised the question, whether the members of each group may not be mere varieties of one species – and which speculates steadily in the direction of the ultimate unity of matter, of a sort of prototype or simple element which may be to the ordinary species of matter what the *protozoa* or component cells of an organism are to the higher sorts of animals and plants – the mind of such an age cannot be expected to let the old belief about species pass unquestioned. It will raise the question, how the diverse sorts of plants and animals came to be as they are and where they are, and will allow that the whole inquiry transcends its powers only when all endeavors have failed. Granting the origin to be supernatural, or miraculous even, will not arrest the inquiry. All real origination, the philosophers will say, is supernatural; their very question is, whether we have yet gone back to the origin, and can affirm that the present forms of plants and animals are the primordial, the miraculously created ones. And even if they admit that, they will still inquire into the order of the phenomena, into the form of the miracle. You might as well expect the child to grow up content with what it is told about the advent of its infant brother. Indeed, to learn that the new-comer is the gift of God, far from lulling inquiry, only stimulates speculation as to how the precious gift was bestowed. That questioning child is father to the man – is philosopher in short-clothes.

Since, then, questions about the origin of species will be raised, and have been raised – and since the theorizings, however different in particulars, all proceed upon the notion that one species of plant or animal is somehow derived from another, that the different sorts which now flourish are lineal (or unlineal) descendants of other and earlier sorts – it now concerns us to ask, What are the grounds in Nature, the admitted facts, which suggest hypotheses of derivation, in some shape or other? Reasons there must be, and plausible ones, for the persistent recurrence of theories upon this genetic basis. A study of Darwin's book, and a general glance at the present state of the natural sciences, enable us to gather the following as perhaps the most suggestive and influential. We can only enumerate them here, without much indication of their particular bearing. There is:

1. The general fact of variability; the patent fact, that all species vary more or less; that domesticated plants and animals, being in conditions favorable to the production and preservation of varieties, are apt to vary widely; and that by interbreeding, any variety may be fixed into a race, that is, into a variety which comes true from seed. Many such races, it is allowed, differ from each other in structure and appearance as widely as do many admitted species; and it is practically very difficult, perhaps impossible, to draw a clear line between races and species. Witness the human races, for instance. Wild species also vary, perhaps about as widely as those of

domestication, though in different ways. Some of them appear to vary little, others moderately, others immoderately, to the great bewilderment of systematic botanists and zoologists, and their increasing disagreement as to whether various forms shall be held to be original species or marked varieties. Moreover, the degree to which the descendants of the same stock, varying in different directions, may at length diverge is unknown. All we know is, that varieties are themselves variable, and that very diverse forms have been educed from one stock.

2. Species of the same genus are not distinguished from each other by equal amounts of difference. There is diversity in this respect analogous to that of the varieties of a polymorphous species, some of them slight, others extreme. And in large genera the unequal resemblance shows itself in the clustering of the species around several types or central species, like satellites around their respective planets. Obviously suggestive this of the hypothesis that they were satellites, not thrown off by revolution, like the moons of Jupiter, Saturn, and our own solitary moon, but gradually and peacefully detached by divergent variation. That such closely related species may be only varieties of higher grade, earlier origin, or more favored evolution, is not a very violent supposition. Anyhow, it was a supposition sure to be made.

3. The actual geographical distribution of species upon the earth's surface tends to suggest the same notion. For, as a general thing, all or most of the species of a peculiar genus or other type are grouped in the same country, or occupy continuous, proximate, or accessible areas. So well does this rule hold, so general is the implication that kindred species are or were associated geographically, that most trustworthy naturalists, quite free from hypotheses of transmutation, are constantly inferring former geographical continuity between parts of the world now widely disjoined, in order to account thereby for the generic similarities among their inhabitants. Yet no scientific explanation has been offered to account for the geographical association of kindred species, except the hypothesis of a common origin.

4. Here the fact of the antiquity of creation, and in particular of the present kinds of the earth's inhabitants, or of a large part of them, comes in to rebut the objection, that there has not been time enough for any marked diversification of living things through divergent variation, – not time enough for varieties to have diverged into what we call species.

So long as the existing species of plants and animals were thought to have originated a few thousand years ago and without predecessors, there was no room for a theory of derivation of one sort from another, nor time enough even to account for the establishment of the races which are generally believed to have diverged from a common stock. Not that five or six thousand years was a short allowance for this; but because some of our familiar domesticated varieties of grain, of fowls, and of other animals, were pictured and mummified by the old Egyptians more than half that number of years ago, if not much earlier. Indeed, perhaps the strongest argument for the original plurality of human species was drawn from the identification of some of the present races of men upon these early historical monuments and records.

But this very extension of the current chronology, if we may rely upon the archaeologists, removes the difficulty by opening up a longer vista. So does the discovery in Europe of remains and implements of pre-historic races of men to whom the use of metals was unknown – men of the *stone age*, as the Scandinavian archaeologists designate them. And now, "axes and knives of flint, evidently wrought by human skill, are found in beds of the drift at Amiens (also in other places, both in France and England) associated with the bones of extinct species of animals." These implements, indeed, were noticed twenty years ago; at a place in Suffolk they have been exhumed from time to time for more than a century; but the full confirmation, the recognition of the age of the deposit in which the implements occur, their abundance, and the appreciation of their bearings upon most interesting questions, belong to the present time. To complete the connection of these primitive people with the fossil ages, the French geologists, we are told, have now "found these axes in Picardy associated with remains of *Elephas primigenius, Rhinoceros tichorhinus, Equus fossilis,* and an extinct species of *Bos*."[2] In plain language, these workers in flint lived in the time of the mammoth, of a rhinoceros now extinct, and along with horses and cattle unlike any now existing – specifically different, as naturalists say, from those with which man is now associated. Their connection with existing human races may perhaps be traced through the intervening people of the stone age, who were succeeded by the people of the bronze age, and these by workers in iron.[3] Now, various evidence carries back the existence of many of the present lower species of animals, and probably of a larger number of plants, to the same drift period. All agree that this was very many thousand years ago. Agassiz tells us that the same species of polyps which are now building coral walls around the present peninsula of Florida actually made that peninsula, and have been building there for centuries which must be reckoned by thousands.

5. The overlapping of existing and extinct species, and the seemingly gradual transition of the life of the drift period into that of the present, may be turned to the same account. Mammoths, mastodons, and Irish elks, now extinct, must have lived down to human, if not almost to historic times. Perhaps the last dodo did not long outlive his huge New Zealand kindred. The auroch, once the companion of mammoths, still survives, but apparently owes his present and precarious existence to man's care. Now, nothing that we know of forbids the hypothesis that some new species have been independently and supernaturally created within the period which other species have survived. It may even be believed that man was created in the days

² See Correspondence of M. Nicklès, in *American Journal of Science and Arts*, for March, 1860.
³ See Morlet, *Some General Views on Archaeology*, in *American Journal of Science and Arts*, for January, 1860, translated from *Bulletin de in Société Vaudoise*, 1859.

of the mammoth, became extinct, and was recreated at a later date. But why not say the same of the auroch, contemporary both of the old man and of the new? Still it is more natural, if not inevitable, to infer, that, if the aurochs of that olden time were the ancestors of the aurochs of the Lithuanian forests, so likewise were the men of that age if men they were – the ancestors of the present human races. Then, whoever concludes that these primitive makers of rude flint axes and knives were the ancestors of the better workmen of the succeeding Stone Age, and these again of the succeeding artificers in brass and iron, will also be likely to suppose that the *Equus* and *Bos* of that time were the remote progenitors of our own horses and cattle. In all candor we must at least concede that such considerations suggest a genetic descent from the drift period down to the present, and allow time enough – if time is of any account – for variation and natural selection to work out some appreciable results in the way of divergence into races or even into so-called species. Whatever might have been thought, when geological time was supposed to be separated from the present era by a clear line, it is certain that a gradual replacement of old forms by new ones is strongly suggestive of some mode of origination which may still be operative. When species, like individuals, were found to die out one by one, and apparently to come in one by one, a theory for what Owen sonorously calls "the continuous operation of the ordained becoming of living things" could not be far off.

That all such theories should take the form of a derivation of the new from the old seems to be inevitable, perhaps from our inability to conceive of any other line of secondary causes, in this connection. Owen himself is apparently in travail with some transmutation theory of his own conceiving, which may yet see the light, although Darwin's came first to the birth. Different as the two theories will probably be in particulars, they cannot fail to exhibit that fundamental resemblance in this respect which betokens a community of origin, a common foundation on the general facts and the obvious suggestions of modern science. Indeed – to turn the point of a taking simile directed against Darwin – the difference between the Darwinian and the Owenian hypotheses may, after all, be only that between homoeopathic and heroic doses of the same drug.

    If theories of derivation could only stop here, content with explaining the diversification and succession of species between the tertiary period and the present time, through natural agencies or secondary causes still in operation, we fancy they would not be generally or violently objected to by the *savans* of the present day. But it is hard, if not impossible, to find a stopping-place. Some of the facts or accepted conclusions already referred to, and several others, of a more general character, which must be taken into the account, impel the theory onward with accumulated force. *Vires* (not to say *virus*) *acquirit eundo*. The theory hitches on wonderfully well to Lyell's uniformitarian theory in geology – that the thing that has been is the thing that is and shall be – that the natural operations now going on will account for all geological changes in a quiet and easy way, only give the time enough, so connecting the present and the proximate with the farthest past by almost imperceptible gradations – a view which finds large and increasing, if not general, acceptance in physical geology, and of which Darwin's theory is the natural complement.

So the Darwinian theory, once getting a foothold, marches boldly on, follows the supposed near ancestors of our present species farther and yet farther back into the dim past, and ends with an analogical inference which "makes the whole world kin." As we said at the beginning, this upshot discomposes us. Several features of the theory have an uncanny look. They may prove to be innocent: but their first aspect is suspicious, and high authorities pronounce the whole thing to be positively mischievous.

In this dilemma we are going to take advice. Following the bent of our prejudices, and hoping to fortify these by new and strong arguments, we are going now to read the principal reviews which undertake to demolish the theory; – with what result our readers shall be duly informed.

Meanwhile, we call attention to the fact, that the Appletons have just brought out a second and revised edition of Mr. Darwin's book, with numerous corrections, important additions, and a preface, all prepared by the author for this edition, in advance of a new English edition.

Asa Gray, "Sequoia and Its History" (excerpt), *American Naturalist* (1872)

An address by Prof. Asa Gray, President of the American Association for the Advancement of Science, Delivered at the Meeting held at Dubuque, Iowa, August, 1872

THE session being now happily inaugurated, your presiding officer of the last year has only one duty to perform before he surrenders his chair to his successor. If allowed to borrow a simile from the language of my own profession, I might liken the President of this association to a biennial plant. He flourishes for the year in which he comes into existence, and performs his appropriate functions as presiding officer. When the second year comes round he is expected to blossom out in an address and disappear. Each President, as he retires, is naturally expected to contribute something from his own investigations or his own line of study, usually to discuss some particular scientific topic.

Now, although I have cultivated the field of North American Botany, with some assiduity, for more than forty years, have reviewed our vegetable hosts, and assigned to no small number of them their names and their place in the ranks, yet, so far as our own wide country is concerned, I have been to a great extent a closet botanist. Until this summer I had not seen the Mississippi, nor set foot upon a prairie.

To gratify a natural interest, and to gain some title for addressing a body of practical naturalists and explorers, I have made a pilgrimage across the continent. I have sought and viewed in their native haunts many a plant and flower which for me had long bloomed unseen, or only in the *hortus siccus*. I have been able to see for myself what species and what forms constitute the main features of the vegetation of each successive region, and record – as the vegetation unerringly does – the permanent characteristics of its climate.

Passing on from the eastern district, marked by its equably distributed rainfall, and therefore naturally forest-clad, I have seen the trees diminish in number, give place to wide prairies, restrict their growth to the borders of streams, and then disappear from the boundless drier plains; have seen grassy plains change into a brown and sere desert – desert in the common sense, but hardly anywhere botanically so; have seen a fair growth of coniferous trees adorning the more favored slopes of a mountain range high enough to compel summer showers; have traversed that broad and bare elevated region shut off on both sides by high mountains from the moisture supplied by either ocean, and longitudinally intersected by sierras which seemingly remain as naked as they were born; and have reached at length the westward slopes of the high mountain barrier which, refreshed by the Pacific, bears the noble forests of the Sierra Nevada and the Coast Range, and among them trees which are the wonder of the world. As I stood in their shade, in the groves of Mariposa and Calaveras, and again under the canopy of the commoner Redwood, raised on columns of such majestic height and ample girth, it occurred to me that I could not do better than to share with you, upon this occasion, some of the thoughts which possessed my mind. In their development they may, perhaps, lead us up to questions of considerable scientific interest.

I shall not detain you with any remarks (which would now be trite) upon the size or longevity of these far-famed Sequoia trees, or of the sugar pines, incense-cedar and firs associated with them, of which even the prodigious bulk of the dominating Sequoia does not sensibly diminish the grandeur. Although no account and no photographic representation of either species of the far-famed Sequoia trees gives any adequate impression of their singular majesty – still less of their beauty – yet my interest in them did not culminate merely or mainly in consideration of their size and age. Other trees, in other parts of the world, may claim to be older. Certain Australian gum trees (Eucalypti) are said to be taller. Some, we are told, rise so high that they might even cast a flicker of shadow upon the summit of the pyramid of Cheops. Yet the oldest of them doubtless grew from seed which was shed long after the names of the pyramid builders had been forgotten. So far as we can judge from the actual counting of the layers of several trees, no Sequoia now alive can sensibly antedate the Christian era.

Nor was I much impressed with an attraction of man's adding. That the more remarkable of these trees should bear distinguishing appellations seems proper enough: but the tablets of personal names which are affixed to many of them in the most visited groves – as if the memory of more or less notable people of our day might be made more enduring by the juxtaposition – do suggest some incongruity. When we consider that a hand's breadth at the circumference of any one of the venerable trunks so placarded has recorded in annual lines the lifetime of the individual thus associated with it, one may question whether the next hand's breadth may not measure the fame of some of the names thus ticketed for adventitious immortality. Whether it be the man or the tree that is honored in the connection, probably either would live as long, in face and in memory, without it.

One notable thing about these Sequoia trees is their *isolation*. Most of the trees associated with them are of peculiar species, and some of them are nearly as local. Yet every pine, fir, and cypress in California is in some sort familiar, because it has near relatives in other parts of the world. But the redwoods have none. The redwood – including in that name the two species of "big-trees" – belongs to the general cypress family, but is *sui generis*. Thus isolated systematically, and extremely isolated geographically, and so wonderful in size and port, they more than other trees suggest questions.

Were they created thus local and lonely, denizens of California only; one in limited numbers in a few choice spots on the Sierra Nevada, the other along the coast range from the Bay of Monterey to the frontiers of Oregon? Are they veritable Melchizedecs, without pedigree or early relationship, and possibly fated to be without descent?

Or are they now coming upon the stage (or rather were they coming but for man's interference) to play a part in the future?

Or, are they remnants, sole and scanty survivors of a race that has played a grander part in the past, but is now verging to extinction? Have they had a career, and can that career be ascertained or surmised, so that we may at least guess whence they came, and how, and when?

Time was, and not long ago, when such questions as these were regarded as useless and vain – when student of natural history, unmindful of what the name denotes, were content with a knowledge of things as they now are, but gave little heed as to how they came to be so. Now, such questions are held to be legitimate, and perhaps not wholly unanswerable. It cannot now be said that these trees inhabit their present restricted areas simply because they are there placed in the climate and soil of all the world most congenial to them. These must indeed be congenial, or they would not survive. But when we see how Australian Eucalyptus trees thrive upon the Californian coast, and how these very redwoods flourish upon another continent; how the so-called wild oat (*Avena sterilis* of the Old World) has taken full possession of California; how that cattle and horses, introduced by the Spaniard, have spread as widely and made themselves as much at home on the plains of La Plata as on those of Tartary, and that the cardoon-thistle seeds and others they brought with them, have multiplied there into numbers probably much exceeding those extant in their native lands; indeed, when we contemplate our own race, and our own particular stock, taking such recent but dominating possession of this New World; when we consider how the indigenous flora of islands generally succumbs to the foreigners which come in the train of man; and that most weeds (*i.e.*, the prepotent plants in open soil) of all temperate climates are not "to the manor born," but are self-invited intruders; – we must needs abandon the notion of any primordial and absolute adaptation of plants and animals to their habitats, which may stand in lieu of explanation, and so preclude our inquiring any further. The harmony of Nature and its admirable perfection need not be regarded as inflexible and changeless. Nor need Nature be likened to a statue, or a cast in rigid bronze, but rather to an organism, with play and adaptability of parts, and life and even soul informing the whole. Under the former view Nature would be "the faultless monster which the world ne'er saw," but inscrutable as the Sphinx, whom it were vain, or worse, to question of the whence and whither. Under the other, the perfection of nature, if relative, is multifarious and ever renewed; and much that is enigmatical now may find explanation in some record of the past.

That the two species, of redwood we are contemplating originated as they are and where they are, arid for the part they are now playing, is, to say the least, not a scientific supposition, nor in any sense a probable one. Nor is it more likely that they are destined to play a conspicuous part in the future, or that they would have done so, even if the Indian's fires and the white man's axe had spared them. The redwood of the coast (*Sequoia semper-virens*) had the stronger hold upon existence, forming as it did large forests throughout a narrow belt about three hundred miles in length, and being so tenacious of life that every large stump sprouts into a copse. But it does not pass the Bay of Monterey, nor cross the line of Oregon, although so grandly developed not far below it. The more remarkable *Sequoia gigantea* of the Sierra exists in numbers so limited that the separate groves may be reckoned upon the fingers, and the trees of most of them have been counted, except near their southern limit, where they are said to be more copious. A species limited in individuals holds its existence by a precarious tenure; and this has a foothold only in a few sheltered spots, of a happy mean in temperature and locally favored with moisture in summer. Even there, for some reason or other, the pines with which they are associated (*Pinus Lambertiana* and *P. ponderosa*), the firs (*Abies grandis* and *A. amabilis*) and even the incense-cedar (*Libocedrus decurrens*) possess a great advantage, and, though they strive in vain to emulate their size, wholly overpower the Sequoias in numbers. "To him that hath shall be given." The force of numbers eventually wins. At least in the commonly visited groves *Sequoia gigantea* is invested in its last stronghold, can neither advance into more exposed positions above, nor fall back into drier and barer ground below, nor hold its own in the long run where it is, under present conditions; and a little further drying of the climate, which must once have been much moister than now, would precipitate its doom. Whatever the individual longevity, certain if not speedy is the decline of a race in which a high death-rate afflicts the young. Seedlings of the big trees occur not rarely, indeed, but in meagre proportion to those of associated trees; and small indeed is the chance that any of these will attain to "the days of the years of their fathers." "Few and evil" are the days of all the forest likely to be, while man, both barbarian and civilized, torments them with fires, fatal at once to seedlings, and at length to the aged also. The forests of California, proud as the State may be of them, are already too scanty and insufficient for her uses. Two lines, such as may be drawn with one sweep of a small brush over the map, would cover them all. The coast redwood – the most important tree in California – although a million times more numerous than its relative of the Sierra, is too good to live long. Such is its value for lumber and its accessibility, that, judging the future by the past, it is not likely, in its primeval growth, to outlast its rarer fellow-species.

Happily man preserves and disseminates as well as destroys. The species will probably be indefinitely preserved to science, and for ornamental and other uses, in its own and other lands; and the more remarkable individuals of the present day are likely to be sedulously cared for, all the more so as they become scarce.

Our third question remains to be answered: Have these famous Sequoias played in former times and upon a larger stage a more imposing part, of which the present is but the epilogue? We cannot gaze high up the huge and venerable trunks, which one crosses the continent to behold, without wishing that these patriarchs of the grove were able, like the long-lived antediluvians of scripture, to hand down to us, through a few generations, the traditions of centuries, and so tell us somewhat of the history of their race. Fifteen hundred annual layers have been counted, or satisfactorily made out, upon one or two fallen trunks. It is probable that close to the heart of some of the living trees may be found the circle that records the year of our Saviour's nativity. A few generations of such trees might carry the history a long way back. But the ground they stand upon, and the marks of very recent geological change and vicissitude in the region around, testify that not very many such generations can have flourished just there, at least in an unbroken series. When their site was covered by glaciers, these Sequoias must have occupied other stations, if, as there is reason to believe, they then existed in the land.

I have said that the redwoods have no near relatives in the country of their abode, and none of their genus anywhere else. Perhaps something may be learned of their genealogy by inquiring of such relatives as they have. There are only two of any particular nearness of kin; and they are far away. One is the bald cypress, our southern cypress, Taxodium, inhabiting the swamps of the Atlantic coast from Maryland to Texas, thence extending into Mexico. It is well known as one of the largest trees of our Atlantic forest-district, and, although it never (except perhaps in Mexico, and in rare instances) attains the portliness of its western relatives, yet it may equal them in longevity. The other relative is Glyptostrobus, a sort of modified Taxodium, being about as much like our bald cypress as one species of redwood is like the other.

Now species of the same type, especially when few, and the type peculiar, are, in a general way, associated geographically, *i.e.*, inhabit the same country, or (in a large sense) the same region. Where it is not so, where near relatives are separated, there is usually something to be explained. Here is an instance. These four trees, sole representatives of their tribe, dwell almost in three separate quarters of the world: the two redwoods in California, the bald cypress in Atlantic North America, its near relative, Glyptostrobus, in China.

It was not always so. In the tertiary period, the geological botanists assure us, our own very Taxodium, or bald cypress, and a Glyptostrobus, exceedingly like the present Chinese tree, and more than one Sequoia, co-existed in a fourth quarter of the globe, *viz.*, in Europe! This brings up the question: Is it possible to bridge over these four wide intervals of space and the much vaster interval of time, so as to bring these extraordinarily separated relatives into connection. The evidence which may be brought to bear upon this question is various and widely scattered. I bespeak your patience while I endeavor to bring together, in an abstract, the most important points of it.

Some interesting facts may come out by comparing generally the botany of the three remote regions, each of which is the sole home of one of these three genera, *i.e.*, Sequoia in California, Taxodium in the Atlantic United States, and Glyptostrobus in China, which compose the whole of the peculiar tribe under consideration.

Note then, first, that there is another set of three or four peculiar trees, in this case of the yew family, which has just the same peculiar distribution, and which therefore may have the same explanation, whatever that explanation be. The genus Torreya, which commemorates our botanical Nestor and a former president of this association, Dr. Torrey, was founded upon a tree rather lately discovered (that is, about thirty-five years ago) in northern Florida. It is a noble, yew-like tree, and very local, being known only for a few miles along the shores of a single river. It seems as if it had somehow been crowded down out of the Alleghanies into its present limited southern quarters; for in cultivation it evinces a northern hardiness. Now another species of Torreya is a characteristic tree of Japan; and the same, or one very like it indeed, inhabits the Himalayas – belongs, therefore, to the Eastern Asiatic temperate region, of which China is a part, and Japan, as we shall see, the portion most interesting to us. There is only one more species of Torreya, and that is a companion of the redwoods in California. It is the tree locally known under the name of the California nutmeg. In this case the three are near brethren, species of the same genus, known nowhere else than in these three habitats.

Moreover, the Torreya of Florida has growing with it a yew tree; and the trees of that grove are the only yew trees of Eastern America; for the yew of our northern woods is a decumbent shrub. The only other yew trees in America grow with the redwoods and the other Torreya in California, and more plentifully farther north, in Oregon. A yew tree equally accompanies the Torreya of Japan and the Himalayas, and this is apparently the same as the common yew of Europe.

So we have three groups of trees of the great coniferous order which agree in this peculiar geographical distribution; the redwoods and their relatives, which differ widely enough to be termed a different genus in each region; the Torreyas, more nearly akin, merely a different species in each region; the yews, perhaps all of the same species, perhaps not quite that, for opinions differ and can hardly be brought to any decisive test. The yews of the Old World, from Japan to Western Europe, are considered the same; the very local one in Florida is slightly different; that of California and Oregon differs a very little more; but all of them are within the limits of variation of many a species. However that may be, it appears to me that these several instances all raise the same question, only with a different degree of emphasis, and, if to be explained at all, will have the same kind of explanation. But the value of the explanation will be in proportion to the number of facts it will explain.

★   ★   ★

These singular relations attracted my curiosity early in the course of my botanical studies, when comparatively few of them were known, and my serious attention in later years, when I had numerous and new Japanese plants to study in the collections made (by Messrs. Williams and Morrow) during Commodore Perry's visit in 1853, and, especially, by Mr. Charles Wright, in Commodore Rodgers' expedition in 1855. I then discussed this subject somewhat fully, and tabulated the facts within my reach.[1]

This was before Heer had developed the rich fossil botany of the Arctic zone, before the immense antiquity of existing species of plants was recognized, and before the publication of Darwin's now famous volume on the "Origin of Species" had introduced and familiarized the scientific world with those now current ideas respecting the history and vicissitudes of species with which I attempted to deal in a moderate and feeble way.

My speculation was based upon the former glaciation of the northern temperate zone, and the inference of a warmer period preceding and perhaps following. I considered that our own present vegetation, or its proximate ancestry, must have occupied the arctic and subarctic regions in pliocene times, and that it had been gradually pushed southward as the temperature lowered and the glaciation advanced, even beyond its present habitation; that plants of the same stock and kindred, probably ranging round the arctic zone as the present arctic species do, made their forced migration southward upon widely different longitudes, and receded more or less as the climate grew warmer; that the general difference of climate which marks the eastern and the western sides of the continents – the one extreme, the other mean – was doubtless even then established, so that the same species and the same sorts of species would be likely to secure and retain foothold in the similar climates of Japan and the Atlantic United States, but not in intermediate regions of different distribution of heat and moisture; so that different species of the same genus, as in Torreya, or different genera of the same group, as redwood, Taxodium and Glyptostrobus, or different associations of forest trees, might establish themselves each in the region best suited to its particular requirements, while they would fail to do so in any other. These views implied that the sources of our actual vegetation and the explanation of these peculiarities were to be sought in, and presupposed, an ancestry in pliocene or still earlier times, occupying the high northern regions. And it was thought that the occurrence of peculiarly North American genera in Europe

[1] Mem. Amer. Acad. vol 6.

in the tertiary period (such as Taxodium, Carya, Liquidambar, Sassafras, Negundo, etc.) might be best explained on the assumption of early interchange and diffusion through North Asia, rather than by that of the fabled Atlantis.

The hypothesis supposed a gradual modification of species in different directions under altering conditions, at least to the extent of producing varieties, subspecies and representative species, as they may be variously regarded; likewise the single and local origination of each type, which is now almost universally taken for granted.

The remarkable facts in regard to the Eastern American and Asiatic floras which these speculations were to explain, have since increased in number, more especially through the admirable collections of Dr. Maximowitz in Japan and adjacent countries, and the critical comparisons he has made and is still engaged upon.

I am bound to state that, in a recent general work[2] by a distinguished European botanist, Prof. Grisebach of Göttingen, these facts have been emptied of all special significance, and the relations between the Japanese and the Atlantic United States floras declared to be no more intimate than might be expected from the situation: climate, and present opportunity of interchange. This extraordinary conclusion is reached by regarding as distinct species all the plants common to both countries between which any differences have been discerned, although such differences would probably count for little if the two inhabited the same country, thus transferring many of my list of identical to that of representative species; and, then, by simply eliminating from consideration the whole array of representative species, *i.e.*, all cases in which the Japanese and the American plant are not exactly alike. As if, by pronouncing the cabalistic word species the question were settled, or rather the greater part of it remanded out of the domain of science; – as if, while complete identity of forms implied community of origin, anything short of it carried no presumption of the kind; so leaving all these singular duplicates to be wondered at, indeed, but wholly beyond the reach of inquiry!

Now the only known cause of such likeness is inheritance; and as all transmission of likeness is with some difference in individuals, and as changed conditions have resulted, as is well known, in very considerable differences, it seems to me that, if the high antiquity of our actual vegetation could be rendered probable, not to say certain, and the former habitation of any of our species or of very near relatives of them in high northern regions could be ascertained, my whole case would be made out. The needful facts, of which I was ignorant when my essay was published, have now been for some years made known, – thanks, mainly, to the researches of Heer upon ample collections of arctic fossil plants. These are confirmed and extended by new investigations, by Heer and Lesquereux, the results of which have been indicated to me by the latter.

The Taxodium, which everywhere abounds in the miocene formations in Europe, has been specially identified, first by Goeppert, then by Heer, with our common cypress of the Southern States. It has been found, fossil, in Spitzbergen, Greenland and Alaska – in the latter country along with the remains of another form, distinguishable,

[2] Die VegeLation der Erde nach ihrer klimatischen Anordnung. 1871.

but very like the common species; and this has been identified by Lesquereux in the miocene of the Rocky Mountains. So there is one species of tree which has come down essentially unchanged from the tertiary period, which for a long while inhabited both Europe and North America, and also, at some part of the period, the region which geographically connects the two (once doubtless much more closely than now), but has survived only in the Atlantic United States and Mexico.

The same Sequoia which abounds in the same miocene formations in Northern Europe has been abundantly found in those of Iceland, Spitzbergen, Greenland, Mackenzie River and Alaska. It is named *S. Langsdorfii*, but is pronounced to be very much like *S. sempervirens*, our living redwood of the Californian coast, and to be the ancient representative of it. Fossil specimens of a similar, if not the same, species have recently been detected in the Rocky Mountains by Hayden, and determined by our eminent paleontological botanist, Lesquereux; and he assures me that he has the common redwood itself from Oregon in a deposit of tertiary age. Another *Sequoia* (*S. Sternbergii*) discovered in miocene deposits in Greenland, is pronounced to be the representative of *S. gigantea*, the big tree of the Californian Sierra. If the Taxodium of the tertiary time in Europe and throughout the Arctic regions is the ancestor of our present bald cypress – which is assumed in regarding them as specifically identical – then I think we may, with our present light, fairly assume that the two redwoods of California are the direct or collateral descendants of the two ancient species which so closely resemble them.

The forests of the Arctic zone in tertiary times contained at least three other species of Sequoia, as determined by their remains, one of which, from Spitzbergen, also much resembles the common redwood of California. Another, "which appears to have been the commonest coniferous tree on Disco," was common in England and some other parts of Europe. So the Sequoias, now remarkable for their restricted station and numbers, as well as for their extraordinary size, are of an ancient stock; their ancestors and kindred formed a large part of the forests which flourished throughout the polar regions, now desolate and ice-clad, and which extended into low latitudes in Europe. On this continent one species, at least, had reached to the vicinity of its present habitat before the glaciation of the region. Among the fossil specimens already found in California, but which our trustworthy palaeontological botanist has not yet had time to examine, we may expect to find evidence of the early arrival of these two redwoods upon the ground which they now, after much vicissitude, scantily occupy.

Differences of climate, or circumstances of migration, or both, must have determined the survival of Sequoia upon the Pacific, and of Taxodium upon the Atlantic coast. And still the redwoods will not stand in the east, nor could our Taxodium find a congenial station in California.

★   ★   ★

Concluding, then, as we must, that our existing vegetation is a continuation of that of the tertiary period, may we suppose that it absolutely originated then? Evidently not. The preceding cretaceous period has furnished to Carruthers in Europe a fossil fruit like that of the *Sequoia gigantea* of the famous groves, associated with pines of the same character as those that accompany the present tree; has furnished to Heer, from Greenland, two more Sequoias, one of them identical with a tertiary species, and one nearly allied to *Sequoia Langsdorfii*, which in turn is a probable ancestor of the common Californian redwood; has furnished to Lesquereux in North America the remains of another ancient Sequoia, a Glyptostrobus, a Liquidambar which well represents our sweet-gum tree, oaks analogous to living ones, leaves of a plane tree, which are also in the tertiary and are scarcely distinguishable from our own *Platanus occidentalis*, of a magnolia and tulip-tree, and "of a sassafras undistinguishable from our living species." I need not continue the enumeration. Suffice it to say that the facts justify the conclusion which Lesquereux – a very scrupulous investigator – has already announced: "that the essential types of our actual flora are marked in the cretaceous period, and have come to us after passing, without notable changes, through the tertiary formations of our continent."

According to these views, as regards plants at least, the adaptation to successive times and changed conditions has been maintained, not by absolute renewals, but by gradual modifications. I, for one, cannot doubt that the present existing species are the lineal successors of those that garnished the earth in the old time before them, and that they were as well adapted to their surroundings then, as those which flourish and bloom around us are to their conditions now. Order and exquisite adaptation did not wait for man's coming, nor were they ever stereotyped. Organic nature – by which I mean the system and totality of living things, and their adaptation to each other and to the world – with all its apparent and indeed real stability, should be likened, not to the ocean, which varies only by tidal oscillations from a fixed level to which it is always returning, but rather to a river, so vast that we can neither discern its shores nor reach its sources, whose onward flow is not less actual because too slow to be observed by the *ephemeroe* which hover over its surface, or are borne upon its bosom.

Such ideas as these, though still repugnant to some, and not long since to many, have so possessed the minds of the naturalists of the present day, that hardly a discourse can be pronounced or an investigation prosecuted without reference to them. I suppose that the views here taken are little, if at all, in advance of the average scientific mind of the day. I cannot regard them as less noble than those which they are succeeding.

An able philosophical writer, Miss Frances Power Cobbe, has recently and truthfully said:[3]

> It is a singular fact, that when we can find out how anything is done, our first conclusion seems to be that God did not do it. No matter how wonderful, how beautiful, how intimately complex and delicate has been the machinery which has worked, perhaps for centuries, perhaps for millions of ages, to bring about some beneficent result, if we can but catch a glimpse of the wheels its divine character disappears.

[3] Darwinism in Morals; in Theological Review, April, 1871.

I agree with the writer that this first conclusion is premature and unworthy; I will add deplorable. Through what faults or infirmities of dogmatism on the one hand and skepticism on the other it came to be so thought, we need not here consider. Let us hope, and I confidently expect, that it is not to last; – that the religious faith which survived without a shock the notion of the fixity of the earth itself, may equally outlast the notion of the absolute fixity of the species which inhabit it; – that, in the future even more than in the past, faith in an order, which is the basis of science, will not (as it cannot reasonably) be dissevered from faith in an Ordainer, which is the basis of religion.

JAMES DWIGHT DANA (1813–1895)

Geologist, mineralogist, zoologist.
Born in Utica, New York. Died in New Haven, Connecticut.

Dana attended Yale for a year and trained under his future father-in-law, Benjamin Silliman. After Yale, Dana served as a mathematics instructor for the U.S. Navy and, while aboard a ship in the Mediterranean witnessed the eruption of Vesuvius, which was later published in Silliman's journal. In 1834 he returned to Yale where he worked on a new mineral classification based on chemistry and crystallography. This system was the basis for his *System of Mineralogy*. In 1838, Dana was invited to accompany the Wilkes Exploration Expedition (1838–42), which charted the Pacific islands. The expedition was a significant undertaking and compelled Dana to think globally.

One of Dana's main gifts to American science was his aptitude for grand synthesis, particularly on geological questions. The most significant of his ideas were the place of geosynclines in orogeny, the age progression of volcanic chains in the Pacific based on erosion and reference to reefs, the distinction of continents from ocean basins and the permanence of both, and the concentric accretion of mountains around the ancient interior of the North American continent. The Wilkes expedition reports would take him over a decade to complete and the specimens and notes he collected would continue to be the raw material for his science the rest of his life. On returning from the expedition he married Henrietta Silliman and resettled in New Haven and Yale. In 1850, on the resignation of Silliman, Dana was appointed Silliman Professor of Natural History and Geology at Yale, a position he held until 1892. He co-edited the *American Journal of Science* with his father-in-law until Silliman's death in 1864, and then continued as editor until his son, Edward Salisbury Dana took the helm in 1895. Dana was awarded the Wollaston medal by the Geological Society in 1874 and the Copley medal by the Royal Society of London in 1877. He was a dogged worker and at 82 was working on revising the fourth edition of his *Manual of Geology*. A deeply devout Christian, Dana saw order and plan in the direction of evolution, not chance, yet remained open and pliant enough as a scientific thinker – in contrast to his friend and colleague Agassiz – to allow the pages of the journal he edited and the minds of the students he molded to remain open to new ideas.

James D. Dana, "On the Origin of the Geographical Distribution of Crustacea" (excerpt), *Annals and Magazine of Natural History* (1856)

THE origin of the existing distribution of species in this department of zoology deserves attentive consideration. Two great causes are admitted by all, and the important question is, how far the influence of each has extended. The first is, *original local creations*; the second, *migration*.

Under the first head, we may refer much that we have already said on the influence of temperature, and the restriction of species to particular temperature regions. It is not doubted that the species have been created in regions for which they are especially fitted; that their fitness for these regions involves an adaptation of structure thereto, and upon this adaptation, their characteristics as species depend. These characteristics are of no climatal origin. They are the impress of the Creator's hand, when the species had their first existence in those regions calculated to respond to their necessities.

The following questions come under this general head:—

1. Have there been local centres of creation, from which groups of species have gone forth by migration?
2. Have genera only and not species, or have species, been repeated by creation in distinct and distant regions?
3. How closely may we recognize in climatal and other physical conditions, the predisposing cause of the existence of specific genera or species?

With regard to the *second* head, migration, we should remember, that Crustacea are almost wholly maritime or marine; that marine waters are continuous the globe around; and that no sea-shore species in zoology are better fitted than crabs for migration. They may cling to any floating log and range the seas wherever the currents drift the rude craft, while the fish of the sea-shores will only wander over their accustomed haunts. Hence it is, that among the Pacific Islands the fishes of each group of islands are mostly peculiar to the group, while the Crustacea are much more generally diffused.

A direction and also a limit to this migration exist, (1) in the currents of the ocean, and (2) in the temperature of its different regions. Through the torrid zone, the currents flow mainly *from the east* towards the west; yet they are reversed in some parts during a certain portion of the year. But this reversed current in the Pacific never reaches the American continent, and hence it could never promote migration to its shores. Again, beyond 30° or 35° of north or south latitude, the general course of the waters is *from the west*, and the currents are nearly uniform and constant. Here is a means of eastward migration in the middle and higher temperate regions. But the temperate regions in these latitudes are more numerous than in the tropics, and species might readily be wafted to uncongenial climates, which would be their destruction; in fact they could hardly escape this. Moreover, such seas are more boisterous than those nearer the equator. Again, these waters are almost entirely bare for very long distances, and not dotted closely with islands like the equatorial Pacific.

In the northern hemisphere, on the eastern coasts especially, there are warm currents from the south and cold currents from the north. The former overlie the latter to a great extent in the summer, and may aid southern species in northward migrations. Cape Hatteras is nearly the termination of the summer line of 70° (see Maury's Chart), a temperature which belongs to the subtorrid region in winter. On the China coast, at Macao, there is a temperature of 83° in July, and in the Yellow Sea, of 78° to 80°. But such northward migrations as are thus favoured, are only for the season; the cold currents of the winter months destroy all such adventurers, except the individuals of some hardier species that belong to the seas or have a wide range in distribution. Sea-shore Crustacea are not in themselves migratory, and are thus unlike many species of fish. Even the swimming Portunidæ are not known voluntarily to change their latitudes with the season.

The following is a brief recapitulation of the more prominent facts bearing on these points:—

1. The distribution of individuals of many species through twelve thousand miles in the torrid zone of the Oriental seas.
2. The very sparing distribution of Oriental species in Occidental seas.
3. The almost total absence of Oriental species from the west coast of America.
4. The world-wide distribution within certain latitudes of the species we have called cosmopolites.
5. The occurrence of closely allied genera at the Hawaiian Islands and in the Japan seas.
6. The occurrence of the same subtorrid species at the Hawaiian Islands and at Port Natal, South Africa, and not in the torrid zone intermediate, as *Kraussia rugulosa* and *Galene natalensis*.
7. The occurrence of identical species in the Japan seas and at Port Natal.
8. The occurrence of the same species (*Plagusia tomentosa*) in South Africa, New Zealand, and Valparaiso; and the occurrence of a second species (*Cancer Edwardsii* (?)) at New Zealand and Valparaiso.
9. The occurrence of closely allied species (as species of *Amphoroidea* and *Ozius*) in New South Wales and Chili.
10. The occurrence of the same species in the Japan seas and the Mediterranean, and of several identical genera.
11. The occurrence of a large number of identical species in the British seas and the Mediterranean; and also in these seas and about the Canary Islands.
12. The occurrence of closely allied, if not identical, species (as of *Palæmon*) in New Zealand and the British seas; and also of certain genera that are elsewhere peculiarly British, or common only to Britain and America.
13. An identity in certain species of Eastern and Western America.

The following are the conclusions to which we are led by the facts:—

I.   The migration of species from island to island through the tropical Pacific and East Indies may be a possibility; and the same species may thus reach even to Port Natal in South Africa. The currents of the oceans favour it, the temperature of the waters is congenial through all this range, and the habits of many Crustacea, although they are not voluntarily migratory, seem to admit of it. The species which actually have so wide a range are not Maioids (which are to a considerable extent deep-water species), but those of the shores; and some, as *Thalamita admete*, are swimming species.

II.   The fact, that very few of the Oriental species occur in the Occidental seas, may be explained on the same ground, by the barrier which the cold waters of Cape Horn and the South Atlantic present to the passage of tropical species around the Cape westward, or to their migration along the coasts.

    Moreover, the diffusion of Pacific tropical species to the Western American coast is prevented, as already observed, by the westward direction of the tropical currents, and the cold waters that bathe the greater part of this coast.

III.   When we compare the seas of Southern Japan and Port Natal, and find species common to the two that are not now existing in the Indian Ocean or East Indies, we hesitate as to migration being a sufficient cause of the distribution. It may however be said, that driftings of such species westward through the Indian Ocean may have occasionally taken place, but that only those individuals that were carried during the season quite through to the *subtorrid* region of the South Indian Ocean (Port Natal, &c.), survived and reproduced; the others, if continuing to live, soon running out under the excessive heat of the intermediate equatorial regions. That they would thus run out in many instances is beyond question; but whether this view will actually account for the resemblance in species pointed out, is open to doubt.

IV.   When, further, we find an identity of species between the Hawaiian Islands and Port Natal – half the circumference of the globe, or twelve thousand miles, apart – and the species, as *Galene natalensis*, not a species found in any part of the torrid region, and represented by another species only in Japan, we may well question whether we can meet the difficulty by appealing to migration. It may however be said, that we are not as yet thoroughly acquainted with the species of the tropics, and that facts may hereafter be discovered that will favour this view. The identical species are of so peculiar a character that we deem this improbable.

V.   The existence of the *Plagusia tomentosa* at the southern extremity of Africa, in New Zealand, and on the Chilian coasts, may perhaps be due to migration, and especially as it is a southern species, and each of these localities is within the subtemperate region. We are not ready however to assert, that such journeys as this range of migration implies are possible. The oceanic currents of this region are in the right direction to carry the species eastward, except that there is no passage into this western current from Cape Horn, through the Lagulhas current, which flows the other way. It appears to be rather a violent assumption, that an individual or more of this species could reach the western current from the coast on which it might have lived; or could have survived the boisterous passage, and finally have had a safe landing on the foreign shore. The distance from New Zealand to South America is five thousand miles, and there is at present not an island between.

VI. Part of the difficulty in the way of a transfer of species between distant meridians might be overcome, if we could assume that the intermediate seas had been occupied by land or islands during any part of the recent epoch. In the case just alluded to, it is possible that such a chain of interrupted communication once existed; and this bare possibility weakens the force of the argument used above against migration. Yet as it is wholly an assumption, we cannot rely upon it for evidence that migration has actually taken place.

VII. The existence of the same species on the east and west coasts of America affords another problem, which migration cannot meet, without sinking the Isthmus of Darien or Central America, to afford a passage across. We know of no evidence whatever that this portion of the continent has been beneath the ocean during the recent epoch. An argument against such a supposition might be drawn from the very small number of species that are identical on the two sides, and the character of these species. *Libinia spinosa* occurs at Brazil and Chili, and has not been found in the West Indies. *Leptopodia sagittaria*, another Maioid, occurs at Valparaiso, the West Indies, and the Canaries.

VIII. The large number of similar species common to the Mediterranean and British seas may be due to migration, as there is a continuous line of coast and no intermediate temperature rendering such a transfer impossible; and the passage farther south to the Canaries of several of the species is not beyond what this cause might accomplish. Still, it cannot be asserted that in all instances the distribution here is owing to migration; nor will it be admitted unless other facts throw the weight of probability on that side.

IX. But when we find the same temperate zone species occurring in distant provinces, these provinces having between them no water-communication except through the torrid or frigid zone, and offering no ground for the supposition that such a communication has existed during the recent epoch, we are led to deny the agency of voluntary or involuntary migration in producing this dissemination. An example of this, beyond all dispute, is that of the Mediterranean Sea and Japan. No water communication for the passage of species can be imagined. An opening into the Red Sea is the only possible point of intercommunication between the two kingdoms; but this opens into the torrid zone, in no part of which are the species found. The two regions have their peculiarities and their striking resemblances; and we are forced to attribute them to original creation, and not intercommunication.

X. The resemblances found are not merely in the existence of a few identical species. There are genera common to the two seas that occur nowhere else in the Oriental kingdom, as *Latreillia, Ephyra, Sicyonia*, &c.; and species where not identical having an exceedingly close resemblance.

Now this *resemblance* in genera and species (without exact identity in the latter) is not explained by supposing a possible intercommunication. But we may reasonably account for it on the ground of a similarity in the temperature and other physical conditions of the seas; and the well-known principle of "like causes, like effects," forces itself upon the mind as fully meeting the case. Mere intercommunication could not produce the resemblance; for just this similarity of physical condition would still be necessary. And where such a similarity exists, creative power may multiply analogous species; we should almost say, *must*; for, as species are made for the circumstances in which they are to live, identical circumstances will necessarily imply identity of genera in a given class, and even of specific structure or of subgenera.

If then the similarity in the characters of these regions is the occasion of the identity of genera, and of the very close likeness in certain species (so close that an identity is sometimes strongly suspected where not admitted), we must conclude that there is a possibility of actual identity of species, through original creation. This, in fact, becomes the only admissible view, and the actually identical species between Japan and the Mediterranean are examples.

XI. When we find a like resemblance of genera and species between temperate-zone provinces in opposite hemispheres that are almost exact antipodes, as in the case of Great Britain and New Zealand, we have no choice of hypotheses left. We must appeal directly to creative agency for the peopling of the New Zealand seas as well as the British, and see, in both, like wisdom, and a like adaptedness of life to physical nature. The *Palæmon affinis* of the New Zealand seas is hardly distinguishable from the common *P. squilla* of Europe, and is one example of this resemblance. It may not be an identity; and on this account it is a still better proof of our principle, because there is no occasion to suspect migration or any other kind of transfer. It is a creation of species in these distant provinces, which are almost identical, owing to the physical resemblances of the seas; and it shows at least, that a very close approximation to identity may be consistent with Divine Wisdom.

The resemblance of the New Zealand and British seas has been remarked upon as extending also to the occurrence in both of the genera *Portunus* and *Cancer*. It is certainly a wonderful fact that New Zealand should have a closer resemblance in its Crustacea to Great Britain, its antipode, than to any other part of the world—a resemblance running parallel, as we cannot fail to observe, with its geographical form, its insular position, and its situation among the temperate regions of the ocean. Under such circumstances, there must be many other more intimate resemblances, among which we may yet distinguish the special cause which led to the planting of peculiar British genera in this antipodal land.

The close resemblance in species and genera from Britain and New Zealand, and from Japan and the Mediterranean, and the actual identity in some species among the latter, prove therefore that, as regards the species of two distant regions, identity as well as resemblance may be attributable to independent creations, these resemblances being in direct accordance with the physical resemblances of the regions. As this conclusion cannot be avoided, we are compelled in all cases to try the hypothesis of migration by considering something beside the mere possibility of its having taken place under certain assumed conditions. The possibility of independent creations is as important a consideration. After all the means of communication between distant provinces have been devised or suggested, the principle still arises, that it is in accordance with Divine Wisdom to create similar and identical species in different regions where the physical circumstances are alike; and we must determine by special and thorough investigation, whether one or the other cause was the actual origin of the distribution in each particular case. Thus it must be with reference to the wide distribution of species in the Oriental tropics, as well as in the European temperate regions, and the temperate zone of the South Pacific and Indian Oceans.

XII. With respect to the creation of identical species in distant regions, we would again point to its direct dependence on a near identity of physical condition. Although we cannot admit that circumstances or physical forces have ever created a species (as like can only beget like, and physical force must result simply in physical force), and while we see in all nature the free act of the Divine Being, we may still believe the connexion between the calling into existence of a species and the physical circumstances surrounding it, to be as intimate nearly as cause and effect. The Creator has, in infinite skill, adapted each species to its place, and the whole into a system of admirable harmony and perfection. In His wisdom, any difference of physical condition and kind of food at hand, is sufficient to require some modification of the intimate structure of species, and this difference is expressed in the form of the body or members, so as to produce an exactness of adaptation, which we are far from fully perceiving or comprehending with our present knowledge of the relations of species to their habitats.

When therefore we find the same species in regions of unlike physical character, as, for example, in the seas of the Canaries and Great Britain – regions physically so unlike – we have strong reason for attributing the diffusion of the species to migration. The difference between the Mediterranean and Great Britain may require the same conclusion for the species common to these seas. They are so far different, that we doubt whether species *created* independently in the two could have been identical, or even have had that resemblance that exists between varieties; for this resemblance is usually of the most trivial kind, and affects only the least essential of the parts of a species.

The continental species of Crustacea from the interior of different continents are not in any case known to be identical; and it is well understood that the zoological provinces and districts of the land are of far more limited extent than those of the ocean. The physical differences of the former are far more striking than those of the latter. As we have observed elsewhere, the varieties of climate are greater; the elevation above the sea may vary widely; and numberless are the diversities of soil and its conditions, and the circumstances above and within it. Hence, as the creation of each species has had reference most intimately to each and all of these conditions, as well as to other prospective ends, an identity between distant continental regions is seldom to be found, and the characteristic groups of genera are very widely diverse. Comparatively few genera of Insects have as wide a range as those of Crustacea; and species, with rare exceptions, have very narrow limits. Where the range of a species in this class is great, we should in general look to migration as the cause, rather than original creation; but the considerations bearing on both should be attentively studied, before either is admitted as the true explanation.

Throughout the warmer tropical oceans, a resemblance in the physical conditions of distant provinces is far more common and more exact than in the temperate zone; and hence it would seem that we cannot safely appeal to actual differences as an argument against the creation of a species in more than one place in the tropics. The species spread over the Oriental torrid zone may hence be supposed to owe their distribution to independent creations of the same species in different places, as well as to migration. Yet we may in this underrate the exactness of physical identity required in regions for independent creations of the same species. We know that for some chemical compounds, the condition of physical forces for their formation is exceedingly delicate; and much more should we infer that, when the creation of a living germ was concerned, a close exactness in the conditions would be required in order that the creation should be repeated in another place. Infinite power, it is true, may create in any place; but the creation will have reference to the forces of matter, the material employed in the creation. The few species common to the Oriental and Occidental torrid seas seem to be evidence on this point. The fact that the Oriental species have so rarely been repeated in the Occidental seas, when the conditions seem to be the same, favours the view that migration has been the main source of the diffusion in the Oriental tropics.

As we descend in the order of Invertebrates, the species are less detailed in structure, with fewer specific parts and greater simplicity of functions, and they therefore admit of a wider range of physical condition; the same argument against multiplication by independent creations in regions for the most part different, does not, therefore, so strongly hold. As we pass, on the contrary, to the highest groups in Zoology, the argument receives far greater weight; and at the same time there are capabilities of migration increasing generally in direct ratio as we ascend, which are calculated to promote the diffusion of species, and remove the necessity of independent creations.

Migration cannot therefore be set aside. It is an actual fact in nature, interfering much with the simplicity which zoological life in its diffusion would otherwise present to us. Where it ends, and where independent creations have taken place, is the great problem for our study. This question has its bearings on all departments of Zoology; but in few has migration had the same extended influence as in that of Crustacea. Mollusks, if we except oceanic species, are no travellers, and keep mostly to narrow limits.

XIII. There is evidence, in the exceedingly small number of torrid-zone species identical in the Atlantic and Indian Oceans, that there has been no water-communication across from one to the other in the torrid zone, during the period since existing species of Crustacea were first on the globe.

XIV. As to zoological centres of diffusion for groups of species, we can point out none. Each species of Crustacea may have had its place of origin and single centre of diffusion in many and perhaps the majority of cases. But we have no reason to say that certain regions were without life, and were peopled by migration from specific centres specially selected for this end. If such centres had an existence, there is at present no means by which they may be ascertained. The particular temperature region in which a species originated may be ascertained by observing which is most favourable to its development: we should thus conclude that the *Ranina dentata*, for example, was created in the subtorrid region, and not the torrid, as it attains its largest size in the latter. By pursuing this course with reference to each species, we may find some that are especially fitted for almost every different locality. Hence we might show, as far as reason and observation can do it, that all regions have had their own special creations.

The world, throughout all its epochs in past history, has been furnished with life in accordance with the times and seasons, each species being adapted to its age, its place, and its fellow species of life.

James D. Dana, "On Cephalization" (excerpt), *New Englander* (1863)

In presenting to our readers a new word, and a scientific principle yet unrecognized except in the writings of a single author,[1] we may reasonably be asked for a fuller exposition of the term and of its bearings than is given in our recent paper on "Man's Zoological Postion;"[2] and, accordingly, we here offer the following thoughts on the subject.

---

[1] Report, by Dames D. Dana, on Crustacea (being one of the Reports of the Exploring Expedition under Captain Wilkes), 1853, p. 1395. – *American Journal of Science*, 2nd series, Vol. XXII, p. 14, 1856; Vol. XXV, p. 213, 1858; Vol. XXXV, p. 65, Jan. 1863; Vol. XXXVI, p. 1, July, 1863.

[2] This volume, p. 282.

The importance of the head to an animal all understand. It makes the great difference between an animal and a plant. The former may be correctly described as a *fore-and-aft* structure. The former has more or less of will emanating from its head-extremity, producing voluntary action; and an animal is therefore, typically, a *forward-moving*, or a "go-ahead" being; while a plant simply stands and grows.[3] An animal is cognizant of existences about him, and, however minutes or simple, it knows enough to steer clear of obstacles, in its head-forward progress, or to attempt it at least; but a plant is, utterly, a non-percipient, unknowing thing.

The head of an animal is the seat of power. It contains not merely the principal nervous mass (the brain, in the higher tribes, and a ganglion or mass corresponding to a brain, in the lower) but also the various organs of the senses, as of sight, hearing, smell, taste, and also the mouth with its parts or appliances.

The *anterior portion* of the structure properly includes all of the body that is devoted to the special service of the head. In a Crab, it comprises not only the organs of the senses and a pair of jaws, but also, following these, *five* pairs of jointed organs called *maxillae* and *maxilla-feet* (a little like short feet in structure), that cover the mouth and serve to put into it the food; and in an Insect, it comprises *two* pairs of such *maxillae*, besides the pair of jaws.

The *posterior portion* of the body stands in direct opposition to the anterior. The kind of opposition may be partly understood from the structure of a plant, in which there is an analogous oppositeness in its extremities – the root end tending downward, whatever obstacles it may encounter, the leaf-end as strongly in the opposite direction; it being remembered that in an animal the opposite extremities are those of a *fore-and-aft* structure.

The functions of the *posterior* portion are, first, *digestion*, which is performed by the various viscera contained within this part of the structure, and is the means of supplying the material for flesh and bone, and involves arrangements for the removal of the refuse material of the food, etc.; and secondly, *locomotion*, the function of the legs in most animals, of legs and wings in birds and insects, of fins in fishes.

Thus the *anterior* and *posterior* portions of the system have their diverse duties. It is obvious, that any animal, as an oyster, for example, whose body is almost wholly a visceral or gastric mass, and which, therefore, has its *posterior portions very large*, and its anterior very small, must be of *very* low grade. This much of the principle of cephalization requires no depth of philosophy to comprehend or apply.

An important part of this *posterior* extremity, in many animals, is the *tail*, which, in Vertebrate species, is not merely a posterior elongation of the body, but also of the bony structure of the body; for the tail, however flexible, has a series of bones running

---

[3] Some kinds of animals, as *Polyps*, are fixed, like plants. But these are not true representations of the animal idea or type. They are animals in having each a mouth and a stomach, muscles and sensation; but they are given up to a vegetative style of growth. Animal life exists in these species under the forms of the vegetable type, and not that of the animal.

the greater part of its length, and this series of bones is a direct continuation of that which makes up the back-bone of the animal. It may be only a switch for switching off insects. But in whales and fishes, this part of the body has great magnitude, and takes the principal part (a few fishes excluded) in the duty of locomotion.

As the head is the seat of power in an animal, the part that gives honor to the whole, it is natural, that among species rank should be marked by means of variations in the structure of the head; and not only by variations in structure, but also in the extent to which the rest of the body directly contributes, by its members, to the uses or purposes of the head. *Cephalization* is, then, simply the degree of head-domination in the structure.

James D. Dana, "On some Results of the Earth's Contraction from cooling, including a discussion of the Origin of Mountains, and the nature of the Earth's Interior" (excerpt), *American Journal of Science* (1873)

WHILE mountains and mountain chains all over the world, and low lands, also, have undergone uplifts, in the course of their long history, that are not explained on the idea that all mountain elevating is simply what may come from plication or crushing, the *component parts* of mountain chains, or those simple mountain or mountain ranges that are *the product of one process of making* – may have received, *at the time of their original making*, no elevation beyond that resulting from plication.

This leads us to a grand distinction in orography, hitherto neglected, which is fundamental and of the highest interest in dynamical geology; a distinction between –

1. A simple or *individual* mountain mass or range, which is the result of *one process of making*, like an individual in any process of evolution, and which may be distinguished as a *monogenetic* range, being *one in genesis*; and
2. A composite or *polygenetic* range or chain, made up of two or more monogenetic ranges combined.

The Appalachian chain – the mountain region along the Atlantic border of North America – is a *polygenetic* chain; it consists, like the Rocky and other mountain chains, of several *monogenetic* ranges, the more important of which are: 1. The Highland range (including the Blue Ridge or parts of it, and the Adirondacks also, if these belong to the same process of making) pre-Silurian in formation; 2. The Green Mountain range, in western New England and eastern New York, completed essentially after the Lower Silurian era or during its closing period; 3. The Alleghany range, extending from southern New York southwestward to Alabama, and completed immediately after the Carboniferous age.

The making of the Alleghany range was carried forward at first through a long-continued subsidence – a *geosynclinal*[1] (not a *true* synclinal, since the rocks of the bending crust may have had in them many true or simple synclinals as well as anticlinals), and a consequent accumulation of sediments, which occupied the whole of Paleozoic time; and it was completed, finally in great breakings, faultings and foldings or plications of the strata, along with other results of disturbance. The folds are in several parallel lines, and rise in succession along the chain, one and another dying out after a course each of 10 to 150 miles; and some of them, if the position of the parts which remain after long denudation be taken as evidence, must have had, it has been stated, an altitude of many thousand feet; and there were also faultings of 8,000 to 10,000 feet, or, according to Lesley, of 20,000 feet.[2] This is one example of a *monogenetic* range.

The Green Mountains are another example in which the history was of the same kind: first, a slow subsidence or geosynclinals, carried forward in this case during the Lower Silurian era or the larger part of it; and, accompanying it, the deposition of sediments to a thickness equal to the depth of the subsidence; finally, as a result of the subsidence and as the climax in the effects of the pressure producing it, an epoch of plication, crushing, etc. between the sides of the trough.

In the Alleghany range the effect of heat were mostly confused to solidification; the reddening of such sandstones and shaly sandstones as contained a little iron in some form;[3] the coking of the mineral coal; and probably, on the western outskirts where the movements were small, the distillation of mineral oil, through the heating of shales or limestones containing carbohydrogen material, and its condensation in cavities among overlying strata; with also some metamorphism to the eastward; while in the making of the Green Mountains, there was metamorphism over the eastern, middle, and southern portions, and imperfect metamorphism over most of the western side to almost none in some western parts.

Another example is offered by the Triassico-Jurassic region of the Connecticut valley. The process included the same stages in kind as in the preceding cases. It began in a geosynclinals of probably 4,000 feet, this much being registered by the thickness of the deposits; but it *stopped short of metamorphism*, the sandstones being only reddened and partially solidified; and *short of plication or crushing*, the strata being only tilted in a monoclinal manner 15° to 25°; it ended in numerous great

---

[1] From the Greek γη, *earth*, and *synclinal*, it being a bend in the earth's crust.

[2] See an admirable paper on these mountains by Professors W. B. and H. D. Rogers, in the Trans. Assoc. Geol. and Nat., 1840–42. J. P. Lesley gives other facts in his "Manual of Coal and its Topography," and in many memoirs in the Proceedings of the American Philosophical Society. A brief account is contained in the author's Manual of Geology.

[3] Oxide of iron produced by a wet process at a temperature even as low as 212° F. is the red oxide $Fe_2 O_3$, or at least hjas a red powder. (Am. Jour. Sci., II, xliv, 292.)

longitudinal fractures, as a final catastrophe from the subsidence, out of which issued the trap (dolerite) that now makes Mt. Holyoke, Mt. Tom, and many other ridges along a range of 100 miles.[4]

These examples exhibit the characteristics of a large class of mountain masses or ranges. A geosynclinal accompanied by sedimentary depositions, and ending in a catastrophe of plications and solidification, are the essential steps, while metamorphism and igneous ejections are incidental results. The process is one that produces final stability in the mass and its annexation generally to the more stable part of the continent, though not stable against future oscillations of level *of wider range*, nor against denudation.

It is apparent that in such a process of formation elevation by direct uplift of the underlying crust has no necessary place. The attending plications may make elevations on a vast scale and so also may the shoves upward along the lines of fracture, and crushing may sometimes add to the effect; but elevation from an upward movement of the downward bent crust is only an incidental concomitant, if it occur at all.

We perceive thus where the truth lies in Professor LeConte's important principle. It should have in view along *monogenetic* mountains and these only *at the time of their making*. It will then read, plication and shovings along fractures being made more prominent than crushing:

Plication, shoving along fractures and crushing are the true sources of the elevation that takes place *during the making* of geosynclinals monogenetic mountains.

And the statement of Professor Hall may be made right if we recognize the same distinction, and, also, reverse the order and causal relation of the two events, accumulation and subsidence; and so make it read:

Regions of monogenetic mountains were, previous, and preparatory, to the making of the mountains, areas each of a slowly progressing geosynclinals, and, *consequently*, of thick accumulations of sediments.

The prominence and importance in orography of the mountain individualities described above as originating through a geosynclinals make it desirable that they should have a distinctive name; and I therefore propose to call a mountain range of this kind a *synclinorium*, from *synclinal* and the Greek *opos, mountain*.

---

[4] This history is precisely that which I have given in my Manual of Geology, though without recognizing the parallelism in stages with the history of the Alleghanies.

DANIEL KIRKWOOD (1814–1895)

Astronomer, mathematician.
Born in Harford County, Maryland. Died in Riverside, California.

Kirkwood was educated at a country school near his family's farm. Uninterested in farming, Kirkwood, at the age of 19, decided to teach. He found a position teaching at a country school not unlike the one he had attended. When a student asked to be taught algebra, which Kirkwood did not know, Kirkwood sat down with the student and together they worked their way through an elementary textbook. Kirkwood had found his passion for mathematics and decided to return to school himself, graduating from York County Academy in Pennsylvania in 1838 and then returning to teach mathematics at various secondary academies over the next decade. Throughout this time, Kirkwood read in mathematics and astronomy and published seven scholarly papers on astronomy. In 1849 he became an overnight sensation when his paper on a mathematical analogy between the rotations of the planets and their orbits was read by renowned astronomer Sears Cook Walker at the second meeting of the American Association for the Advancement of Science. It was Walker who first dubbed Kirkwood the "American Kepler." In 1851 he was offered a professorship in mathematics at Delaware College and then in 1856 at Indiana University, where he would remain for the next thirty years. His most significant contribution to astronomy is his study of the orbits of asteroids. The paper anthologized here presents his discovery of gaps in the distances of asteroids from the sun, called "Kirkwood Gaps" in his honor. He was also the first to suggest that meteor showers were the result of broken up comets.

Daniel Kirkwood, "On Comets and Meteors,"
Proceedings of the American Philosophical Society (1869)

THE comets which passed their perihelia in August, 1862, and January, 1866, will ever be memorable in the annals of science, as having led to the discovery of the intimate relationship between comets and meteors. These various bodies found revolving about the sun in very eccentric orbits may all be regarded as similar in their nature and origin, differing mainly in the accidents of magnitude and density. The recent researches, moreover, of Hoek, Leverrier and Schiaparelli, have led to the conclusion that such objects exist in great numbers in the interstellar spaces; that in consequence of the sun's progressive motion they are sometimes drawn towards the centre of our system; and that if undisturbed by any of the large planets they again pass off in parabolas or hyperbolas. When, however, as must sometimes be the case, they approach near Jupiter, Saturn, Uranus or Neptune, their orbits may be transformed into ellipses. Such, doubtless, has been the origin of the periodicity of the August and November meteors, as well as of numerous comets. In the present paper it is proposed to consider the probable consequences of the sun's motion through regions of space in which cosmical matter is widely diffused; to compare these theoretical deductions with the observed phenomena of comets, aerolites and falling stars; and thus, if possible, explain a variety of facts in regard to those bodies, which have hitherto received no satisfactory explanation.

1. As comets now moving in elliptic orbits owe their periodicity to the disturbing action of the major planets, and as this planetary influence is sometimes sufficient, especially in the case of Jupiter and Saturn, to change the direction of cometary motion, the great majority of periodic comets should move in the same direction with the planets. Now, of the comets known to be elliptical, 70 percent have direct motion. In this respect, therefore, theory and observation are in striking harmony.

2. When the relative positions of a comet and the disturbing planet are such as to give the transformed orbit of the former a small perihelion distance, the comet must return to the point at which it received its greatest perturbation; in other words, to the orbit of the planet. The aphelia of the comets of short period ought therefore to be found, for the most part, in the vicinity of the orbits of the major planets. The actual distances of these aphelia are as follows:

I. *Comets whose Aphelion Distances are Nearly Equal to 5.20, the Radius of Jupiter's Orbit.*

	Comets	Aph. Dist.		Comets	Aph. Dist.
1.	Encke's	4.09	7.	1766 II	5.47
2.	1819 IV	4.81	8.	1819 III	5.55
3.	De Vico's	5.02	9.	Brorsen's	5.64
4.	Pigott's (1743)	5.28	10.	D'Arrest's	5.75
5.	1867 I1	5.29	11.	Faye's	5.93
6.	1743 I	5.32	12.	Biela's	6.19

II. *Comets whose Aphelion Distances are Nearly Equal to 9.54, the Radius of Saturn's Orbit.*

	Comets.	Aph. Dist
1.	Peters' (1846 VI)	9.45
2.	Tuttle's (1856 I)	10.42

III. *Comets whose Aphelion Distances are Nearly Equal to* 19.18, *the Radius of Uranus's Orbit.*

Comets.	Aph. Dist.
1. 1867 I	19.28
2. Nov. Meteors	19.65
3. 1866 I	19.92

IV. *Comets whose Aphelion Distances are Nearly Equal to* 30.04, *the Radius of Neptune's Orbit.*

Comets	Aph. Dist.	Comets	Aph. Dist.
1. Westphal's (1852 IV)	31.97	4. De Vico's (1846 IV)	34.35
2. Pons' (1812)	33.41	5. Brorsen's (1847 V)	35.07
3. Olbers' (1815)	34.05	6. Halley's	35.37

The coincidences here pointed out (some of which have been noticed by others) appear, then, to be necessary consequences of the motion of the solar system through spaces occupied by meteoric nebulae. Hence the observed facts receive an obvious explanation.

In regard to comets of long period we have only to remark that, for any thing we know to the contrary, there may be causes of perturbation far exterior to the orbit of Neptune.

3. From what we observe in regard to the *larger* bodies of the universe – a clustering tendency being everywhere apparent, – it seems highly improbable that *cometic and meteoric* matter should be uniformly diffused through space. We would expect, on the contrary, to find it collected in cosmical clouds, similar to the visible nebule. Now, this, in fact, is precisely what has been observed in regard both to comets and meteors. In 150 years, from 1600 to 1750, 16 comets were visible to the naked eye;[1] of which 8 appeared in the 25 years from 1664 to 1689. Again, during 60 years, from 1750 to 1810, only 5 comets were visible to the naked eye, while in the next 50 years there were double that number. The probable cause of such variations is sufficiently obvious. As the sun in his progressive motion approaches a cometary group, the latter must, by reason of his attraction, move toward the centre of our system, the nearer members with greater velocity than the more remote. Those of the same cluster would enter the solar domain at periods not very distant from each other; the forms of their orbits depending upon their original relative positions with reference to the sun's course, and also on planetary perturbation. It is evident also that the

---

[1] See Humboldt's Cosmos, vol. IV. p. 538. The writer called attention to this variation as long since as 1861.

passage of the solar system through a region of space comparatively destitute of cometic clusters would be indicated by a corresponding paucity of comets. By the examination, moreover, of any complete table of falling stars we shall find a still more marked variation in the frequency of meteoric showers.

Previous to 1833, the periodicity of shooting stars had not been suspected. Hence the showers seen up to that date were observed *accidentally*. Since the great display of that year, however, they have been regularly *looked for*, especially at the November and August epochs. Consequently the numbers recently observed cannot properly be compared with those of former periods. Now, according to the Catalogue of Quetelet, 244 meteoric showers were observed from the Christian era to 1833. These were distributed as follows:

Centuries	No. of Showers	Centuries	No. of Showers
0 to 100	5	1000 to 1100	22
100 to 200	0	1100 to 1200	12
200 to 300	3	1200 to 1300	3
300 to 400	1	1300 to 1400	4
400 to 500	1	1400 to 1500	4
500 to 600	20	1500 to 1600	7
600 to 700	1	1600 to 1700	7
700 to 800	14	1700 to 1800	24
800 to 900	37	1800 to 1833	48
900 to 1000	31		

A remarkable secular variation in the number of showers is obvious from the foregoing table. During the 5 centuries from 700 to 1200, 116 displays are recorded; while in the 5 succeeding, from 1200 to 1700, the number is only 25. It will also be observed that another period of abundance commenced with the 18th century. A catalogue of meteoric stonefalls indicates also a corresponding increase in the number of aerolites, which cannot be wholly accounted for by the increased number of observers. Now, there are two obvious methods by which these variations may be explained. Either (1) the orbits of the meteoric rings which intersect the earth's path were so changed by perturbation towards the close of the 12th century as to prevent the appulse of the meteoric groups with the earth's atmosphere; or (2) the nebulous matter is very unequally diffused through the sidereal spaces. That the former has not been the principal cause is rendered extremely probable by the fact that the number of epochs of periodical showers was no greater during the cycle of abundance than in that of paucity. We conclude, therefore, that during the interval from 700 to 1200 the solar system was passing through, or near, a meteoric cloud of very great extent; that from 1200 to 1700 it was traversing a region comparatively destitute of such matter; and that about the commencement of the 18th century it again entered a similar nebula of unknown extent.

The fact that the August meteors, which have been so often subsequently observed, were *first* noticed in 811, renders it probable that the cluster was introduced into the planetary system not long previous to the year 800. It may be also worthy of remark that the elements of the comet of 770 A.D., are not very different front those of the August meteors and the 3d comet of 1862.[2]

Adopting Struve's estimate of the sun's orbital velocity, we find the diameter of the nebula traversed in 500 years to be 14 times that of Neptune's orbit.

It is remarkable that with the exception of Mars the perihelia of the orbits of all the principal planets fall in the same semi-circle of longitude − a fact which can hardly be regarded as accidental. Now, if the orbits were originally circular, the motion of the solar system through a nebulous mass not of uniform density would have the obvious effect of compelling the planets to deviate from their primitive orbits and move in ellipses of various eccentricities. It is easy to perceive, moreover, that the original perihelion points of all the orbits would be on that side of the system which had passed through the rarer portion of the nebulous mass. We have thus a possible cause of the eccentricity of the planetary orbits, as well as of the observed distribution of their perihelia.[3]

4. The particles of a cometic mass, being at unequal distances from the sun, will tend to move at different rates and in somewhat different orbits. This tendency will gradually overcome the feeble attractive force between the particles themselves. The most distant parts will thus become separated from the nucleus, and move in independent orbits. The motion of such meteoric matter will be in the same plane with that of the parent comet; the orbit of the former, however, being generally exterior to that of the latter. The connection recently discovered between comets and meteors, and especially the fact that the period of the November group is somewhat greater than that of the comet of 1866, are in striking harmony with the views here presented.

5. Owing to this loss of matter, periodic comets must become less brilliant, other things being equal, at each successive return; − a fact observed in regard to the comets of Halley and Biela.

6. The line of apsides of a large proportion of comets will be approximately coincident with the solar orbit. The point towards which the sun is moving is in longitude about 260°. The quadrants bisected by this point and that directly opposite extend from 215° to 305°, and from 35° to 125°. The number of cometary perihelia found in these

[2] The interval between the perihelion passage of 770 and that of 1862 is equal to 9 periods of 121.36 years. Oppolzer's determination of the period of 1862 II. is 121.5 years. Hind remarks that the elements of the comet of 770 are "rather uncertain," but says "that the general character of the orbit is decided." It may be worthy of remark that a great meteoric shower, the exact date of which has not been preserved, occurred in 770.

[3] This suggestion is due to R. A. Proctor, F.R.A.S., the distinguished author of "Saturn and its System."

quadrants up to July, 1868 (periodic comets being counted but once) was 159, or 62 per cent.; in the other two quadrants, 98, or 38 per cent.

This tendency of the perihelia to crowd together in two opposite regions has been noticed by different writers.

7. Comets whose positions before entering our system were very remote from the solar orbit must have *overtaken* the sun in its progressive motion; hence their perihelia must fall for the most part, in the vicinity of the point towards which the sun is moving; and they must in general have very small perihelion distances. Now, what are the observed facts in regard to the longitudes of the perihelia of the comets which have approached within the least distance of the sun's surface? But three have had a perihelion distance less than 0.01. *All* these, it will be seen by the following table, have their perihelia in close proximity to the point referred to:

I. *Comets whose Perihelion Distances are Less than 0.01.*

Perihelion Passage.	Per. Dist.	Long of Per.
1. 1668, Feb. 28d. 13h.	0.0047	277° 2′
2. 1680, Dec. 17 23	0.0062	262 4
3. 1843, Feb. 27 9	0.0055	278 39

In table II all but the last have their perihelia in the same quadrant.

II. *Comets whose Perihelion Distances are Greater than 0.01 and Less than 0.05*

Perihelion Passage.	Per. Dist.	Long. of Per.
1. 1689, Nov. 29d 4h	0.0189	269° 41′
2. 1816, March 1 8	0.0485	267 35
3. 1826, Nov. 18 9	0.0268	315 31
4. 1847, March 30 6	0.0425	276 2
5. 1865, Jan. 14 7	0.0260	141 15

The perihelion of the first comet in table III is remote from the direction of the sun's motion; that of the second is distant but 14°, and of the third, 21°.

III. *Comets whose Perihelion Distances are Greater than 0.05 and Less than 0.1.*

Perihelion Passage.	Per. Dist.	Long. of Per.
1. 1593, July 18d. 13h.	0.0891	176° 19′
2. 1780, Sept. 30 22	0.0963	246 35
3. 1821, March 21 12	0.0918	239 29

With greater perihelion distances the tendency of the perihelia to crowd together around the point indicated is less distinctly marked.

8. Few comets of small perihelion distance should have their perihelia in the vicinity of longitude 80°, the point opposite that towards which the sun is moving. Accordingly we find, by examining a table of cometary elements, that with a perihelion distance less than 0.1, there is not a single perihelion between 35° and 125°; between 0.1 and 0.2, but 3; and between 0.2 and 0.3 only 1.

BENJAMIN PEIRCE (1809–1880)

Mathematician, astronomer.
Born in Salem, Massachusetts. Died in Cambridge, Massachusetts.

AFTER graduating from Harvard College in 1829, Peirce taught briefly at the Round Hill School, an experimental secondary school in Northampton. He soon returned to Harvard as a math tutor and assisted Nathaniel Bowditch in his translation of Laplace's *Mecanique celeste*. In 1833 Peirce was hired by Harvard as a professor in mathematics, a position that came to include astronomy after 1842. He would hold this position until his death. He was never a widely popular teacher, having little patience for small talent, but he could be quite stimulating for the exceptional student. As a colleague, Julian Lowell Coolidge, put it, "His great natural mathematical talent and originality of thought, combined with a total inability to put anything clearly, produced upon his contemporaries a feeling of awe that amounted almost to dread" (Coolidge, p. 246). Peirce investigated the orbit of Neptune in relation to Uranus, wrote on complex associative algebras, number theory, and the nature and necessity of mathematics. He produced a number of textbooks marked by the same terse brevity that lesser students found unwelcoming. In addition to his duties at Harvard he was consulting astronomer to the *American Nautical Almanac*, for which he prepared new tables of the moon in 1852 and superintendent of the Coast Survey from 1867 to 1874. In his lifetime he was probably best known for his book *System of Analytical Mechanics* but his real brilliance could be better seen in his *Linear Associative Algebra* (the opening of which is anthologized here). Peirce had only a few copies of the latter lithographed for friends and colleagues in 1870, but his son Charles Sanders Peirce had the work reprinted in the *American Journal of Mathematics* in 1882. The sons of Benjamin Peirce were another great legacy: James Mills, a Harvard mathematician; Benjamin Mills, a mining engineer; Herbert Henry Davis, a diplomat; and the brilliant polymath Charles Sanders, who is anthologized in Part III.

Benjamin Peirce, "Linear Associative Algebra," A Memoir read before the National Academy of Sciences in Washington, 1870 (excerpt), *American Journal of Mathematics* (1881)[1]

1. Mathematics is the science which draws necessary conclusions.
    This definition of mathematics is wider than that which is ordinarily given, and by which its range is limited to quantitative research. The ordinary definition,

[1] With notes and addenda, by C. S. Peirce, son of the author.

like those of other sciences, is objective; whereas this is subjective. Recent investigations, of which quaternions is the most noteworthy instance, make it manifest that the old definition is too restricted. The sphere of mathematics is here extended, in accordance with the derivation of its name, to all demonstrative research, so as to include all knowledge strictly capable of dogmatic teaching. Mathematics is not the discoverer of laws, for it is not induction; neither is it the framer of theories, for it is not hypothesis; but it is the judge over both, and it is the arbiter to which each must refer its claims; and neither law can rule nor theory explain without the sanction of mathematics. It deduces from a law all its consequences, and develops them into the suitable form for comparison with observation, and thereby measures the strength of the argument from observation in favor of a proposed law or of a proposed form of application of a law.

Mathematics, under this definition, belongs to every enquiry, moral as well as physical. Even the rules of logic, by which it is rigidly bound, could not be deduced without its aid. The laws of argument admit of simple statement, but they must be curiously transposed before they can be applied to the living speech and verified by observation. In its pure and simple form the syllogism cannot be directly compared with all experience, or it would not have required an Aristotle to discover it. It must be transmuted into all the possible shapes in which reasoning loves to clothe itself. The transmutation is the mathematical process in the establishment of the law. Of some sciences, it is so large a portion that they have been quite abandoned to the mathematician, – which may not have been altogether to the advantage of philosophy. Such is the case with geometry and analytic mechanics. But in many other sciences, as in all those of mental philosophy and most of the branches of natural history, the deductions are so immediate and of such simple construction, that it is of no practical use to separate the mathematical portion and subject it to isolated discussion.

2. The branches of mathematics are as various as the sciences to which they belong, and each subject of physical enquiry has its appropriate mathematics. In every form of material manifestation, there is a corresponding form of human thought, so that the human mind is as wide in its range of thought as the physical universe in which it thinks. The two are wonderfully matched. But where there is a great diversity of physical appearance, there is often a close resemblance in the processes of deduction. It is important, therefore, to separate the intellectual work from the external form. Symbols must be adopted which may serve for the embodiment of forms of argument, without being trammeled by the conditions of external representation or special interpretation. The words of common language are usually unfit for this purpose, so that other symbols must be adopted, and mathematics treated by such symbols is called *algebra*. Algebra, then, is formal mathematics.

3. All relations are either qualitative or quantitative. Qualitative relations can be considered by themselves without regard to quantity. The algebra of such enquiries may be called logical algebra, of which a fine example is given by Boole.

Quantitative relations may also be considered by themselves without regard to quality. They belong to arithmetic, and the corresponding algebra is the common or arithmetical algebra.

In all other algebras both relations must be combined, and the algebra must conform to the character of the relations.

4. The symbols of an algebra, with the laws of combination, constitute its *language*; the methods of using the symbols in the drawing of inferences is its *art*; and their interpretation is its *scientific application*. This three-fold analysis of algebra is adopted from President Hill, of Harvard University, and is made the basis of a division into books.

Book I[2]

The Language of Algebra

5. The language of algebra has its alphabet, vocabulary, and grammar.
6. The symbols of algebra are of two kinds: one class represents its fundamental conceptions and may be called its *letters*, and the other represent the relations or modes of combination of the letters and are called the signs.
7. The *alphabet* of an algebra consists of its letters; the *vocabulary* defines its signs and the elementary combinations of its letters: and the *grammar* gives the rules of composition by which the letters and signs are united into a complete and consistent system.

The Alphabet

8. Algebras may be distinguished from each other by the number of their independent fundamental conceptions, or of the letters of their alphabet. Thus an algebra which has only one letter in its alphabet is a *single* algebra; one which has two letters is a *double* algebra; one of three letters a *triple* algebra; one of four letters a *quadruple* algebra, and so on.

This artificial division of the algebras is cold and uninstructive like the artificial Linnean system of botany. But it is useful in a preliminary investigation of algebras, until a sufficient variety is obtained to afford the material for a natural classification.

Each fundamental conception may be called a *unit*; and thus each unit has its corresponding letter, and the two words, unit and letter, may often be used indiscriminately in place of each other, when it cannot cause confusion.

9. The present investigation, not usually extending beyond the sextuple algebra, limits the demand of the algebra for the most part to six letters; and the six letters, i, j, k, l, m and n, will be restricted to this use except in special cases.

---

[2] Only this book was ever written. [C. S. P.]

10. *For any given letter another may be substituted*, provided a new letter represents a combination of the original letters of which the replaced letter is a necessary component.

   For example, any combination of two letters, which is entirely dependent for its value upon both of its components, such as their sum, difference, or product, may be substituted for either of them.

   This *principle of the substitution of letters* is radically important, and is a leading element of originality in the present investigation; and without it, such an investigation would have been impossible. It enables the geometer to analyse an algebra, reduce it to its simplest and characteristic forms, and compare it with other algebras. It involves in its principle a corresponding substitution of *units* of which it is in reality the formal representative.

   There is, however, no danger in working with the symbols, irrespective of the ideas attached to them, and the consideration of the change of the original conceptions may be safely reserved for the *book of interpretation*.

11. In making the substitution of letters, the original letter will be preserved with the distinction of a subscript number.

   Thus, for the letter i there may successively be substituted i_1, i_2, i_3, etc. In the final forms, the subscript numbers can be omitted, and they may be omitted at any period of the investigation, when it will not produce confusion.

   It will be practically found that these subscript numbers need scarcely ever be written. They pass through the mind, as a sure ideal protection from erroneous substitution, but disappear from the writing with the same facility with which those evanescent chemical compounds, which are essential to the theory of transformation, escape the eye of the observer.

12. A *pure* algebra is one in which every letter is connected by some indissoluble relation with every other letter.

13. When the letters of an algebra can be separated into two groups, which are mutually independent, it is a *mixed algebra*. It is mixed even when there are letters common to the two groups, provided those which are not common to the two groups are mutually independent. Were an algebra employed for the simultaneous discussion of distinct classes of phenomena, such as those of sound and light, and were the peculiar units of each class to have their appropriate letters, but were there no recognized dependence of the phenomena upon each other, so that the phenomena of each class might have been submitted to independent research, the one algebra would be actually a mixture of two algebras, one appropriate to sound, the other to light.

   It may be farther observed that when, in such a case as this, the component algebras are identical in form, they are reduced to the case of one algebra with two diverse interpretations.

EDWARD DRINKER COPE (1840–1897)

Paleontologist, herpetologist, ichthyologist.
Born in Philadelphia, Pennsylvania. Died in Philadelphia, Pennsylvania.

COPE came from a wealthy Quaker family. He was educated as a youth through a combination of Friends' Schools and private tutors. Although he never graduated from a four-year college, he did attend the University of Pennsylvania for a year around 1858, studying anatomy and working for Joseph Leidy. The following year, at the age of 19, Cope wrote his first paper, on Salamandridae, and read it before the Academy of Natural Sciences at Philadelphia. He had by this time decided that natural history was his calling. In 1863when he went on a tour of Europe, it was to visit the great museums, take classes at the renowned German universities, and talk with European scientists. When he returned in 1864, he took a position as professor of Natural Sciences at Haverford College. Throughout this period he continued to write almost exclusively on amphibians and reptiles, but in 1864 he began to expand his focus. The following year he became curator to the Philadelphia Academy of Natural Sciences and the collections there only further widened his interests. He left Haverford in 1867 and by 1871 was beginning the exploration phase of his career throughout the West, on expeditions of his own design and then from 1874–77 on three separate U.S. Geological Survey trips.

In 1875 Cope's father died and left him a substantial inheritance. In 1878 he purchased the rights to the *American Naturalist*, which became something of a neo-Lamarckian publication under Cope's editorship. Cope was an incredibly prodigious scientist; although his prodigiousness is at times noted as a fault as much as a virtue, he was undoubtedly one of the great naturalists of the second half of the century. His biographers always stress his love of herpetology even though much of his fame came from his paleontology. It is for Cope that *Copeia*, the journal of the American Society of Ichthyologists and Herpetologists is named. While his nemesis Marsh may have beat him in the number of fossil dinosaurs identified, Cope was far ahead in total number of papers published, over 1,200 on extant and extinct fish, reptiles, amphibians, and mammals. Cope accepted a professorship at the University of Pennsylvania in 1889, where he remained until his death.

E. D. Cope, "The Laws of Organic Development" (excerpt), *American Naturalist* (1871)

II. On the Laws of Evolution

Wallace and Darwin have propounded as the cause of modification in descent their law of natural selection. This law has been epitomized by Spencer as the "preservation of the fittest." This neat expression no doubt covers the case, but it leaves the origin of the fittest entirely untouched. Darwin assumes a "tendency to variation" in nature, and it is plainly necessary to do this in order that materials for the exercise of a selection should exist. Darwin and Wallace's law is, then, only restrictive, directive, conservative, or destructive of something already created. Let us, then, seek for the originative laws by which these subjects are furnished – in other words, for the causes of the origin of the fittest.

The origin of new structures which distinguish one generation from those which have preceded it, I have stated to take place under the law of *acceleration*. As growth (creation) of parts usually ceases with maturity, it is entirely plain that the process of acceleration is limited to the period of infancy and youth in all animals. It is also plain that the question of growth is one of nutrition, or of the construction of organs and tissues out of protoplasm.

The construction of the animal types is restricted to two kinds of increase – the addition of identical segments and the addition of identical cells. The first is probably to be referred to the last, but the laws which give rise to it cannot be here explained. Certain it is that segmentation is not only produced by addition of identical parts, but also by subdivision of a homogeneous part. In reducing the vertebrate or most complex animal to its simplest expression, we find that all its specialized parts are but modifications of the segment, either simply or as sub-segments of compound but identical segments. Gegenbaur has pointed out that the most complex limb with hand or foot is constructed, first, of a single longitudinal series of identical segments, from each of which a similar segment diverges, the whole forming parallel series, not only in the oblique transverse, but generally in the longitudinal sense. Thus the limb of the Lepidosiren represents the simple type, that of the Icthyosaurus a first modification. In the latter the first segment only (femur or humerus) is specialized, the other pieces being undistinguishable. In the Plesiosaurian paddle the separate parts are distinguished; the ulna and radius well marked, the carpal pieces hexagonal, the phalanges well marked, etc.

As regards the whole skeleton the same position may be safely assumed. Though Huxley may reject Owen's theory of the vertebrate character of the segments of the cranium, because they are so very different from the segments in other parts of the column, the question rests entirely on the definition of a vertebra. If a vertebra be a segment of the skeleton, of course the skull is composed of vertebrae; if not, then the cranium may be said to be formed of "sclerotomes," or some other name may be used. Certain it is, however, that the parts of the segments of the cranium may be now more or less completely parallelized or homologised with each other, and that as we descend the scale of vertebrated animals, the resemblance of these segments to vertebrae increases, and the constituent segments of each become more similar. In the types where the greatest resemblance is seen, segmentation of either is incomplete, for they retain the original cartilaginous basis. Other animals which present cavities or parts of a solid support are still more easily reduced to a simple basis of segments, arranged either longitudinally (worm) or centrifugally (star-fish, etc.)

Each segment – and this term includes not only the parts of a complex whole but parts always subdivided, as the jaw of a whale or the sac-body of a mollusk, – is constructed, as is well know by cell division. In the growing foetus the first cell divides its nucleus and then its whole outline, and this process repeated millions of times produces, according to the cell theory, all the tissues of the animal organism or their bases from first to last. That the ultimate or histological elements of all organs are produced originally by repetitive growth of simple, nucleated cells with various modifications of exactitude of repetition in the more complex, is taught by the cell theory. The formation of some of the tissues is as follows:

First Change – Formation of simple nucleated cells from homogeneous protoplasm or the cytoblastema.

Second – Formation of new cells by division of body and nucleus of the old.

Third – Formation of tissues by accumulation of cells with or without addition of intercellular cytoblastema.

A. In connective tissue by slight alteration of cells and addition of cytoblastema.
B. In blood, by addition of fluid cytoblastema (fibrin) to free cells (lymph corpuscles) by loss of membrane, and by cell development of nucleus.
C. In muscles by simple confluence of cells, end to end and mingling of contents (Kölliker).
D. Of cartilage by formation of cells in cytoblast which break up, their contents being added to cytoblast; this occurring several times, the result being an extensive cytoblast with few and small cells (Vogt). The process is here an attempt at development with only partial success, the result being a tissue of small vitality.

Even in repair-nutrition recourse is had to the nucleated cell. For Cohnheim first shows that if the corner of a frog's eye be scarified, repair is immediately set on foot by the transportation thither of white or lymph or nucleated corpuscles from the neighboring lymph heart. This he ascertained by introducing aniline dye into the latter. Repeated experiments have shown that this is the history in great part of the construction of new tissue in the adult man.

Now, it is well know that the circulating fluid of the foetus contains for a period only these nucleated cells as corpuscles, and that the lower vertebrates have a greater proportion of these corpuscles than the higher, whence probably the greater facility for repair or reconstruction of lost limbs or parts enjoined by them. The invertebrates possess only nucleated blood corpuscles.

What is the relation of cell division to the forces of nature, and to which of them as a cause is it to be referred, if to any? The animal organism transfers the chemism of the food (protoplasm) to correlated amounts of heat, motion, electricity, light (phosphorescence), and nerve force. But cell division is an affection of protoplasm distinct from any of these. Addition to homogeneous lumps or parts of protoplasm (as in that lowest animal, *Protamoeba* of Haeckel) may be an exhibition of mere molecular force, or addition as is seen in the crystal, but cell division is certainly something distinct. It looks to me like an exhibition of another force, and though this is still an open question, it may be called for the present *growth force*. It is correlated to the other forces, for its exhibitions cease unless the protoplasm exhibiting it be fed. It is potential in the protoplasm of both protoplasmic animal mass and protoplasmic food, and becomes energetic on the union of the two. So long as cell division continues it is energetic; when cells burst and discharge the contained cytoblastema, as in the formation of cartilage, it becomes again potential.

The size of a part is then dependent on the amount of cell division or growth force, which as given it origin, and the number of segments is due to the same cause. The whole question, then, of the creation of animal and vegetable types is reduced to one of the amount and location of growth force.

Before discussing the influences which have increased and located growth force, it will be necessary to point out the mode in which these influences must necessarily have affected growth. Acceleration is only possible during the period of growth in animals, and during that time most of them are removed from the influence of physical or biological causes either through their hidden lives or incapacity for the energetic performance of life functions. These influences must, then, have operated on the parents, been rendered potential in their reproductive cells, and become energetic in the growing foetus of the next generation. However little we may understand this mysterious process, it is nevertheless a fact. Says Murphy, "There is no action which may not become habitual, and there is no habit that may not be inherited." Materialized, this may be rendered – there is no act which does not direct growth force, and therefore there is no determination of grown force which may not become habitual; there is, then, no habitual determination of growth force which may not be inherited; and of course in a growing foetus becomes at once energetic in the production of new structure in the direction inherited, which is acceleration.

III. The Influences Directing Growth Force

Up to this point we have followed paths more or less distinctly traced in the field of nature. The positions taken appear to me either to have been demonstrated or to have a great balance of probability in their favor. In the closing part of these remarks I shall indulge in more of hypothesis than heretofore.

What are the influences locating growth force? First, physical and chemical causes; second, use; third, effort. I leave the first, as not especially prominent in the economy of type growth among animals, and confine myself to the two following. The effects of use are well known. We cannot use a muscle without increasing its bulk; we cannot use the teeth in mastication without inducing a renewed deposit of dentine within the pulp-cavity to meet the encroachments of attrition. The hands of the laborer are always larger than those of men of other pursuits. Pathology furnishes us with a host of hypertrophies, exostoses, etc., produced by excessive use, or necessity for increased means of performing excessive work. The tendency, then, induced by use in the parent is to add segments or cells to the organ used. Use thus determines the locality of new repetitions of parts already existing, and determines an increase of growth force at the same time, by the increase of food always accompanying increase of work done, in every animal.

But supposing there be no part or organ to use. Such must have been the condition of every animal prior to the appearance of an additional digit or limb or other useful element. It appears to me that the cause of the determination of growth force is not merely the irritation of the part or organ used by contact with the objects of its use. This would seem to be the remote cause of the deposit of dentine in the used tooth, in the thickening epidermis of the hand of the laborer, in the wandering of the lymph-cells to the scarified cornea of the frog in Cohnheim's experiment. You cannot rub the sclerotica of the eye without producing an expansion of the capillary arteries and corresponding increase in the amount of nutritive fluid. But the case may be different in the muscles and other organs (as the pigment cells of reptiles and fishes) which are under the control of the volition of the animal. Here, and in many other instances which might be cited, it cannot be asserted that the nutrition of use is not under the direct control of the will through the mediation of nerve force. Therefore I am disposed to believe that growth force may be, by the volition of the animal, as readily determined to a locality where an executive organ does not exist, as to the first segment or cell of such an organ already commenced, and that therefore effort is in the order of time, the first factor in acceleration.

Effort and use have, however, very various stimuli to their exertion.

Use of a part by an animal is either compulsory or optional. In either case the use may be followed by an increase of nutrition under the influence of reflex force or of direct volition.

A compulsory use would naturally occur in new situations which take place apart from the control of the animal, where no alternatives are presented. Such a case would arise in a submergence of land where land animals might be imprisoned on an island or in swamps surrounded by water, and compelled to assume a more or less aquatic life. Another case which has also probably often occurred, would be when the enemies of a species might so increase as to compel a large number of the latter to combat who would previously have escaped it.

In these cases the structure produced would be necessarily adaptive. But the effect would be most frequently to destroy or injure the animals (retard them) thus brought into new situations and compelled to an additional struggle for existence, as has, no doubt, been the case in geologic history. Preservation, with modifications would only ensue where the changes should be introduced very gradually. This mode is always a consequence of the optional use. The cases here included are those where choice selects from several alternatives, thus exercising its influence on structure. Choice will be influenced by the emotions, the imagination, and by intelligence.

As examples of intelligent selection the modified organisms of the varieties of bees and ants must be regarded as striking examples of its exercise. Had all in the hive or hill been modified alike, as soldiers, queens, etc., the origin of the structures might have been thought to be compulsory; but varied and adapted as the different forms are to the wants of a community, the influence of intelligence is too obvious to be denied. The structural results are obtained in this case by a shorter road than by inheritance.

The selection of food offers an opportunity for the exercise of intelligence, and the adoption of means for obtaining it still greater ones. It is here that intelligent selection proves its supremacy as a guide of use, and consequently of structure, to all the other agencies here proposed. The preference for vegetable or for animal food determined by the choice of individual animals among the omnivores, which were, no doubt, according to the palaeontological record the predecessors of our herbivores, and perhaps of carnivores also, must have determined their course of life and thus of all their parts into those totally distinct directions. The choice of food under ground, on the ground, or in the trees would necessarily direct the uses of organs in those directions respectively.

Intelligence is a conservative principle and always will direct effort and use into lines which will be beneficial to its possessor. Thus we have the source of the fittest – *i.e.*, addition of parts by increase and location of growth force directed by the will – the will being under the influence of various kinds of compulsory choice in the lower, and intelligent option among higher animals. Thus intelligent choice may be regarded as the originator of the fittest, while natural selection is the tribunal to which all the results of accelerated growth are submitted. This preserves or destroys them, and determines the new points of departure on which accelerated growth shall build.

Acceleration under the influence of effort accounts for the existence of rudimental characters. Many other characters will follow at a distance, the modifications proceeding in accordance with the laws here proposed, and retardation is accounted for by complementary or absolute loss of growth force.

OTHNIEL CHARLES MARSH (1831–1899)

Palaeontologist.
Born in Lockport, New York. Died in New Haven, Connecticut.

MARSH was the nephew of George Peabody, the financier and philanthropist, and it was Peabody who financed Marsh's education: Philips Exeter Academy, Yale College, two years more at Yale's Sheffield Scientific School, and then several years of study at German universities in Berlin and Breslau. In 1867, Marsh convinced his uncle to finance a museum of natural history at Yale, which would become the Peabody Museum of Natural History. The same year, Marsh became professor of palaeontology at Yale, the first in the United States and only the second such professor in the world. In 1869, Peabody died and left Marsh a considerable inheritance, which allowed him to forgo his Yale salary and teaching. Marsh then focused on building the fossil and archaeological collections for the Peabody Museum and writing up his findings. Marsh began seriously collecting in the late 1860s and found a large variety of horse like bones in Nebraska and the Dakotas. By the mid-seventies, Marsh had one of the largest collections in the world of early mammal fossils, particularly equine ones, having spent, quite literally, a small fortune on them. It is perhaps his work on the fossil horses anthologized here that is most memorable (the Bone Wars not withstanding; see the introduction to Part II for more on Marsh, Cope and the Bone Wars), because he had enough fossils to be able to track the change or evolution of the early horse in America. The evidence of evolution he presented was the most complete to date. His later work on bird fossils with teeth would also prove probative. Thomas Henry Huxley visited the Yale collection to see the complete horse fossils and believed they demonstrated clearly the line of descent of an extant animal. Darwin too expressed a desire to see the fossils that so clearly demonstrated his theory but never did. Marsh was an affirmed evolutionist and considered Huxley a friend and mentor. Marsh was active in the AAAS, serving as vice president, and was in charge of the vertebrate palaeontology section of the U.S. Geological Survey. In May 1871 he discovered the first pterodactyl remains in North America. He wrote important monographs on *Odontornithes*, *Dinocerata*, as well as a general monograph on the North American dinosaurs.

O. C. Marsh, "Fossil Horses in America," *American Naturalist* (1874)

IT is a well-known fact that the Spanish discoverers of America found no horses on this continent, and that the modern horse (*Equus caballus* Linn.) was subsequently introduced from the old world. It is, however, not so generally known that these animals had formerly been abundant here, and that long before, in Tertiary time, near relatives of the horse, and probably his ancestors, existed in the far west in countless numbers, and in a marvellous variety of forms. The remains of equine mammals, now know from the Tertiary and Quaternary deposits of this country, already represent more than double the number of genera and species hitherto found in the strata of the eastern hemisphere, and hence afford most important aid in tracing out the genealogy of the horses still existing.

The animals of this group which lived in this country during the three divisions of the Tertiary period were especially numerous in the Rocky Mountain regions, and their remains are well preserved in the old lake basins which then covered so much of that country. The most ancient of these lakes – which extended over a considerable part of the present territories of Wyoming and Utah – remained so long in Eocene times that the mud and sand, slowly deposited in it, accumulated to more than a mile in vertical thickness. In these deposits, vast numbers of tropical animals were entombed, and here the oldest equine remains occur, four species of which have been described. These belong to the genus *Orohippus* Marsh, and are all of diminutive size, hardly larger than a fox. The skeleton of these animals resembled that of the horse in many respects, much more indeed than any other existing species, but instead of the single toe on each foot, so characteristic of all modern equines, the various species of *Orohippus* had four toes before and three behind, all of which reached the ground. The skull, too, was proportionately shorter, and the orbit was not enclosed behind by a bridge of bone. There were forty-four teeth in all, and the premolars were smaller than the molars. The crowns of these teeth were very short. The canine teeth were developed in both sexes, and the incisors did not have the "mark" which indicates the age of the modern horse. The radius and ulna were separate, and the latter was entire throughout its whole length. The tibia and fibula were distinct. In the fore foot, all the digits except the pollex, or first, were well developed, as shown in the accompanying figure (73) of the left fore foot of *Orohippus agilis* Marsh. The third digit is the largest, and its close resemblance to that of the horse is clearly marked. The terminal phalanx, or coffin bone, was a shallow median groove in front, as in many species of this group in the later Tertiary. The fourth digit exceeds the second in size, and the fifth is much the shortest of all. Its metacarpal bone is considerably curved outward. In the hind foot of this genus, there are but three digits. The fourth metatarsal is much larger than the second.

The only species of *Orohippus* at present known are from the Eocene of Wyoming and Utah, and are as follows: – *Orohippus gracilis* Marsh, *O. pumilus* Marsh, *O. agilis* Marsh, and *O. major* Marsh.[1]

In the middle of Tertiary, or Miocene, there were two other lakes on either side of the great Eocene basin. The largest of these was east of the Rocky Mountains, extending over portions of what are now Dakota, Nebraska and Colorado. The clays deposited in this lake form the "*Mauvaises terres*," or "Bad lands," of that region, and

[1] American Journal of Science, Vol. vii, p. 247, March, 1874.

Fig. 73.

Orohippus
agilis.

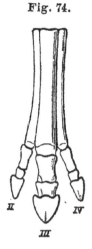

Fig. 74.

Miohippus
annectens.

are well known for their fossil treasures. The other Miocene lake was west of the Blue mountains, where eastern Oregon now is, but its extent is unknown, as this whole region has since been covered with a vast sheet of basalt, a thousand or more feet in thickness, and the original lake sediments are only to be seen where this lava has been washed away. In both of these ancient lake basins, many remains of animals allied to the horse are found, showing that during the Miocene this group of mammals were well represented.

In the western, or Oregon basin, the genus *Miohippus* Marsh first makes its appearance. It resembles *Orohippus* of the Eocene in its general characters, especially in the shape of the skull, number and form of teeth, and separate ulna; but it had only three toes in the fore foot, as well as behind, and the fibula was coössified with the tibia at its lower end. In this genus, all the toes reached the ground, as shown in the accompanying figure of the left fore foot of *Miohipus annectens* Marsh, the type species (Fig. 74). In the same deposits, the genus *Anchitherium* Meyer occurs, represented by a single species, *A. anceps* Marsh. This genus is closely allied to *Miohippus*, but differs in having a deep depression in the skull in front of the orbit. The radius and ulna are united, and the outer toes are reduced in size. In the eastern basin, *Anchitherium Bairdi* Leidy is abundant, and with it is found a smaller species, *A. celer* Marsh. The animals of these two genera are all larger than the species of *Orohippus* from the Eocene, some of them exceeding a sheep in size. The Miocene species known with certainty are as follows: – *Miohippus annectens* Marsh, *Miohippus Condoni* (*Anchitherium Condoni* Leidy) and *Anchetherium anceps* Marsh, from Oregon; and *A. Bairdi*, Leidy, and *A. celer* Marsh, from the eastern basin.

During the Pliocene, or later Tertiary, a great development of the horse family took place, and vast numbers of these animals left their remains in the lake deposits of that epoch. The largest of these lakes had the Rocky Mountains for its western border, and extended from Dakota to Texas, its northern part covering the bed of the older Miocene basin. Another Pliocene lake, of unknown limits, extended over the older Tertiary strata of eastern Oregon, and evidence of still others may be seen in Idaho, Nevada and California. In all of these basins, equine remains of various kinds have been found, but the most important localities are the region of the Niobrara river east of the mountains, and the valley of the John Day river in Oregon.

The equine genera of the Pliocene which appear to be most nearly related to their predecessors from older strata are, *Anchipus* Leidy, *Hipparion* Christol, and *Protohippus* Leidy, all three-toed forms, but with the outer digits reduced to much the same proportions as the posterior hooflets of the modern deer and ox. The genus *Pliohippus* Marsh, from the same deposits, had feet like those of the recent horse. Other genera, less know, which have been proposed, are *Parahippus*, *Merychippus*, and *Hypohippus* of Leidy, to whose researches we are so largely indebted for our present knowledge of this group. Of these Pliocene genera, more than twenty species have been described from American strata, all apparently larger than their Miocene relatives, but all smaller than the present horse, and many of them approaching the ass in size. Among the more characteristic of these species may be mentioned, *Anchippus Texanus* Leidy, from Texas; *A. brevidens* Marsh, from Oregon; *Hipparion occidentale* Leidy, and *H. speciosum* Leidy, from Nebraska; *Protohippus perditus* Leidy, from the Niobrara; *P. parvulus* Marsh, from Nebraska, the smallest Pliocene species; *Parahippus cognatus* Leidy, and *Pliohippus pernix* Marsh, from the Niobrara.

In the upper Pliocene, or more probably in the transition beds above, there first appears a true *Equus*, and in the Quaternary deposits, remains of this genus are not uncommon. Five or six species are known from the United States, and several others from Central and South America. The latest extinct species appears to have been *Equus fraternus* Leidy, which cannot be distinguished anatomically from the existing horse. These later extinct horses are all larger than the Pliocene Equines, and some of them even exceeded in size the living species.

The large number of equine mammals now know from the Tertiary deposits of this country, and their regular distribution through the subdivisions of this formation, afford a good opportunity to ascertain the probable lineal descent of the modern horse. The American representative of the latter is the extinct *Equus fraternus* Leidy, a species almost, if not entirely, identical with the old world *Equus caballus* Linn., to which our recent horse belongs. Huxley has traced successfully the later genealogy of the horse through European extinct forms,[2] but the line in America was probably a more direct one, and the record is more complete. Taking, then, as the extremes of a series, *Orohippus agilis* Marsh, from the Eocene, and *Equus fraternus* Leidy, from the Quaternary, intermediate forms may be intercalated with considerable certainty from the thirty or more well marked species that lived in the intervening periods. The

[2] Anniversary Address, Geological Society of London, 1870.

Fig. 75.

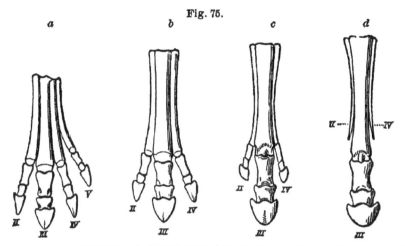

a, Orohippus (Eocene); *b*, Miohippus (Miocene); *c*, Hipparion (Pliocene); *d*, Equus (Quaternary).

natural line of descent would seem to be through the following genera: – *Orohippus*, of the Eocene; *Miohippus* and *Anchitherium*, of the Miocene; *Anchippus*, *Hipparion*, *Protohippus* and *Pliohippus*, of the Pliocene; and *Equus*, Quaternary and recent.

The most marked changes undergone by the successive equine genera are as follows: 1st, increase in size; 2d, increase in speed, through concentration of limb bones; 3d, elongation of head and neck, and modifications of skull. The increase in size is remarkable. The Eocene *Orohippus* was about the size of a fox. *Miohippus* and *Anchitherium*, from the Miocene, were about as large as a sheep. *Hipparion* and *Pliohippus*, of the Pliocene, equaled the ass in height: while the size of the Quaternary *Equus* was fully up to that of the modern horse.

The increase of speed was equally marked, and was a direct result of the gradual modification of the limbs. The latter were slowly concentrated, by the reduction of their lateral elements and enlargement of the axial one, until the force exerted by each limb came to act directly through its axis, in the line of motion. This concentration is well seen, *e.g.*, in the fore limb. There was, 1st, a change in the scapula and humerus, especially in the latter, which facilitated motion in one line only; 2d, an expansion of the radius, and reduction of the ulna, until the former alone remained entire, and effective; 3d, a shortening of all the carpal bones, and enlargement of the median ones, ensuring a firmer wrist; 4th, an increase in size of the third digit, at the expense of those on each side, until the former alone supported the limb. The latter change is clearly shown in the above diagram (Fig. 75), which represents the fore feet of four typical genera in the equine series, taken in succession from each of the geological periods in which this group of mammals is known to have lived.

The ancient *Orohippus* had all four digits of the fore feet well developed. In *Miohippus*, of the next period, the fifth toe has disappeared, or is only represented by a rudiment, and the limb is supported by the second, third and fourth, the middle one being the largest.

Hipparion, of the later Tertiary, still has three digits, but the third is much stouter, and the outer ones have ceased to be of use, as they do not touch the ground. In *Equus*, the last of the series, the lateral hoofs are gone, and the digits themselves are represented only by the rudimentary splint bones.[3] The middle, or third digit, supports the limb, and its size has increased accordingly. The corresponding changes in the posterior limb of these genera are very similar, but not so striking, as the oldest type (*Orohippus*) had but three toes behind. An earlier ancestor of the group, perhaps in the lowest Eocene, probably had four toes on this foot, and five in front. Such a predecessor is as clearly indicated by the feet of *Orohippus*, as the latter is by its Miocene relative. A still older ancestor, possibly in the Cretaceous, doubtless had five toes in each foot, the typical number in mammals. This reduction in the number of toes may, perhaps, have been due to elevation of the region inhabited, which gradually led the animals to live on higher ground, instead of the soft lowlands where a polydactyl foot would be an advantage.

The gradual elongation of the head and neck, which took place in the successive genera of this group during the Tertiary period, was a less fundamental change than that which resulted in the reduction of the limbs. The process may be said to have already began in *Orohippus*, if we compare that form with other most nearly allied mammals. The diastema, or "place for the bit," was well developed in both jaws even then, but increased materially in succeeding genera. The number of the teeth remained the same until the Pliocene, when the front lower premolar was lost, and subsequently the corresponding upper tooth ceased to be functionally developed. The next upper premolar, which in *Orohippus* was the smallest of the six posterior teeth, rapidly increased in size, and soon became, as in the horse, the largest of the series. The grinding teeth at first had very short crowns, without cement, and were inserted by distinct roots. In Pliocene species, the molars became longer, and were more or less coated with cement. The modern horse has extremely long grinders, without true roots, and covered with a thick external layer of cement. The canine teeth were very large in *Orohippus*, and in this genus, as well as those from the Middle Tertiary, appear to have been well developed in both sexes. In later forms, these teeth declined in size, especially as the changes in the limbs afforded other facilities for defence, or escape from danger. The incisors in the early forms were small, and without the characteristic "mark" of the modern horse. In the genera from the American Eocene and Miocene, the orbit was not enclosed behind by an entire bridge of bone, and this first makes it appearance in this country in Pliocene forms. The depression in front of the orbit, so characteristic of *Anchitherium* and some of the Pliocene genera, is, strange to say, not seen in *Orohippus*, or the later *Miohippus*, and is wanting, likewise, in existing horses. It is an interesting fact that the peculiarly equine features acquired by *Orohippus* are retained persistently throughout the entire series of succeeding forms. Such, *e.g.*, is the form of the anterior part of the lower jaw, and also the characteristic astragalus, with its narrow, oblique, superior ridges, and its small articular facet for the cuboid.

---

[3] The modern horse occasionally has one of the ancestral hooflets developed, usually on the fore foot.

Such is, in brief, a general outline of the more marked changes that seem to have produced in America the highly specialized modern *Equus* from his diminutive, four-toed predecessor, the Eocene *Orohippus*. The line of descent appears to have been direct, and the remains now known supply every important intermediate form. It is, of course, impossible to say with certainty through which of the three-toed genera of the Pliocene that lived together, the succession came. It is not impossible that the later species, which appear generically identical, are the descendants of more distinct Pliocene types, as the persistent tendency in all the earlier forms was in the same direction. Considering the remarkable development of the group through the entire Tertiary period, and its existence even later, it seems very strange that none of the species should have survived, and that we are indebted for our present horse to the old world.

O. C. Marsh, "Odontornithes, or Birds with Teeth,"[1]
American Naturalist (1875)

REMAINS of birds are among the rarest fossils, and few have been discovered except in the more recent formations. With the exception of *Archaeopteryx* from the Jurassic, and a single species from the Cretaceous, no birds are known in the old world below the Tertiary. In this country numerous remains of birds have been found in the Cretaceous, but there is no satisfactory evidence of their existence in any older formation, the three-toed footprints of the Triassic being probably all made by Dinosaurian reptiles.

The Museum of Yale College contains a large series of remains of birds from the Cretaceous deposits of the Atlantic coast and the Rocky Mountain region, thirteen species of which have already been described by the writer. The most important of these remains, so far as now known, are the *Odontornithes*, or birds with teeth, and it is the object of the present communication to give some of the more marked characters of this group, reserving the full description for a memoir now in course of preparation.

The first species of birds in which teeth were detected was *Ichthyornis dispar* Marsh, described in 1872.[2] Fortunately the type specimen of this remarkable species was in excellent preservation, and the more important portions of both the skull and skeleton were secured. These remains indicate an aquatic bird, fully adult, and about as large as a pigeon.

The skull is of moderate size, and the eyes were placed well forward. The lower jaws are long, rather slender, and the rami were not coössified at the symphysis. In each lower jaw there are twenty-one distinct sockets, and the series extends over the entire upper margin of the dentary bone (Plate II, figures 1 and 2). The teeth in these

[1] Published in part in the *American Journal of Science*, x, November 1875.
[2] *American Journal of Science*, iv, p. 344, and v, p. 74.

sockets are small, compressed and pointed, and all are directed more or less backward. The crowns are covered with nearly smooth enamel. The maxillary teeth appear to have been numerous, and essentially the same as those in the mandible. Whether the premaxillary bones supported teeth, or were covered with a horny beak, cannot be determined from the present specimen.

The scapular arch and the bones of the wings and legs all conform closely to the true avian type. The sternum has a prominent keel, and elongated grooves for the expanded coracoids. The wings were very large in proportion to the legs, and the humerus had an extended radial crest. The metacarpals are coössified, as in recent birds, thus differing widely from those of *Archaeopteryx*. The bones of the posterior extremities are slender, and resemble those of some aquatic birds. The centra of the vertebrae are all biconcave, the concavities at each end being distinct, and nearly equal (Plate II, figures 3 and 4). The sacrum is elongated, and made up of a large number of coössified vertebrae. Whether the tail was elongated or not cannot at present be decided.

The jaws and teeth of this species show it to have been carnivorous, and it was probably aquatic. Its powerful wings indicate that it was capable of prolonged flight.

Another Cretaceous bird (*Apatornis celer* Marsh), belonging apparently to the same order as *Ichthyornis*, was found by the writer in 1872 in the same geological horizon in Kansas. The remains preserved indicate an individual about the same size as *Ichthyornis dispar*, but of more slender proportions. The vertebrae are biconcave, and there were probably teeth.

The most interesting bird with teeth yet discovered is perhaps *Hesperornis regalis*, a gigantic diver, also from the Cretaceous of Kansas, and discovered by the writer in 1870. The type specimen, which was found by the writer in 1871, and described soon after, consisted mainly of vertebrae and the nearly complete posterior limbs, all in excellent preservation.[3]

A nearly perfect skeleton of this species was obtained in Western Kansas by Mr. T. H. Russell and the writer in November 1872, during the exploration of the Yale College party, and several other less perfect specimens have since been secured, and are now in the Yale Museum. These various remains apparently all belong to one species.

The skull of *Hesperornis* has the same general form as that in *Colymbus torquatus* Brün., but there is a more prominent median crest between the orbits, and the beak is less pointed. The brain cavity was quite small. The maxillary bones are massive, and have throughout their length a deep inferior groove which was thickly set with sharp, pointed teeth. These teeth had no true sockets, but between their bases there are slight projections from the sides of the grooves. (Plate III, figure 2). The teeth have pointed crowns, covered with enamel, and supported on stout fangs. (Plate III, figure 1*a*). In form of crown and base, they most resemble the teeth of Mosasauroid reptiles. The method of replacement, also, was the same, as some of the teeth preserved have the crowns of the successional teeth implanted in cavities in their fangs. The maxillary grooves do not extend into the premaxillaries, and the latter do not appear to have

---

[3] *American Journal of Science*, iii, p. 360, May 1872.

supported teeth. The external appearance, moreover, of the premaxillaries seems to indicate that these bones were covered with a horny bill, as in modern birds.

The lower jaws are long, and slender, and the rami were united in front only by cartilage. The dentary bone has a deep groove throughout its entire length, and in this, teeth were thickly planted, as in the jaws of *Ichthyosaurus*. The lower teeth are similar to those above, and all were more or less recurved (Plate III, fig. 2).

The scapular arch of *Hesperornis* presents many features of interest. The sternum is think and weak, and entirely without a keel. In front, it resembles the sternum of *Apteryx*, but there are two very deep posterior emarginations, as in the Penguins. The scapula and coracoid are very small. The wing bones are diminutive, and the wings were rudimentary, and useless as organs of either flight or swimming.

The vertebrae in the cervical and dorsal regions are of the true ornithic type, the articular faces of the centra being quite as in modern birds (Plate III, figure 3 and 4). The sacrum is elongated, and resembles that in recent diving birds. The last sacral vertebra is quite small. The caudal vertebrae, which are about twelve in number, are very peculiar, and indicate a structure not before seen in birds. The anterior caudals are short, with high neural spines and moderate transverse processes. The middle and posterior caudals have very long and horizontally expanded transverse processes, which restrict lateral motion, but clearly indicate that the tail was moved vertically, probably in diving. The last three or four caudal vertebrae are firmly coössified, forming a flat terminal mass, analogous to, but quite unlike, the "ploughshare" bone of modern birds. The anterior two at least of these caudals have expanded transverse processes.

The pelvic bones, although avian in type, are peculiar, and present some well marked reptilian features. A resemblance to the corresponding bones of a Cassowary is at once evident, especially in a side view, as the ilium, ischium, and pubis all have their posterior extremities separate. The two latter are slender, and also free, back of their union with the ilium at the acetabulum. The ischium is spatulate at its distal end, and the pubis rodlike. The acetabulum differs from that in all known birds, in being closed internally by bone, except a foramen, that perforates the inner wall.

The femur is unusually short and stout, much flattened antero-posteriorly, and the shaft curved forward. It somewhat resembles in form the femur of *Colymbus torquatus* Brün., but the great trochanter is proportionally much less developed in a fore-and-aft direction, and the shaft is much more flattened. The tibia is straight and elongated. Its proximal end has a moderately developed cnemial process, with an obtuse apex. The epi-cnemial ridge is prominent, and continued distally about one-half the length of the shaft. The distal end of the tibia has on its anterior face no ossified supratendinal bridge, differing in this respect from nearly all known aquatic birds. The fibula is well developed, and resembles that of the Divers. The patella is large, as in *Podiceps*, and in position extends far above the elevated rotular process of the tibia.

The tarso-metatarsal bone is much compressed transversely, and resembles in its main features that of *Colymbus*. On its anterior face there is a deep groove between the third and fourth metatarsal elements, bounded on its outer margin by a prominent rounded ridge, which expands distally into the free articular end of the fourth metatarsal. This extremity projects far beyond the other two, and is double the size of either, thus showing a marked difference from any known recent or fossil bird. There is a shallow groove, also, between the second and third metatarsals. The second metatarsal is much shorter than the third or fourth, and its trochlear end resembles in shape and size that of the former. The existence of a hallux is indicated by an elongated oval indentation on the inner margin above the articular face of the second metatarsal. The free extremities of the metatarsals have the same oblique arrangement as in the *Colymbidae*, to facilitate the forward stroke of the foot through the water. There are no canals or even grooves for tendons on the posterior face of the proximal end, as in the Divers and most other birds; but below this, there is broad, shallow depression, extending rather more than half way to the distal extremity.

The phalanges are shorter than in most swimming birds. Those of the large, external toe are very peculiar, although an approach to the same structure is seen in the genus *Podiceps*. On the outer, inferior margin, they are all deeply excavated. The first, second, and third have, at their distal ends, a single, oblique, articular face on the inner half of the extremity, and the outer portion is produced into an elongated, obtuse process, which fits into a corresponding cavity in the adjoining phalanx. This peculiar articulation prevents flexion except in one direction, and greatly increases the strength of the joints. The terminal phalanx of this toe was much compressed. The third, or middle toe, was greatly inferior to the fourth in size, and had slender, compressed phalanges, which correspond essentially in their main features with those of modern Divers.

The remains preserved of *Hesperornis regalis* show that this species was larger than any known aquatic bird. All the specimens discovered are in the Yale College Museum, and agree essentially in size, the length from the apex of the bill to the end of the toes being between five and six feet. The habits of this gigantic bird are clearly indicated in the skeleton, almost every part of which has now been found. The rudimentary wings prove that flight was impossible, while the powerful swimming legs and feet were peculiarly adapted to rapid motion through the water. The tail appears to have been much expanded horizontally, as in the Beaver, and doubtless was an efficient aid in diving, perhaps compensating in part for want of wings, which the Penguins use with so much effect in swimming under water. That *Hesperornis* was carnivorous is clearly proven by its teeth; and its food was probably fishes.

The zoological position of *Hesperornis* is evidently in the *Odontornithes*; but the insertion of the teeth in grooves, the absence of a keel on the sternum, and the wide difference in the vertebrae require that it be placed in a distinct order, which may be called *Odontolcae*, in allusion to the position of the teeth in grooves.

The two orders of birds with teeth would then be distinguished as follows:—

Sub-Class, Odontornithes (or Aves Dentatae)

A. Teeth in sockets. Vertebrae biconcave. Sternum with keel. Wings well developed.

Order, Odontotormae.[4]

B. Teeth in grooves. Vertebrae as in recent birds. Sternum without keel. Wings rudimentary.

Order, Odontolcae.

In comparing *Ichthyornis* and *Hesperornis*, it will be noticed that the combination of characters in each is very remarkable, and quite the reverse of what would naturally be expected. The former has teeth in distinct sockets, with biconcave vertebrae; while the latter has teeth in grooves, and yet vertebrae similar to those of modern birds. In point of size, and means of locomotion, the two present the most marked contrast. The fact that two birds, so entirely different, living together during the Cretaceous, should have been recovered in such perfect preservation, suggests what we may yet hope to learn of life in that period.

The geological horizon of all the *Odontornithes* now known is the Upper Cretaceous. The associated vertebrate fossils are mainly Mosasauroid reptiles and Pterodactyls.

Explanation of Plates

Plate II. – *Ichthyornis dispar* Marsh. Twice natural size.
      Figure 1. Left lower jaw; side view.
      Figure 2. Left lower jaw; top view.
      Figure 3. Cervical vertebra; side view.
      Figure 4. Same vertebra; front view.

Plate III. – *Hesperornis regalis* Marsh.
      Figure 1. Left lower jaw; side view; half natural size.
      Figure 1*a*. Tooth; four times natural size.
      Figure 2. Left lower jaw; top view; half natural size.
      Figure 3. Dorsal vertebra; side view; natural size.
      Figure 4. Same vertebra; front view; natural size.

---

[4] The name *Ichthyornithes*, first proposed for this order by the writer, proves to be preoccupied, and *Odontotormae* may be substituted. The name *Ichthyornidae* may be retained for the family. – O. C. M.

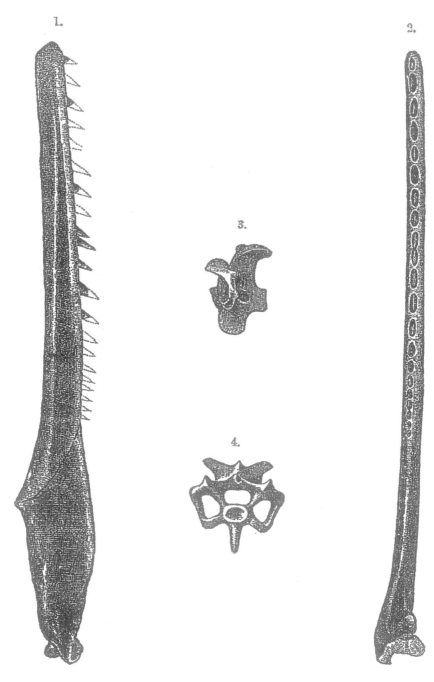

Ichthyornis Dispar

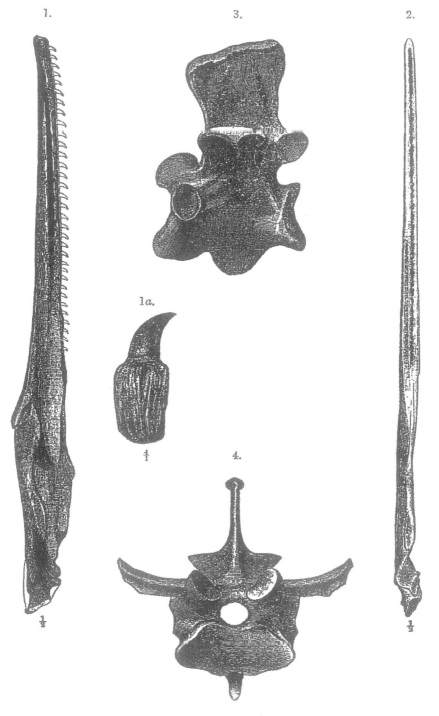

Hesperornis Regalis

CHAUNCEY WRIGHT (1830–1875)

Mathematician, philosophy of science.
Born in Northampton, Massachusetts. Died in Cambridge, Massachusetts.

Wright was born in modest circumstances and went to Harvard thanks to the generosity of a wealthy Northampton benefactor. He was not a particularly distinguished student in the classic Harvard curriculum of the time, but he did excel in mathematics (enough so that even Benjamin Peirce took note) and pursued two years of advanced study in natural history as a supplement to the traditional coursework. He was perhaps at his best, however, in informal philosophical discussions, where he was at once brilliant, challenging, and generous. Throughout his adult life, he attracted exceptional intellects for regular conversations in what came to be known as the Metaphysical Club, the core of which included C. S. Peirce, William James (and on occasion his younger brother Henry), and Oliver Wendell Holmes, Jr. This discussion group was in many ways the emotional core of his life. After graduating from Harvard, he took a job with the Nautical Almanac office in Cambridge as a computer (that is, Benjamin Peirce secured the job for him since he was simply too radical a thinker for the usual Harvard venues of college or clergy). His computing job perfectly fit his needs. While there he could condense a year's worth of calculations into three months of the year and spend the other nine reading, thinking, and discussing the nature of science, philosophy, and metaphysics. He wrote the occasional article or review, mostly on philosophical issues and science, but early on in mathematics too. His biggest influences were John Stuart Mill and Darwin, and as a modern and untraditional thinker, he was unsurpassed among his peers. It might even be argued that he out thought Darwin in that he saw, even before Darwin did fully, that variation could be "accidental" without violating causal laws. Darwin was so taken by Wright's defense in "Genesis of Species" that he arranged to have the whole book reprinted and circulated in England. Above all else, Wright was a champion of scientific empiricism, naturalism, and the clear separation of theology and metaphysics from science. He did not discount metaphysics in life just in science, and his rigorous defense of the Maginot line of the mind put him ahead, and perhaps beyond, the thinking of his time. He was particularly critical of the American addiction to an old school, Scottish common-sense view of realism (which he once termed a "dogmatic coma") because it inevitably collapsed back upon the "wonders of God's world." Wright's brand of realism relied on no authority beyond "our experience of the world and of our own thoughts" (Wright, *The Evolutionary Philosophy*, p. 338). He would be a principal force behind the pragmatism of Peirce, James, and Dewey, and would also influence the fields of psychology and evolutionary epistemology with his essay, "The Evolution of Self-Consciousness," which applied natural selective theory to human psychological development. Wright struggled with depression all his life.

A heavy smoker, and on occasion a heavy drinker, Wright died of a heart attack at 45. His good friend Charles Norton gathered all of Wright's essays into a book which he then published in 1877. The opening biography of Wright by Norton is an excellent place to start for a better understanding of this most gifted and elusive American thinker.

Chauncey Wright, "The Genesis of Species" (excerpt), *North American Review* (1871)

Art. III. – 1. *Contributions to the Theory of Natural Selection*. By Alfred Russel
Wallace. London and New York: Macmillan & Co. 1870. 8vo. Pp. 384.

2. *On the Genesis of Species*. By St. George Mivart, F. R. S. London and New
York: Macmillan & Co. 1871. 8vo. Pp. 314.

3. *The Descent of Man, and Selection in Relation to Sex*. By Charles Darwin,
M. A., F. R. S., etc. In two volumes. New York: D. Appleton & Co. 1871.
8vo. Pp. 409 and 436.

4. *On the Origin of Species by Means of Natural Selection, or the Preservation of
Favored Races in the Struggle for Life*. By Charles Darwin, M. A., F. R. S., etc. Fifth
Edition, with Additions and Corrections. New York: D. Appleton & Co. 1871.
8vo. Pp. 447.

IT is now nearly twelve years since the discussion of that "mystery of mysteries," the origin of species, was reopened by the publication of the first edition of Mr. Darwin's most remarkable work. Again and again in the history of scientific debate this question had been discussed, and, after exciting a short-lived interest, had been condemned by cautious and conservative thinkers to the limbo of insoluble problems or to the realm of religious mystery. They had, therefore, sufficient grounds, *a priori*, for anticipating that a similar fate would attend this new revival of the question, and that, in a few years, no more would be heard of the matter; that the same condemnation awaited this movement which had overwhelmed the venturesome speculations of Lamarck and of the author of the "Vestiges of Creation." This not unnatural anticipation has been, however, most signally disappointed. Every year has increased the interest felt in the question, and at the present moment the list of publications which we place at the head of this article testifies to the firm hold which the subject has acquired in this short period on the speculative interests of all inquisitive minds. But what can we say has really been accomplished by this debate; and what reasons have we for believing that the judgment of conservative thinkers will not, in the main, be proved right after all, though present indications are against them? One permanent consequence, at least, will remain, in the great additions to our knowledge of natural history, and of general physiology, or theoretical biology, which the discussion has produced; though the greater part of this positive contribution to science is still to be credited directly to Mr. Darwin's works, and even to his original researches. But, besides this, an advantage has been gained which cannot be too highly estimated. Orthodoxy has been won over to the doctrine of evolution. In asserting this result, however, we are obliged to make what will appear to many persons important qualifications and explanations. We do not mean that the heads of leading religious bodies, even in the most enlightened communities, are yet willing to withdraw the dogma that the origin of species is a special religious mystery, or even to assent to the hypothesis of evolution as a legitimate question for scientific inquiry. We mean only, that many eminent students of science, who claim to be orthodox, and who are certainly actuated as much by a spirit of reverence as by scientific inquisitiveness, have found means of reconciling the general doctrine of evolution with the dogmas they regard as essential to religion. Even to those whose interest in the question is mainly scientific this result is a welcome one, as opening the way for a freer discussion of subordinate questions, less trammeled by the religious prejudices which have so often been serious obstacles to the progress of scientific researches.

But again, in congratulating ourselves on this result, we are obliged to limit it to the doctrine of evolution in its most general form, the theory common to Lamarck's zoölogical philosophy, to the views of the author of the "Vestiges of Creation," to the general conclusions of Mr. Darwin's and Mr. Wallace's theory of Natural Selection, to Mr. Spencer's general doctrine of evolution, and to a number of minor explanations of the processes by which races of animals and plants have been derived by descent from different ancestral forms. What is no longer regarded with suspicion as secretly hostile to religious belief by many truly religious thinkers is that which is denoted in common by the various names "transmutation," "development," "derivation," "evolution," and "descent with modification." These terms are synonymous in their primary and general signification, but refer secondarily to various hypotheses of the processes of derivation. But there is a choice among them on historical grounds, and with reference to associations, which are of some importance from a theological point of view. "Transmutation" and "development" are under ban. "Derivation" is, perhaps, the most innocent word; though "evolution" will probably prevail, since, spite of its etymological implication, it has lately become most acceptable, not only to the theological critics of the theory, but to its scientific advocates; although, from the neutral ground of experimental science, "descent with modification" is the most pertinent and least exceptional name.

While the general doctrine of evolution has thus been successfully redeemed from theological condemnation, this is not yet true of the subordinate hypothesis of Natural Selection, to the partial success of which this change of opinion is, in great measure, due. It is, at first sight, a paradox that the views most peculiar to the eminent naturalist, whose work has been chiefly instrumental in effecting this change of opinion, should still be rejected or regarded with suspicion by those who have nevertheless been led by him to adopt the general hypothesis, – an hypothesis which his explanations have done so much to render credible. It would seem, at first sight, that Mr. Darwin has won a victory, not for himself, but for Lamarck. Transmutation, it would seem, has been accepted, but Natural Selection, its explanation, is still rejected by many converts to the general theory, both on religious and scientific grounds. But too much weight might easily be attributed to the deductive or explanatory part of the evidence, on which the doctrine of evolution has come to rest. In the half-century preceding the publication of the "Origin of Species," inductive evidence on the subject has accumulated, greatly outweighing all that was previously known; and the "Origin of Species" is not less remarkable as a compend and discussion of this evidence than for the ingenuity of its explanations. It is not, therefore, to what is now known as "Darwinism" that the prevalence of the doctrine of evolution is to be attributed, or only indirectly. Still, most of this effect is due to Mr. Darwin's work, and something undoubtedly to the indirect influence of reasonings that are regarded with distrust by those who accept their conclusions; for opinions are contagious, even where their reasons are resisted.

The most effective general criticism of the theory of Natural Selection which has yet appeared, or one which, at least, is likely to exert the greatest influence in overcoming the remaining prejudice against the general doctrine of evolution, is the work of Mr. St. George Mivart "On the Genesis of Species." Though, as we shall show in the course of this article, the work falls far short of what we might have expected from an author of Mr. Mivart's attainments as a naturalist, yet his position before the religious world, and his unquestionable familiarity with the theological bearings of his subject, will undoubtedly gain for him and for the doctrine of evolution a hearing and a credit, which the mere student of science might be denied. His work is mainly a critique of "Darwinism"; that is, of the theories peculiar to Mr. Darwin and the "Darwinians," as distinguished from the believers in the general doctrine of evolution which our author accepts. He also puts forward an hypothesis in opposition to Mr. Darwin's doctrine of the predominant influence of Natural Selection in the generation of organic species, and their relation to the conditions of their existence. On this hypothesis, called "Specific Genesis," an organism, though at any one time a fixed and determinate species, approximately adapted to surrounding conditions of existence, is potentially, and by innate potential combinations of organs and faculties, adapted to many other conditions of existence. It passes, according to the hypothesis, from one form to another of specific "manifestation," abruptly and discontinuously in conformity to the emergencies of its outward life; but in any condition to which it is tolerably adapted it retains a stable form, subject to variation only within determinate limits, like oscillations in a stable equilibrium. For this conception our author is indebted to Mr. Galton, who, in his work on "Hereditary Genius," "compares the development of species with a many-faceted spheroid tumbling over from one facet or stable equilibrium to another. The existence of internal conditions in animals,"

Mr. Mivart adds (p. 111), "corresponding with such facets is denied by pure Darwinians, but it is contended in this work that something may also be said for their existence." There are many facts of variation, numerous cases of abrupt changes in individuals both of natural and domesticated species, which, of course, no Darwinian or physiologist denies, and of which Natural Selection professes to offer no direct explanation. The causes of these phenomena, and their relations to external conditions of existence, are matters quite independent of the principle of Natural Selection, except so far as they may directly affect the animal's or plant's well-being, with the origin of which this principle is alone concerned. General physiology has classified some of these sudden variations under such names as "reversion" and "atavism," or returns more or less complete to ancestral forms. Others have been connected together under the law of "correlated or concomitant variations," changes that, when they take place, though not known to be physically dependent on each other, yet usually or often occur together. Some cases of this law have been referred to the higher, more fundamental laws of homological variations, or variations occurring together on account of the relationship of homology, or due to similarities and physical relations between parts of organisms, in tissues, organic connections, and modes of growth. Other variations are explained by the laws and causes that determine monstrous growths. Others again are quite inexplicable at yet, or cannot yet be referred to any general law or any known antecedents. These comprise, indeed, the most common cases. The almost universal prevalence of well-marked phenomena of variation in species, the absolutely universal fact that no two individual organisms are exactly alike, and that the description of a species is necessarily abstract and in many respects by means of averages, – these facts have received no particular explanations, and might indeed be taken as ultimate facts or highest laws in themselves, were it not that in biological speculations such an assumption would be likely to be misunderstood, as denying the existence of any real determining causes and more ultimate laws, as well as denying any known antecedents or regularities in such phenomena. No physical naturalist would for a moment be liable to such a misunderstanding, but would, on the contrary, be more likely to be off his guard against the possibility of it in minds otherwise trained and habituated to a different kind of studies. Mr. Darwin has undoubtedly erred in this respect. He has not in his works repeated with sufficient frequency his faith in the universality of the law of causation, in the phenomena of general physiology or theoretical biology, as well as in all the rest of physical nature. He has not said often enough, it would appear, that in referring any effect to "accident," he only means that its causes are like particular phases of the weather, or like innumerable phenomena in the concrete course of nature generally, which are quite beyond the power of finite minds to anticipate or to account for in detail, though none the less really determinate or due to regular causes. That he has committed this error appears from the fact that his critic, Mr. Mivart, has made the mistake, which nullifies nearly the whole of his criticism, of supposing that "the theory of Natural Selection may (though it need not) be taken in such a way as to lead men to regard the present organic world as formed, so to speak, *accidentally*, beautiful and wonderful as is confessedly the hap-hazard result" (p. 33). Mr. Mivart, like many another writer, seems to forget the age of the world in which he lives and for which he writes, – the age of "experimental philosophy," the very stand-point of which, its fundamental assumption, is the universality of physical causation. This is so familiar to minds bred in physical studies, that they rarely imagine that they may be mistaken for disciples of Democritus, or for believers in "the fortuitous concourse of atoms," in the sense, at least, which theology has attached to this phrase. If they assent to the truth that may have been meant by the phrase, they would not for a moment suppose that the atoms move fortuitously, but only that their conjunctions, constituting the actual concrete orders of events, could not be anticipated except by a knowledge of the natures and regular histories of each and all of them, – such knowledge as belongs only to omniscience. The very hope of experimental philosophy, its expectation of constructing the sciences into a true philosophy of nature, is based on the induction, or, if you please, the *a priori* presumption, that physical causation is universal; that the constitution of nature is written in its actual manifestations, and needs only to be deciphered by experimental and inductive research; that it is not a latent invisible writing, to be brought out by the magic of mental anticipation or metaphysical meditation. Or, as Bacon said, it is not by the "anticipations of the mind," but by the "interpretation of nature," that natural philosophy is to be constituted; and this is to presume that the order of nature is decipherable, or that causation is everywhere either manifest or hidden, but never absent.

Mr. Mivart does not wholly reject the process of Natural Selection, or disallow it as a real cause in nature, but he reduces it to "a subordinate role" in his view of the derivation of species. It serves to perfect the imperfect adaptations and to meet within certain limits unfavorable changes in the conditions of existence. The "accidents" which Natural Selection acts upon are allowed to serve in a subordinate capacity and in subjection to a foreordained, particular, divine order, or to act like other agencies dependent on an evil principle, which are compelled to turn evil into good. Indeed, the only difference on purely scientific grounds, and irrespective of theological considerations, between Mr. Mivart's views and Mr. Darwin's is in regard to the *extent* to which the process of Natural Selection has been effective in the modifications of species. Mr. Darwin himself, from the very nature of the process, has never supposed for it, as a cause, any other than a co-ordinate place among other causes of change, though he attributes to it a superintendent, directive, and controlling agency among them. The student of the theory would gather quite a different impression of the theory from Mr. Mivart's account of it, which attributes to "Darwinians" the absurd conception of this cause as acting "alone" to produce the changes and stabilities of species; whereas, from the very nature of the process, other causes of change, whether of a known or as yet unknown nature, are presupposed by it. Even Mr. Galton's and our author's hypothetical "facets," or internal conditions of abrupt changes and successions of stable equilibriums, might be among these causes, if there were any good inductive grounds for supposing their existence. Reversional and correlated variations are, indeed, due to such internal conditions and to laws of inheritance, which have been ascertained inductively as at least laws of phenomena, but of which the causes, or the antecedent conditions in the organism, are unknown. Mr. Darwin continually refers to variations as arising from unknown causes, but these are always such, so far as observation can determine their relations to the organism's conditions of existence, that they are far from accounting for, or bearing any relations to, the adaptive characters of the organism. It is solely upon and with reference to such adaptive characters that the process of Natural Selection has any agency, or could be supposed to be effective. If Mr. Mivart had cited anywhere in his book, as he has not, even a single instance of sudden variation in a whole race, either in a state of nature or under domestication, which is not referable by known physiological laws to the past history of the race on the theory of evolution, and had further shown that such a variation was an adaptive one, he might have weakened the arguments for the agency and extent of the process of Natural Selection. As it is, he has left them quite intact.

Part Three: 1876–1900

Scientists

PART THREE: 1876–1900

SCIENTISTS

Diversity and Differentiation

In 1876 Simon Newcomb (an American polymath who published in mathematics, astronomy, physics, and economics) wrote an essay, "Abstract Science in America, 1776–1876," for an issue of the *North American Review* dedicated to the centennial of the American Revolution. The essay provides a recap of American science – clear, concise, contemporary – and a view of what contemporaries thought about the state of American science:

> [W]e require no increase in the number of our museums, observatories, or laboratories during the present generation. . . . We are deficient in the number of men actively devoted to scientific research of the higher types, in public recognition of the labors of those who are so engaged, in the machinery for making the public acquainted with their labors and their wants, and in the pecuniary means for publishing their researches. Each of these deficiencies is, to a certain extent, both a cause and an effect of the others. . . . The supply of any one of these deficiencies would, to a certain extent, remedy all the others . . . . In other intellectual nations, science has a fostering mother, – in Germany the universities, in France the government, in England the scientific societies; and if science could find one here, it would speedily flourish. The only one it can look to here is the educated public; and if that public would find some way of expressing in a public and official manner its generous appreciation of the labors of American investigators, we should have the best entering wedge for supplying all the wants of our science. (p. 118)

All of the deficiencies Newcomb identifies – numbers of scientists, recognition, and machinery for dissemination of information – would improve by the end of the century. There would be a great boom in the creation of scientific societies and the scientific journals they often found. I have noted some of the more national ones in the timeline to this section (medicine and engineering had a specialty boom of their own) but scores more were started across the country at regional, state, and local levels, including the journals in which Forbes and Cowles first published the essays anthologized here. These societies went a long way toward easing both the dissemination and recognition issues, and many of the national societies began awards and recognitions for excellence within their respective fields.

The numbers game was a twofold problem requiring greater access to advanced education in science and a reason to pursue it. The first of these would be greatly addressed by the Morrill Act of 1862 which established the Land Grant University system for states and territories. A second Morrill Act in 1890 targeted land grant colleges for the devastated South and required each state to show that race was not an admissions criterion or else designate a separate land-grant institution. Over 70 colleges were established through this act, including many of the traditional black colleges. In 1887 the Hatch Act would authorize and fund agricultural experiment stations in connection with the land grant colleges, adding another layer of scientific opportunity. In addition to the land grant colleges, three universities with a strong science bent would be founded in this last quarter thanks to philanthropic bequests: Johns Hopkins (1876), University of Chicago (1890), and Stanford (1891). In the late 1880s two especially influential field/marine stations would be formalized, the Marine Biological Laboratory (1888) at Woods Hole, Massachusetts, and Stanford's Hopkins Marine Station (1892) on the Monterey Peninsula, California. Of course, all of this expansion of scientific education must be tempered by the ever-present resistance to change. Schools like Yale, Harvard, Princeton, and Dartmouth still held a general disdain for science as a subject for gentleman and kept the science schools separate and unequal to their traditional undergraduate curriculum. As one undergraduate would comment about Yale's Sheffield Scientific School in the 1880s, "Sheff did not count, at least not in the affirmative" (Levine, p. 40). As far as job opportunities for those with science training, even though there was a boom in college and university building, the federal government was still the largest employer of scientists with the U.S. Geological Survey, the Department of Agriculture, the Fish Commission, the Weather Service, Naval Observatory, Coast and Geodetic Survey, and numerous lesser offices and departments. The federal government did not, however, dominate the sciences. More and more research and experimentation was coming from the universities and the new trend of industrial laboratories in the wake of Edison's lab at Menlo Park, New Jersey (1876). What becomes obvious when reading biographies of key American figures, research on American studies, or histories of American science is that the energy (good, bad, and chaotic) was coming from a plethora of sources. The same pluralism that expanded and populated Turner's frontier was feeding the new frontiers of science in America. Paradoxically, the same pluralism that often gave an anti-science tinge to much of American individualism also provided the individual space and motivation to pursue scientific and inventive avenues. Newcomb noted the strengths of the German universities, the government support of the French, the scientific societies of the English. By the end of the nineteenth century, as slow and unwieldy as it might at times appear, American science would be working toward its own pluralistic formula incorporating all of the above.

Revisited: Neo-Lamarckians to Post-Darwinians

Charles Darwin's *The Descent of Man* appeared in 1871. The reaction to it was decidedly different from the reaction to the *On the Origin of Species*. There was a much stronger feeling against this book among the religious community and a much more accepting reception among the scientific community. There would follow through the 1870s and 1880s in colleges (predominantly Southern) with religious affiliations and seminaries a certain amount of persecution and dismissal of faculty who were not adequately anti-Darwinian. During this same period the science community seemed to pretty well absorb Darwin's latest work into the ongoing dialogue about evolution. The neo-Lamarckians (a term coined by Henry Fairfield Osborn, a Cope student, in 1884) were a very active force throughout this period and buffered the issues raised in Darwinian evolution by suggesting that there were ways to evolution other than natural selection. Even though the American neo-Lamarckians were ultimately proven wrong, their contributions to American science were substantial. They did not attempt to supplant Darwin – whom all claimed to revere – as much as supplement and refocus. They did not disagree with natural selection per se but did not give it much primacy as the source of change. And they did not find it at all applicable to the "why" of change which was for the American neo-Lamarckian scientists and their philosophical comrades, the Spencerian thinkers and writers of the time, their raison d'etre: progress, progress, progress. Chauncey Wright had argued against this progressive, cosmic bias in discussions of evolution: "Strictly speaking, Natural Selection is not a cause at all, but is the mode of operation of a certain quite limited class of causes" (Wright, *Philosophical Discussions*, p. 109). With Wright's death in 1875 the most forceful and exacting explicator of what was and was not science and evolution was lost to America, and the neo-Lamarckians and Spencerian thinkers (particularly historian John Fiske) could insist that their inheritance of acquired traits and evolutionary progress were not just a cause, but *the* cause.

Europe and Great Britain also had a mix of neo-Lamarckians and neo-Darwinists, but they were more actively concerned about the mechanism of evolution than the American neo-Lamarckians, who thought they pretty much understood it: inheritance of acquired traits by the addition of ontological stages or the degeneration of stages, environment, and the force of (cosmic) progress, which was viewed by Spencerians as linked to the laws of gravitation and conservation of energy and to evolution, particularly through variation and the survival of the fittest. Without a known mechanism for evolution, the force of progress would do.

In the early 1880s, August Weismann (Germany) and Edward Ray Lankester (England) began to argue directly against the idea of the inheritance of acquired traits and to challenge those that believed the idea to prove it through experimentation. Weismann carried out his famous experiment cutting off the tails of 21 generations of mice and finding no change in tail growth, which put a real dent in neo-Lamarckian argument. The American neo-Lamarckians would spend the rest of the century trying to find proof of the inheritance of acquired traits in their own experiments. They would not, but the turn to experimentation would prove instrumental for future generations. In 1900 three different scientists in three different parts of the world rediscovered Mendel's genetic work and the mechanism for natural selection was at last found. Neo-Lamarckian thought in science was effectively dead, and the modern age of post-Darwinian evolution could begin.

Revisited: Utility

A great many studies of American science in the last quarter of the nineteenth century make reference to Henry A. Rowland's speech "A Plea for Pure Science," which he delivered to the Physics Section of the American Association for the Advancement of Science in 1883. Rowland reprimands, complements, and dares his fellow scientists to rise above the mercantile tide, seeming at times to disparage the attention paid to technology, yet when the essay is taken as a whole, Rowland also comes off as sincerely struggling with the complexities of the time. David A. Hounshell's essay "Edison and the Pure Science Ideal in Nineteenth Century America" is very helpful in its methodical look at the "scientific, technical, and social milieu" leading up to Rowland's talk – especially the relationship between the academy, Rowland, and Edison, and provides the layers of connection, tension, reluctance, and compromise between the ever-expanding venues of science at the time: academy, industry, laboratory, the Gilded Age, the Edison Age, the age of ingenuity, the age of robber-barons, late Victorian, early modern. The issues of pure and applied were – then as now – interwoven, needful, symbiotic and potentially predatory. What is particularly meaningful for the fullest view of American science in the nineteenth century is the contrast between the end-of-century rhetoric and that at the start. Utility had been both a source and a shield for science early on, but by the close of the nineteenth century American science was strong enough to step beyond that shield and stand alone. Science for science's sake? Well, yes, and fifteen years after his "A Pleas for Pure Science" speech, Rowland, addressing the newly launched American Physical Society, would sound the same high tones, issuing a similar challenge but with a small, significant difference: "My answer must still be now as it was fifteen years ago, that much of the intellect of the country is still wasted in the pursuit of so-called practical science which ministers to our physical needs and but little thought and money is given to the grander portion of the subject which appeals to our intellect alone. But your presence here gives evidence that such a condition is not to last forever" (p. 5). This call for "intellect alone" was not just a measure of how far American science had to go; it was as well a measure of how far it had come. There was no longer the need to always justify science in terms of utility. That much had changed. The new utility for American science in the twentieth century would prove to be its capacity to allow pure and applied to flourish together or alone as need be.

1877   Asaph Hall discovers the two satellites of Mars.

Charles S. Peirce, "The Fixation of Belief," *Popular Science Monthly*.

Clarence King, "Catastrophishm and Evolution," *The American Naturalist*.

Bell Telephone founded.

An outbreak of locust leads to the founding of the U.S. Entomological Commission.

1878   **J. Willard Gibbs, "On the Equilibrium of Heterogeneous Substances".**

United States Coast Survey becomes the Coast and Geodetic Survey.

Joseph Henry dies.

Thomas J. Burrill finds that bacteria are responsible for "fire blight".

American Microscopical Society established.

American Journal of Mathematics founded at Johns Hopkins University.

Michelson begins speed of light measurements.

1879   American Chemical Journal begins, focus on pure research.

Edison invents light bulb.

George Eastman working on photographic film and cameras.

U.S. Geological Survey authorized, consolidating all the various surveys of the West.

Joseph Leidy indentifies parasitic amoebae.

1880   Alpheus Hyatt, "The Genesis of the Tertiary Species of Planorbis at Steinheim," *Anniversary Memoirs of Boston Society of Natural History*.

Journal *Science* started. This journal, which from the beginning was modelled after *Nature*, the British journal started by Huxley and others, had a very long, slow infancy. Supported at different times by Edison and Bell, it was only after James McKeen Cattell bought the journal and arranged for it to be the journal of the AAAS that it found its footing.

1881   **Charles S. Peirce, "On the Logic of Numbers,"** *American Journal of Mathematics*.

S. P. Langley, "The Bolometer and Radiant Energy," *Proceedings of the American Academy of Arts and Sciences*.

1882   Johann Bernhard Stallo, *Concepts and Theories of Modern Physics* (New York).

Rowland devises machine for producing precision diffraction gratings.

Clarence Edward Dutton publishes *The Tertiary History of the Grand Canyon District*.

American Association for the Advancement of Science increases to nine sections from two, illustrating the increasing specialization of science.

1883   American Society of Naturalists.

American Ornithologists Union.

E. D. Cope published *The Vertebrata of the Tertiary Formations of the West*.

Edison discovered the outflow of electricity from heated metal, to be named the Edison Effect, important in making vacuum tubes for radio and television.

1884   Nikola Tesla invents electric alternator and emigrates to Unites States.

1884–86   Allison Commission looks at role of government and science and endorses the status quo, that is, to continue to support the various scientific roles in federal government but not provide the centralizing department of science suggested by NAS.

1886   Sigma Xi.

1887   **S. P. Langley, "The New Astronomy,"** *The Century, a popular quarterly*.

Albert A. Michelson and Edward W. Morley, "On the Relative Motion of the Earth and the Luminiferous Ether," *American Journal of Science*.

Stephen A. Forbes, "The Lake as a Microcosm," *Bulletin of the Peoria Scientific Association*.

American Physiological Society.

Spencer F. Baird dies, one of the last of the great nineteenth century institution builders. For over a decade he had been Secretary of the Smithsonian during its great museum building phase, Commissioner of U.S. Fish and Fisheries, and a driving force behind the creation of the Marine Biological Laboratory at Woods Hole.

1888   Marine Biological Laboratory, Woods Hole, Massachusetts begins.

American Mathematical Society.

Geological Society of America.

American Association of Anatomists.

National Geographic Society.

1889   Alfred O. Coffin becomes the first African American to receive a PhD in biology.

        Entomological Society of America.

        Department of Agriculture elevated to cabinet status.

1890   Cold Spring Harbor Laboratory (New York) begins.

        American Society of Zoologists founded.

1891   Journal of Comparative Neurology founded.

        Astronomy and Astro-Physics journal founded.

1893   American Fern Society.

        Botanical Society of America.

1894   **C. Hart Merriam, "Law of Temperature Control of the Geographic Distribution of Terrestrial Animals and Plants,"** *National Geographic Magazine.*

1896   Journal of Physical Chemistry founded.

1897   Society for Plant Morphology and Physiology established.

1898   American Bryological and Lichenological Society.

        American Journal of Physiology established.

1899   **Henry Chandler Cowles, "The Ecological Relations of the Vegetation on the Sand Dunes of Lake Michigan,"** *Botanical Gazette.*

        American Astronomical Society.

        Society of American Bacteriologists.

        American Physical Society.

        American Society for Microbiology.

1900   American Roentgen Ray Society.

.

CHARLES SANDERS PEIRCE (1839–1914)

Mathematician, astronomer, chemist, geodesist, philosopher.
Born in Cambridge, Massachusetts. Died in Milford, Pennsylvania.

C. S. PEIRCE, son of mathematician Benjamin Peirce, graduated from Harvard in 1859 and then went on to receive a degree in chemistry from Harvard's Lawrence Scientific School. From 1859 until 1891 he worked for the U.S. Coast Survey (known as the U.S. Coast and Geodetic Survey after 1878). He also worked for a time at a second job teaching logic in the mathematics department at Johns Hopkins University (1879–84). He was part of the core intellectual trust – with Chauncey Wright and William James – that was known as the Metaphysical Club and met regularly with a rotating cast of other bright stars to discuss a wide range of topics. Peirce was, like Chauncey Wright, an extraordinary individual, in both advantageous and disadvantageous ways. He suffered all his life from facial neuralgia, a disorder that results in episodes of severe, nearly crippling pain across various regions of the face. The condition might explain his tendency to be temperamental and unsociable. He may also have been bipolar. His first biographer, Joseph Brent, has conjectured he may have taken ether, morphine, and/or cocaine to try to manage both his mood swings and his neuralgia. His first wife left him in 1875, but their divorce did not become final until 1883. His subsequent cohabitation and eventual marriage with a European Gypsy named Juliette led to his dismissal from the teaching position at Johns Hopkins and may have been part of the reason he was asked to resign from the Survey in 1891. After leaving the Survey, Peirce worked odd jobs but remained in financial difficulty for the rest of his life. He was often kept from abject poverty only by the charity of friends, particularly William James. Peirce had always written the occasional brilliant piece on any number of subjects – logic, mathematics, astronomy, chemistry – but after 1891 he devoted himself to writing full time. The bulk of Peirce's writings would not be published until well after his death, some are still in the process of being edited and published by Peirce scholars to this day. At the time of his death, Peirce had produced over 80,000 sheets of unpublished, handwritten manuscript pages and over 10,000 of published pages. His wife, Juliette, sold his papers to Harvard after his death, but they were largely ignored until the 1930's and only adequately edited and studied beginning in the 1980s. (I offer this observation carefully; because of his role in the thinking of William James and John Dewey, the brilliant essays he did publish in his lifetime, and his centrality in nineteenth century discussions of science and philosophy, he has always been studied. Recently, though, his fuller reverberations are finally becoming accessible.) Today he is increasingly recognized as one of the most important and innovative American thinkers of his or any other period. He is credited with developing the modern study of semiotics and with helping to define the logical basis for modern mathematics. Best known as the founder and source for the pragmatism of William James and John Dewey, he even today is inspiring a whole new generation of scholars, philosophers, logicians, mathematicians.

Charles S. Peirce, "The Fixation of Belief" (excerpt), *Popular Science Monthly* (1877)

I

Few persons care to study logic, because everybody conceives himself to be proficient enough in the art of reasoning already. But I observe that this satisfaction is limited to one's own ratiocination, and does not extend to that of other men.

We come to the full possession of our power of drawing inferences, the last of all our faculties; for it is not so much a natural gift as a long and difficult art. The history of its practice would make a grand subject for a book. The medieval schoolman, following the Romans, made logic the earliest of a boy's studies after grammar, as being very easy. So it was as they understood it. Its fundamental principle, according to them, was, that all knowledge rests either on authority or reason; but that whatever is deduced by reason depends ultimately on a premise derived from authority. Accordingly, as soon as a boy was perfect in the syllogistic procedure, his intellectual kit of tools was held to be complete.

To Roger Bacon, that remarkable mind who in the middle of the thirteenth century was almost a scientific man, the schoolmen's conception of reasoning appeared only an obstacle to truth. He saw that experience alone teaches anything – a proposition which to us seems easy to understand, because a distinct conception of experience has been handed down to us from former generations; which to him likewise seemed perfectly clear, because its difficulties had not yet unfolded themselves. Of all kinds of experience, the best, he thought, was interior illumination, which teaches many things about Nature which the external senses could never discover, such as the transubstantiation of bread.

Four centuries later, the more celebrated Bacon, in the first book of his *Novum Organum*, gave his clear account of experience as something which must be open to verification and reexamination. But, superior as Lord Bacon's conception is to earlier notions, a modern reader who is not in awe of his grandiloquence is chiefly struck by the inadequacy of his view of scientific procedure. That we have only to make some crude experiments, to draw up briefs of the results in certain blank forms, to go through these by rule, checking off everything disproved and setting down the alternatives, and that thus in a few years physical science would be finished up – what an idea! "He wrote on science like a Lord Chancellor," indeed, as Harvey, a genuine man of science said.

The early scientists, Copernicus, Tycho Brahe, Kepler, Galileo, Harvey, and Gilbert, had methods more like those of their modern brethren. Kepler undertook to draw a curve through the places of Mars; and to state the times occupied by the planet in describing the different parts of that curve; but perhaps his greatest service to science was in impressing on men's minds that this was the thing to be done if they wished to improve astronomy, that they were not to content themselves with inquiring whether one system of epicycles was better than another but that they were to sit down to the figures and find out what the curve, in truth, was. He accomplished this by his incomparable energy and courage, blundering along in the most inconceivable way (to us), from one irrational hypothesis to another, until, after trying twenty-two of these, he fell, by the mere exhaustion of his invention, upon the orbit which a mind well furnished with the weapons of modern logic would have tried almost at the outset.

In the same way, every work of science great enough to be well remembered for a few generations affords some exemplification of the defective state of the art of reasoning of the time when it was written; and each chief step in science has been a lesson in logic. It was so when Lavoisier and his contemporaries took up the study of Chemistry. The old chemist's maxim had been, "*Lege, lege, lege, labora, ora, et relege.*" Lavoisier's method was not to read and pray, but to dream that some long and complicated chemical process would have a certain effect, to put it into practice with dull patience, after its inevitable failure, to dream that with some modification it would have another result, and to end by publishing the last dream as a fact: his way was to carry his mind into his laboratory, and literally to make of his alembics and cucurbits instruments of thought, giving a new conception of reasoning as something which was to be done with one's eyes open, in manipulating real things instead of words and fancies.

The Darwinian controversy is, in large part, a question of logic. Mr. Darwin proposed to apply the statistical method to biology. The same thing has been done in a widely different branch of science, the theory of gases. Though unable to say what the movements of any particular molecule of gas would be on a certain hypothesis regarding the constitution of this class of bodies, Clausius and Maxwell were yet able, eight years before the publication of Darwin's immortal work, by the application of the doctrine of probabilities, to predict that in the long run such and such a proportion of the molecules would, under given circumstances, acquire such and such velocities; that there would take place, every second, such and such a relative number of collisions, etc.; and from these propositions were able to deduce certain properties of gases, especially in regard to their heat-relations. In like manner, Darwin, while unable to say what the operation of variation and natural selection in any individual case will be, demonstrates that in the long run they will, or would, adapt animals to their circumstances. Whether or not existing animal forms are due to such action, or what position the theory ought to take, forms the subject of a discussion in which questions of fact and questions of logic are curiously interlaced.

Charles S. Peirce, "How to Make Our Ideas Clear" (excerpt), *Popular Science Monthly* (1878)

WHOEVER has looked into a modern treatise on logic of the common sort, will doubtless remember the two distinctions between *clear* and *obscure* conceptions, and between *distinct* and *confused* conceptions. They have lain in the books now for nigh two centuries, unimproved and unmodified, and are generally reckoned by logicians as among the gems of their doctrine.

A clear idea is defined as one which is so apprehended that it will be recognized wherever it is met with, and so that no other will be mistaken for it. If it fails of this clearness, it is said to be obscure.

This is rather a neat bit of philosophical terminology; yet, since it is clearness that they were defining, I wish the logicians had made their definition a little more plain. Never to fail to recognize an idea, and under no circumstances to mistake another for it, let it come in how recondite a form it may, would indeed imply such prodigious force and clearness of intellect as is seldom met with in this world. On the other hand, merely to have such an acquaintance with the idea as to have become familiar with it, and to have lost all hesitancy in recognizing it in ordinary cases, hardly seems to deserve the name of clearness of apprehension, since after all it only amounts to a subjective feeling of mastery which may be entirely mistaken. I take it, however, that when the logicians speak of "clearness," they mean nothing more than such a familiarity with an idea, since they regard the quality as but a small merit, which needs to be supplemented by another, which they call *distinctness*.

When Descartes set about the reconstruction of philosophy, his first step was to (theoretically) permit skepticism and to discard the practice of the schoolmen of looking to authority as the ultimate source of truth. That done, he sought a more natural fountain of true principles, and thought he found it in the human mind; thus passing, in the directest way, from the method of authority to that of apriority, as described in my first paper. Self-consciousness was to furnish us with our fundamental truths, and to decide what was agreeable to reason. But since, evidently, not all ideas are true, he was led to note, as the first condition of infallibility, that they must be clear. The distinction between an idea *seeming* clear and really being so, never occurred to him. Trusting to introspection, as he did, even for a knowledge of external things, why should he question its testimony in respect to the contents of our own minds? But then, I suppose, seeing men, who seemed to be quite clear and positive, holding opposite opinions upon fundamental principles, he was further led to say that clearness of ideas is not sufficient, but that they need also to be distinct, *i.e.*, to have nothing unclear about them. What he probably meant by this (for he did not explain himself with precision) was, that they must sustain the test of dialectical examination; that they must not only seem clear at the outset, but that discussion must never be able to bring to light points of obscurity connected with them.

Such was the distinction of Descartes, and one sees that it was precisely on the level of his philosophy. It was somewhat developed by Leibnitz. This great and singular genius was as remarkable for what he failed to see as for what he saw. That a piece of mechanism could not do work perpetually without being fed with power in some form, was a thing perfectly apparent to him; yet he did not understand that the machinery of the mind can only transform knowledge, but never originate it, unless it be fed with facts of observation. He thus missed the most essential point of the Cartesian philosophy, which is, that to accept propositions which seem perfectly evident to us is a thing which, whether it be logical or illogical, we cannot help doing. Instead of regarding the matter in this way, he sought to reduce the first principles of science to two classes, those which cannot be denied without self-contradiction, and those which result from the principle of sufficient reason (of which more anon), and was apparently unaware of the great difference between his position and that of Descartes. So he reverted to the old trivialities of logic; and, above all, abstract definitions played a great part in his philosophy. It was quite natural, therefore, that on observing that the method of Descartes labored under the difficulty that we may seem to ourselves to have clear apprehensions of ideas which in truth are very hazy, no better remedy occurred to him than to require an abstract definition of every important term. Accordingly, in adopting the distinction of *clear* and *distinct* notions, he described the latter quality as the clear apprehension of everything contained in the definition; and the books have ever since copied his words. There is no danger that his chimerical scheme will ever again be over-valued. Nothing new can ever be learned by analyzing definitions. Nevertheless, our existing beliefs can be set in order by this process, and order is an essential element of intellectual economy, as of every other. It may be acknowledged, therefore, that the books are right in making familiarity with a notion the first step toward clearness of apprehension, and the defining of it the second. But in omitting all mention of any higher perspicuity of thought, they simply mirror a philosophy which was exploded a hundred years ago. That much-admired "ornament of logic" – the doctrine of clearness and distinctness – maybe pretty enough, but it is high time to relegate to our cabinet of curiosities the antique *bijou*, and to wear about us something better adapted to modern uses.

★   ★   ★

The principles set forth in the first part of this essay lead, at once, to a method of reaching a clearness of thought of higher grade than the "distinctness" of the logicians. It was there noticed that the action of thought is excited by the irritation of doubt, and ceases when belief is attained; so that the production of belief is the sole function of thought. All these words, however, are too strong for my purpose. It is as if I had described the phenomena as they appear under a mental microscope. Doubt and Belief, as the words are commonly employed, relate to religious or other grave discussions. But here I use them to designate the starting of any question, no matter how small or how great, and the resolution of it. If, for instance, in a horse-car, I pull out my purse and find a five-cent nickel and five coppers, I decide, while my hand is going to the purse, in which way I will pay my fare. To call such a question Doubt, and my decision Belief, is certainly to use words very disproportionate to the occasion. To speak of such a doubt as causing an irritation which needs to be appeased, suggests a temper which is uncomfortable to the verge of insanity. Yet, looking at the matter minutely, it must be admitted that, if there is the least hesitation as to whether I shall pay the fiver coppers or the nickel (as there will be sure to be, unless I act from some previously contracted habit in the matter), though irritation is too strong a word, yet I am excited to such small mental activity as may be necessary to deciding how I shall act. Most frequently doubts arise from some indecision, however momentary, in our action. Sometimes it is not so. I have, for example, to wait in a railway-station, and to pass the time I read the advertisements on the walls. I compare the advantages of different trains and different routes which I never expect to take, merely fancying myself to be in a state of hesitancy, because I am bored with having nothing to trouble me. Feigned hesitancy, whether feigned for mere amusement or with a lofty purpose, plays a great part in the production of scientific inquiry. However the doubt may originate, it stimulates the mind to an activity which may be slight or energetic, calm or turbulent. Images pass rapidly through consciousness, one incessantly melting into another, until at last, when all is over – it may be in a fraction of a second, in an hour, or after long years – we find ourselves decided as to how we should act under such circumstances as those which occasioned our hesitation. In other word, we have attained belief.

    In this process we observe two sorts of elements of consciousness, the distinction between which may best be made clear by means of an illustration. In a piece of music there are the separate notes, and there is the air. A single tone may be prolonged for an hour or a day, and it exists as perfectly in each second of that time as in the whole taken together; so that, as long as it is sounding, it might be present to a sense from which everything in the past was as completely absent as the future itself. But it is different with the air, the performance of which occupies a certain time, during the portions of which only portions of it are played. It consists in an orderliness in the succession of sounds which strike the ear at different times; and to perceive it there must be some continuity of consciousness which makes the events of a lapse of time present to us. We certainly only perceive the air by hearing the separate notes; yet we cannot be said to directly hear it, for we hear only what is present at the instant, and an orderliness of succession cannot exist in an instant. These two sorts of objects, what we are *immediately* conscious of and what we are *mediately* conscious of, are found in all consciousness. Some elements (the sensations) are completely present at every instant so long as they last, while others (like thought) are actions having beginning, middle, and end, and consist in a congruence in the succession of sensations which flow through the mind. They cannot be immediately present to us, but must cover some portion of the past or future. Thought is a thread of melody running through the succession of our sensations.

★   ★   ★

And what, then, is belief? It is the demi-cadence which closes a musical phrase in the symphony of our intellectual life. We have seen that it has just three properties. First, it is something that we are aware of; second, it appeases the irritation of doubt; and, third, it involves the establishment in our nature of a rule of action, or, say for short, a *habit*. As it appeases the irritation of doubt, which is the motive for thinking, thought relaxes, and comes to rest for a moment when belief is reached. But, since belief is a rule for action, the application of which involves further doubt and further thought, at the same time that it is a stopping-place, it is also a new starting-place for thought. That is why I have permitted myself to call it thought at rest, although thought is essentially an action. The *final* upshot of thinking is the exercise of volition, and of this thought no longer forms a part; but belief is only a stadium of mental action, an effect upon our nature due to thought, which will influence future thinking.

The essence of belief is the establishment of a habit; and different beliefs are distinguished by the different modes of action to which they give rise. If beliefs do not differ in this respect, if they appease the same doubt by producing the same rule of action, then no mere differences in the manner of consciousness of them can make them different beliefs, any more than playing a tune in different keys is playing different tunes. Imaginary distinctions are often drawn between beliefs which differ only in their mode of expression:—the wrangling which ensues is real enough, however.

★   ★   ★

After this tedious explanation, which I hope, in view of the extraordinary interest of the conception of force, may not have exhausted the reader's patience, we are prepared at last to state the grant fact which this conception embodies. This fact is that if the actual changes of motion which the different particles of bodies experience are each resolved in its appropriate way, each component acceleration is precisely such as is prescribed by a certain law of Nature, according to which bodies, in the relative positions which the bodies in question actually have at the moment, always receive certain accelerations, which, being compounded by geometrical addition, give the acceleration which the body actually experiences.

This is the only fact which the idea of force represents, and whoever will take the trouble clearly to apprehend what this fact is, perfectly comprehends what force is. Whether we ought to say that a force *is* an acceleration, or that it *causes* an acceleration, is a mere question of propriety of language, which has no more to do with our real meaning than the difference between the French idiom "*Il fait froid*" and its English equivalent "*It is cold.*" Yet it is surprising to see how this simple affair has muddled men's minds. In how many profound treatises is not force spoken of as a "mysterious entity," which seems to be only a way of confessing that the author despairs of ever getting a clear notion of what the word mean! In a recent admired work on Analytic Mechanics it is stated that we understand precisely the effect of force, but what force itself is we do not understand! This is simply a self-contradiction. The idea which the word force excites in our minds has no other function than to affect our actions, and these actions can have no reference to force otherwise than through its effects. Consequently, if we know what the effects of force are, we are acquainted with every fact which is implied in saying that a force exists, and there is nothing more to know. The truth is, there is some vague notion afloat that a question may mean something which the mind cannot conceive; and when some hair-splitting philosophers have been confronted with the absurdity of such a view, they have invented an empty distinction between positive and negative conceptions, in the attempt to give their non-idea a form not obviously nonsensical. The nullity of it is sufficiently plain from the considerations given a few pages back; and, apart from those considerations, the quibbling character of the distinction must have struck every mind accustomed to real thinking.

★   ★   ★

[A]ll the followers of science are animated by a cheerful hope that the processes of investigation, if only pushed far enough, will give one certain solution to each question to which they apply it. One man may investigate the velocity of studying the transit of Venus and the aberration of the stars; another by the oppositions of Mars and the eclipses of Jupiter's satellites; a third by the method of Fizeau; a fourth by that of Foucault; a fifth by the motions of the curves of Lissajoux; a sixth, a seventh, an eighth, and a ninth, may follow the different methods of comparing the measures of statical and dynamical electricity. They may at first obtain different results, but, as each perfects his method and his processes, the results are found to move steadily together toward a destined centre. So with all scientific research. Different minds may set out with the most antagonistic views, but the progress of investigation carries them by a force outside of themselves to one and the same conclusion. This activity of thought by which we are carried, not where we wish, but to a fore-ordained goal, is like the operation of destiny. No modification of the point of view taken, no selection of other facts for study, no natural bent of mind even, can enable a man to escape the predestinate opinion. This great hope is embodied in the conception of truth and reality. The opinion which is fated to be ultimately agreed to by all who investigate, is what we mean by the truth, and the object represented in this opinion is the real. That is the way I would explain reality.

But it may be said that this view is directly opposed to the abstract definition which we have give of reality, inasmuch as it makes the character of the real depend on what is ultimately thought about them. But the answer to this is that, on the one hand, reality is independent, not necessarily of thought in general, but only of what you or I or any finite number of men may think about it; and that, on the other hand, though the object of the final opinion depends on what that opinion is, yet what the opinion is does not depend on what you or I or any man thinks. Our perversity and that of others may indefinitely postpone the settlement of opinion; it might even conceivably cause an arbitrary proposition to be universally accepted as long as the human race should last. Yet even that would not change the nature of the belief, which alone could be the result of investigation carried sufficiently far; and if, after the extinction of our race, another should arise with faculties and disposition for investigation, that true opinion must be the one which they would ultimately come to. "Truth crushed to earth shall rise again," and the opinion which would finally result from investigation does not depend on how anybody may actually think. But the reality of that which is real does depend on the real fact that investigation is destined to lead, at last, if continued long enough, to a belief in it.

CLARENCE KING (1842–1901)

Geologist.
Born in Newport, Rhode Island. Died in Phoenix, Arizona.

KING was in the first graduating class of the Sheffield Scientific School at Yale (1862) and studied with James D. Dana. After graduating, he spent the following winter attending Agassiz's lectures at Harvard before joining the California Geological Survey as an unpaid geologist. King was known for being a bright, politic, and charismatic fellow. He lobbied hard for an exploration of the 40th Parallel, and by 1867 the Geological Exploration of the 40th Parallel was a reality and he was its principle geologist. He spent six years in that capacity, producing the genre-bending travel/ adventure/natural history/local-color classic *Mountaineering in the Sierra Nevada* in 1872 and the more traditionally academic *Systematic Geology* in 1878. In 1872 King investigated and uncovered the truth behind an infamous diamond and gemstone hoax, which made him a bit of a celebrity for a time. King was always a man with varied interests. He was drawn to art, literature, travel and elegant living. In the late 1870s he withdrew from science in the hope of making a fortune in mining but never did. He moved in the highest social circles of New York and Washington and was something of a raconteur. He was very good friends with both Henry Adams and John Hay, Lincoln's private secretary. He traveled throughout Europe and was received at the court of England. He always seemed to live above his means. In addition, he lived something of a double life that only his friends Adams and Hay ever knew about. He had a common law wife, Ada Copeland, an African-American woman, and with her four or five children. Ada never knew his real name until he was on his deathbed. After his death, Adams and Hay provided for Ada and his children. It was a shame that King felt the need to pursue wealth over geology because he was not only a bright and innovative thinker but a great facilitator of science, government, and business. As part of the second generation of American neo-Lamarckians, King could utilize both soft and hard views of evolution and offer unique reconsiderations. King didn't think a Lyellian uniformitarianism could explain the sort of geological dynamic he had witnessed in the West and considered himself a catastrophist. His catastrophism was different than Agassiz's, however, in that King didn't think it contrary to evolution. King's catastrophism called for a greater variation in rate and intensity of evolutionary processes — in contrast to the wipe-out scenarios of Agassiz or the gradualism of Lyell. King viewed the shakeup of biological equilibrium caused by catastrophic events as a way to rapid change in "plastic species."

Clarence King, "Catastrophism and Evolution" (excerpt), *The American Naturalist* (1877)

THE earliest geological induction of primeval man is the doctrine of terrestrial catastrophe. This ancient belief has its roots in the actual experience of man, who himself has been witness of certain terrible and destructive exhibitions of sudden, unusual telluric energy. Here in America our own species has seen the vast, massive eruptions of Pliocene basalt, the destructive invasion of northern lands by the slow-marching ice of the glacial period, has struggled with the hardly conceivable floods which marked the recession of the frozen age, has felt the solid earth shudder beneath its feet and the very continent change its configuration. Yet these phenomena are no longer repeated; nothing comparable with them ever now breaks the geological calm.

★   ★   ★

When complete evidence of the antiquity of man in California and the catastrophes he has survived come to be generally understood, there will cease to be any wonder that a theory of the destructive in nature is an early, deeply rooted archaic belief, most powerful in its effect on the imagination. Catastrophe, speaking historically, is both an awful memory of mankind and a very early piece of pure scientific induction. After it came to be woven into the Sanskrit, Hebrew, and Mohammedan cosmogonies, its perpetuation was a matter of course.

From the believers in catastrophe there is, however, a totally different class of minds, whose dominant characteristic is a positive refusal to look further than the present, or to conceive conditions which their senses have never reported. They lack the very mechanism of imagination. They suffer from a species of intellectual near-sightedness too lamentably common among all grades and professions of men. They are bounded – I might almost say imprisoned – by the evident facts and ideas of their own to-day and their own environment. With that sort of detective sharpness of vision which is often characteristic of those who cannot see far beyond their noses, these men have most ably accumulated an impressive array of geological facts relating to the existing operation of natural laws. They have saturated themselves with the present *modus operandi* of geological energy, and culminating in Lyell have founded the British School of Uniformitarianism.

Men are born either catastrophists or uniformitarians. You may divide the race into imaginative people who believe in all sorts of impending crises – physical, social, political – and others who anchor their very souls *in statu quo*. There are men who build arks straight through their natural lives, ready for the first sprinkle, and there are others who do not watch Old Probabilities or even own an umbrella. This fundamental differentiation expresses itself in geology by means of the two historic sects of catastrophists and uniformitarians.

★   ★   ★

Remembering distinctly that uniformitarianism claims one dynamic rate past and present, let us turn to the broader geological features of North America and try to unravel the past enough to test the tenets of the two schools by actual fact. Beneath our American lies buried another distinct continent, – an archaean America. Its original coast-lines we may never be able fully to survey, but its great features, the lofty chains of the mountains which made its bones, were very nearly coextensive with our existing systems, the Appalachians and Cordilleras. The cañon-cutting rivers of the present Western mountains have dug out the peaks and flanks of those underlying, primeval uplifts and developed an astonishing topography: peaks rising in a single sweep thirty thousand feet from their bases, precipices lifting bold, solid fronts ten thousand feet into the air, and profound mountain valleys. The work of erosion which has been carried on by torrents of the Quaternary age – that is to say, within the human period – brings to light buried primeval chains far loftier than any of the present heights of the globe. Man's enthusiastic hand may clear away the shallow dust or rubbish from an Oriental city, and lay bare the stratified graves of perished communities: it is only a mountain torrent which can dig through thousands of feet of solid rock and let in the light of day on the time-strained features of a long-buried continent.

Archaean America was made up of what was originally ocean beds lifted into the air and locally crumpled into vast mountain chains, which were eroded by torrents into true subaerial mountain peaks. This conversion of sea strata into the early continent is the first record of a series of oscillations in which land and sea successively occupied the area of America. In pre-Cambrian time the continent we are considering sank, leaving some of its mountain tops as islands, and the neighboring oceans flowed over it, their bottoms emerging and becoming continents. This is the second of the recorded oscillations of the first magnitude.

★　★　★

Let us pass now to a remarkable chapter of events which closed the Palaeozoic ages. What is now the eastern half of the Mississippi basin had through the coal period often extended itself as a land mass as far west as the Mississippi River, and had as often suffered subsidence and resubmergence. To the west, however, still stretched the open ocean, which since the beginning of the Cambrian, had, with a single exception, never been invaded by land. At the close of the Palaeozoic the two bordering land areas of Atlantis and Pacifis, since the beginning of the Cambrian permanent and perhaps extended continents, began to sink. They rapidly went down, and at last completely disappeared, their places being taken by the present Atlantic and Pacific oceans, while the sea floor of the American ocean, which had been for the most part permanent oceanic area ever since the submergence of the archaean America, emerged and became the new continent of America, which has lasted with local vicissitudes up to the present. The east and west were, indeed, separated by a mediterranean sea, the sole relic of the American ocean, which now occupied a narrow north and south depression.

In that mediterranean sea, we may say that the conditions have been uniformitarian; that is to say, in the great post-Palaeozoic catastrophe that ocean was spared. It remained a body of deep water, its bottom undisturbed by folds or dislocations, and there is no evidence of a cessation of sediments; yet the species which lived there throughout the vast length of the coal period were completely extinguished, and entirely new forms made their appearance. Although spared from the actual physical catastrophe, the effect of the general disturbance of that whole quarter of the globe was thoroughly catastrophic, and exerted a fatal influence upon life far beyond the actual theatre of upheaval.

Passing over the Mesozoic age, which in detail offers much instructive material as to rate of change, we pause only to notice a catastrophe which marked the close of that division of time.

In a quasi-uniformitarian way, 20,000 or 30,000 feet of sediment had accumulated in the Pacific and 14,000 in the mediterranean sea, when these regions, which, during their reception of sediment, had been areas of subsidence, suddenly upheaved, the doming up of the middle of the continent quite obliterating the mediterranean sea and uniting the two land masses into one.

The catastrophe which removed this sea resulted in the folding up of mountain ranges 20,000 and 40,000 feet in height, thereby essentially changing the whole climate of the continent. Of the land life of the Mesozoic age we have abundant remains. Thanks to the palaeontologists, the wonderful reptilian and avian fauna of the Mesozoic age is not familiar to us all. But after the catastrophe and the change of climate which must necessarily have ensued, this fauna totally perished. The rate of this post-Cretaceous change was, in other words, catastrophic.

During the Tertiary, fresh-water lakes of wide extent occupied the western half of the continent. Such was the character of the great post-Cretaceous uplift that there were left broad, deep continental basins above the level of the sea. Into these the early Tertiary rivers found their way, creating extended lakes in which accumulated strata rivaling in importance the deposits of the great oceans. The whole history of the Tertiary is that of the accumulation of thick sedimentary series in fresh-water lakes, accompanied by gradual and periodic subsidence, carried on smoothly and uniformly up to a certain point, and then interrupted by a sudden, mountain-building upheaval, which drained the lakes and created new basins. The five minor catastrophes which have taken place in the western half of America during the Tertiary age have never resulted in those broader changes which mark the close of the Archaean, the Palaeozoic, and the Mesozoic ages. They never broke the grander outline of the continent. They were, however, of such an important scale as to very greatly vary the conditions of half the continent. I may cite the latest important movement, which took place probably within the human epoch, certainly at the close of the great Pliocene lake period of the west. The whole region of the great plains, as far north as we are acquainted with their geology, and southward to the borders of the Gulf, was occupied by a broad lake which existed through the Pliocene period, having always a subtropical climate. In that lake, beds 1,000 to 1,200 feet thick had accumulated, when suddenly the level floor was tilted, causing a difference of height of 7,000 feet between the south and west shores, making the great inclined surface of the present plains, and utterly changing the climate of the whole region. Not a species survived.

★　★　★

If the vicissitudes of our planet have been as marked by catastrophes as I believe, how does that law affect our conceptions of the development of life and the hypothesis of evolution? Man, whatever the drift of life or philosophy, returns with restless eagerness, with pathetic anxiety, to the enigma of his own origins, his own nature, his own destiny. With reverence, with levity, with faith, with doubt, with courage, with cowardice, by ever avenue of approach, in every age, the same old problem is confronted. We pour out our passionate questionings, and hearken lest mute nature may this time answer. But nature yields only one syllable of reply at a time.

Darwin, who in his day has caught the one syllable from nature's lips, advances always with caution, and although he practically rejects does not positively deny the existence of sudden great changes in the earth's history. Huxley, permeated in every fibre by belief in evolution, feels that even to-day catastrophism is not yet wholly out of the possibilities. It is only lesser men who bang all the doors, shut out all doubts, and flaunt their little sign, "Omniscience on draught here." It must be said, however, that biology, as a whole, denies catastrophism in order to save evolution. It is the common mistake of biologists to assume that catastrophes rest for their proof on breaks in the palaeontological record, meaning by that the observed gaps of life or the absence of connecting links of fossils between older and newer sets of successive strata. There never was a more serious error. Catastrophes are far more surely proved by the observed mechanical rupture, displacement, engulfment, crumpling, and crushing of the rocky surface of the globe. Granted that the evidence would have been slightly less perfect had there been no life till the present period, still the reading would have been amply conclusive. The palaeontological record is as imperfect as Darwin pleads, but the dynamic record is vitiated by no such ambiguity.

It is the business of geology to work out the changes of the past configuration of the globe and its climate; to produce a series of maps of the successive stages of the continents and ocean basins, but it is also its business to investigate and fix the rates of change. Geology is not solely a science of ancient configuration. It is also a history of the varying rates and mode of action of terrestrial energy. The development of inorganic environment can and must be solved regardless of biology. It must be based on sound physical principles, and established by irrefragable proof. The evolution of environment, a distinct branch of geology which must soon take form, will, I do not hesitate to assert, be found to depend on a few broad laws, and neither the uniformitarianism of Lyell and Hutton, Darwin and Haeckel, nor the universal catastrophism of Cuvier and the majority of teleologists, will be numbered among these laws. In the dominant philosophy of the modern biologist there is no admission of a middle ground between these two theories, which I, for one, am led to reject. Huxley alone, among prominent evolutionists, opens the door for union of the residua of truth in the two schools, fusing them in his proposed evolutional geology. Looking back over a trail of thirty thousand miles of geological travel, and after as close a research as I am capable, I am impelled to say that his far-sighted view precisely satisfies my interpretation of the broad facts of the American continent.

The admission of even modified catastrophe, namely, suddenly-destructive, but not all-destructive change, is, of course, a downright rejection of strict uniformitarianism. I comprehend the importance of the position, how far-reaching and radical the logical consequences of this belief must be. If true, it is nothing less than an ignited bomb-shell thrown into the camp of the biologists, who have tranquilly built upon uniformitarianism, and the supposed imperfection of the geological record. I quote a few of their characteristic utterances. Lamarck, in his Philosophie Geologique, 1809, says, "The kinds or species of organisms are of unequal age, developed one after another, and show only a relative and temporary persistence. Species arise out of varieties. . . . In the first beginning only the very simplest and lowest animals and plants came into existence; those of a more complex organization only at a later period. The course of the earth's development and that of its organic inhabitants was continuous, not interrupted by violent revolutions. . . . The simplest animals and the simplest plants, which stand at the lowest point in the scale of organization, have originated and still originate by spontaneous generation." Darwin[1] says: "We must be cautious in attempting to correlate as strictly contemporaneous two formation, which include few identical species, by the general succession of their forms of life. As species are produced and exterminated by slowly acting and still acting causes, and not by miraculous acts of creation and by catastrophes. . . . And again, for my part, following out Lyell's metaphor, I look at the natural geological record as a history of the world imperfectly, kept and written in a changing dialect; of this history we possess the last volume alone, relating only to two or three countries. Of this volume, only here and there a short chapter has been preserved; and of each page only here and there a few lines. Each word of the slowly changing language in which the history is written, being more or less different in the successive chapters, may represent the apparently abruptly changed forms of life entombed in our consecutive but widely separated formations. On this view, the difficulties above discussed are greatly diminished, or even disappear."

It is unnecessary to repeat here the well-known views of Lyell. How far biologists have learned to lean on his uniformitarian conclusions may be seen from the following quotation from Haeckel,[2] "He [Lyell] demonstrated that those changes of the earth's surface which are still taking place before our eyes are perfectly sufficient to explain everything we know of the development of the earth's crust in general, and that it is superfluous and useless to seek for mysterious causes in inexplicable revolutions.

---

[1] Origin of Species, p. 522.
[2] History of Creation, vol. I., pages 127–129.

He showed that we need only have recourse to the hypothesis of exceedingly long periods of time, in order to explain the formation of the crust of the earth in the simplest and most natural manner, but the means of the very same causes which are still active. Many geologists had previously imagined that the highest chains of mountains which rise on the surface of the earth could owe their origin only to enormous revolutions transforming a great part of the earth's surface, especially to colossal volcanic eruptions. Such chains of mountains as those of the Alps or the Cordilleras were believed to have arisen direct from the fiery fluid of the interior of the earth through an enormous chasm in the broken crust. Lyell, on the other hand, showed that we can explain the formation of such enormous chains of mountains quite naturally by the same slow and imperceptible risings and depressions of the earth's surface which are still continually taking place, and the causes of which are by no means miraculous. Although these depressions and risings may perhaps amount only to a few inches, or at most a few feet, in the course of a century, still in the course of some millions of years they are perfectly sufficient to raise up the highest chains of mountains without the aid of mysterious and incomprehensible revolutions. . . . We have long known, even from the structure of the stratified crust of the earth alone, that its origin and the formation of neptunic rocks from water must have taken at least several millions of years. From a strictly philosophical point of view, it makes no difference whether we hypothetically assume for this process ten millions or ten thousand billions of years. Before us and behind us lies eternity." This is even bolder than Hutton, who says: "I take things as I find them at present; and from these I reason as regards that which must have been. . . . A theory, therefore, which is limited to the actual constitution of this earth, cannot be allowed to proceed one step beyond the present order of things."

The successive hypotheses which, linked together, form the chain of evolution are, first, the nebular hypothesis; second, spontaneous generation; third, natural selection. It is only with the last that geology has intimate relation. The general theory of a derivative genesis or the descent of all organisms by the various modes of reproduction from one or a few primitive types which came into existence by spontaneous generation was believed long before the Darwinian theory was advanced. Darwin's great contribution was the modus operandi of derivative genesis. It was a mode of accounting for the infinite branching out and differentiation of the complex forms of life from the primitive germs. His theory is natural selection, or the survival of the fittest, a doctrine which, left where Darwin leaves it, has its very roots in uniformitarianism.

Analyzed into its component parts, natural selection resolves, as is well know, into two laws, hereditivity and adaptivity: first, the power on the part of organisms to transmit to offspring their own complex structure down to the minutest details; and, secondly, the power by slight alterations on the part of all individuals to vary slightly in order to bring themselves into harmony with a changed environment. When we bring geology into contact with Darwinism, it is evident that hereditivity is out of the domain of our inquiry; it is not the engine of change, it is the conservator of the past; but the companion law of adaptivity, or the accommodation to circumstances, is one which depends half upon the organism and half upon the environment; half upon the vital interior, half upon the pressure which the environment brings to bear upon it. Now, environment, as conclusively shown by biologists, is a twofold thing, a series of complicated relationships with contemporaneous life, but, besides, with the general inorganic surrounding, involving climate and position upon the globe. Preoccupied with the strictly biological environment, namely, the intricate relation of dependence of any species upon some of its surrounding species, biologists have signally failed to study the power and influence of the inorganic or geologic environment. The actual limits of the influence of physical conditions on life are practically unknown. In America more than in Europe this branch of inquiry has begun to attract notice, but it is yet in its swaddling-clothes. It has lain little and weak from inanition, while the favorite child, Natural Selection has been fed into a plethoric, overgrown monster. Darwin, Wallace, Haeckel, and the other devoted students of natural selection have brought to light the most astonishingly complex struggle for existence, everywhere progressing – the fiercest battle for life and for subsistence, for standing-room, for breath. Some species gain, others lose, some go down to annihilation. In this battle they see adequate cause for all the great, highly organized products of the millions of years since life began. From their logic, you and I are conquerors who have mounted to manhood by treading out the life of infinite generations. We are what we are because this brain and this body form the most effective fighting-machine the dice-box of ages has thrown.

From their conclusions and philosophy let us turn, but with no revolt of prejudice, no rebound of a happier intuition, for this is a question of science. Those who defend the stronghold of natural selection are impregnable to the assaults of feeling. They are dislodged only by the solid projectiles of fact, and to facts cast in the mold of nature they count it no dishonor to surrender. If, as I have said, the evolution and power of environment have been singularly neglected studies, if biologists have allowed the splendor of their achievements within the province of life to blind them to the working of that other and no less important side of the problem, what then is the general relation in time and space of the inorganic environment to life?

Let us first acknowledge frankly that the present and later parts of the Quaternary period are uniformitarian; that the changes going on in organic life now do obey the great law of survival of the fittest, and that if the uniformitarians were true in making of the past a mere infinite projection of the present, then the biologists would have based their theories on a solid foundation, and my protest would have no weight. Let us go further and cordially admit that in all periods of uniformity the progress of life would adjust itself to its surroundings, and the war of competitive extermination become the dominant engine of change and development. This is giving full credit to the greatness of the biological result, and simply asserts that they who achieved it are sound as far as the analogy of present uniformity may be permitted to go. But uniformity has not been the sole law; it has, as we have seen, been often broken by catastrophes, – that is, by accelerated rate of change. Rapid physical change has been it seems to me, the more important of the two conditions of the past, the one whose influence will at last prove to have been the dominant one in life change.

Has environment, with all the catastrophic changes, been merely passive as regards life? It has either had no effect, or has restrained the progress of evolution, or has advanced it, or its influence has been as varied as its own history, – now by the development of favoring conditions accelerating vital progress, now suddenly exterminating on a vast scale, again urging evolution forward, again leaving lapses of calm in which species took the matter into their own hands and worked out their own destiny. It is only through rapid movements of the crusts and sudden climatic changes, due either to terrestrial or cosmical causes, that environment can have seriously interfered with the evolution of life. These effects would, I conceive be, first extermination; secondly, destruction of the biological equilibrium, thus violating natural selection; and thirdly, rapid morphological change on the part of plastic species. When catastrophic change burst in upon the ages of uniformity, and sounded in the ear of every living thing the words "change or die," plasticity became the sole principle of salvation. Plasticity, then is that quality which, in suddenly enforced physical change, is the key to survival and prosperity. And the survival of the plastic, that is of the rapidly and healthily modifiable during periods when terrestrial revolution offers to species the rigorous dilemma of prodigious change or certain death, is a widely different principle from the survival of the fittest in a general biological battle during terrestrial uniformity. In one case it is an accommodation between the individual organism and inorganic environment, in which the most yielding and plastic lives. In the other it is a Malthusian death struggle, in which only the victor survives. At the end of a period of uniformitarian conditions, the Malthusian conqueror, being the fittest, would have won the prize of survival and ascendency. Suppose now an internal of accelerated change. At the end only the most plastic would have deviated from their late forms and reached the point of successful adaptation, which is survival in health. Whatever change takes place by natural selection in uniformitarian ages, according to Darwin, advances by spontaneous, aimless sporting and the survival of those varieties best adapted to surrounding conditions, and of these conditions the biological relations are by far the most important of all. By that means, and by that alone, it is asserted, species came into existence, and inferentially all the other forms from first to last. This is the gospel of chance.

If the out-door facts of American geology shall be admitted to bear me out in my assertion of catastrophes, and if the epochs of maximum vital change do, as I hold, coincide with the epochs of catastrophes, then that coincidence should be directly determinable in the field. I confidently assert that no American geologist will be able to disprove the law that in the past every one of the great breaks in the column of life coincide with datum points of catastrophe. It remains to be determined how far this coincidence is the expression of environmental cause, responded to in terms of vital effect.

From a comparison of the list and character of geological changes in America with those mysterious lines across which no species march, I feel warranted in harboring the belief that catastrophe was an integral part of the cause; changed life, the effect. Biologists are accustomed to explain the cause of a great gap like that which divides the Palaeozoic and Mesozoic life by an admission that the Palaeozoic forms ceased to live, but that the succeeding changed forms at the beginning of the Mesozoic were not the local progeny, greatly modified by catastrophic change, but merely immigrants from some other conveniently assumed country. They succeed in rendering this highly probably, if not certain, in many instances. But they are stopped from always advancing this migration theory. Greek art was fond of decorating the friezes of its sacred edifies with the spirited form of the horse. Times change; around the new temple of evolution the proudest ornament is that strange procession of fossil horse skeletons, among whose captivating splint-bones and general anatomy may be descried the profiles of Huxley and Marsh. Those two authorities, whose knowledge we may not dispute, assert that the American genealogy of the horse is the most perfect demonstrative proof of derivative genesis ever presented. Descent they consider proved, but the fossil jaws are utterly silent as to what the cause of the evolution may have been.

I have studied the country from which these bones came, and am able to make this suggestive geological commentary. Between each two successive forms of the horse there was a catastrophe which seriously altered the climate and configuration of the whole region in which these animals lived. Huxley and Marsh assert that the bones prove descent. My own work proves that each new modification succeeded a catastrophe. And the almost universality of such coincidences is to my mind warrant for the anticipation that not very far in the future it may be seen that the evolution of environment has been the major cause of the evolution of life; that a mere Malthusian struggle was not the author and finisher of evolution; but that He who brought to bear that mysterious energy we call life upon primeval matter bestowed at the same time a power of development by change, arranging that the interaction of energy and matter which made up environment should, from time to time, burst in upon the current of life and sweep it onward and upward to ever higher and better manifestations. Moments of great catastrophe, thus translated into the language of life, become moments of creation, when out of plastic organisms something newer and nobler is called into being.

SAMUEL PIERPONT LANGLEY (1834–1906)

Astronomer, physicist, aviation pioneer.
Born in Roxbury, Massachusetts. Died in Aiken, South Carolina.

LANGLEY graduated from the Boston Latin School and was, thereafter, largely self-taught. He became an assistant at Harvard College Observatory and used the college library to read in astronomy, physics, and engineering. After a couple years he moved on to a job at the U.S. Naval academy as an astronomer and mathematics instructor. In 1867 he became director of the Allegheny Observatory of the Western University of Pennsylvania at Pittsburgh. He would stay there the next twenty years and carry out many of the pioneering studies of the infrared portions of the solar spectrum for which he is remembered as a scientist. He would also invent the bolometer, which detects and measures radiation (visible light, infrared, ultraviolet) in amounts as small as one millionth of an erg. His bolometer enabled him to measure changes in lunar and solar radiation. In 1881 he organized an expedition to Mt. Whitney, California, in order to measure radiation at high altitudes and determine their relative intensity. After this expedition, he became a highly sought after lecturer. *The New Astronomy*, which came out as a book in 1888 and became something of an instant classic, had been serialized in *The Century* for much of 1885 and had made Langley an even more high-profile scientist. The focus of his new astronomy was more on the physics of structure than the traditional inquiries about position, and he was the first scientist to envision astrophysics as a distinct field. In 1887 he became the third Secretary of the Smithsonian Institute and in 1890 founded the Smithsonian Astrophysical Observatory, where he continued his studies of the Sun (including the Sun's effect on our weather and atmosphere that would lead the way to considerations of global warming).Langley received the Henry Draper Medal from the National Academy of Science in 1886 for his work on the Sun. In the 1890s Langley became seriously interested in flight and by 1896 had built a steam-driven aeroplane which gained flight by being catapulted. It flew over the Potomac River for 90 seconds and was the first recorded flight of an unmanned, engine-powered aircraft. This success propelled Langley into further inquiry, and in 1901 he produced a small model of his aerodrome which tested successfully and convinced the government to give him $50,000 for a full-scale model. This was ready for test flight in October 1903. The plane crashed soon after being launched, but not beyond repair, and a second attempt was made that December. This attempt would prove to be a spectacular disaster with the force of the catapult basically collapsing the entire rear of the plane, which sort of folded from the platform straight into the river. Langley's second failure was ridiculed mercilessly in the press and in the congress. It did not help that nine days later the Wright brothers would successfully fly at Kitty Hawk in a plane that cost them $1,000. Although

Langley never quite recovered from the embarrassment and disappointment of his foray into aviation, a decade after his death his name would be attached to two of the corner stones of U.S. aviation and scientific research, Langley Air Force Base and Langley Memorial Aeronautical Laboratory (Langley Research Center).

S. P. Langley, "The Spectrum of an Argand Burner," *Science* (1883)[1]

I have been lately requested to determine the distribution of energy in the spectrum of an argand burner, and have been able to do this by means of the apparatus and methods previously employed at the Allegheny observatory for mapping the invisible spectrum of the sun. The results are curious; and, in the hope that they may also be found useful, I desire to communicate them to the academy. The difficulty is such a determination lies in the mapping of something which is wholly invisible; and it has not been made before, I presume, in spite of its economical importance, because there has been no means known of measuring this invisible energy, except in a rough way, by the thermometer or themopile, by a process which gives incomplete results.

It was my object not merely to indicate how much of the radiation from a gas-burner was visible, and how much was not, but to give a map of its distribution on the normal or wave-length scale, which would enable any one to see the quality and amount of the energy in each part of the light and head region.

The ordinary argand burner, burning common house-gas within a glass chimney, was first placed at the centre of the curvature of a large Rowland concave grating; and, but means of the bolometer, the head was measured at successive points in the spectrum down to a wave-length of about .001 mm., where the overlapping second spectrum began to be sensible. Even in the preliminary determination, it was interesting to observe that the distribution of the head was totally different from that in the sun, and that, instead of growing smaller, it grew greater, as the bolometer passed from the visible to the invisible end. As it was evident that the head was still increasing at the point where the evidence from the grating failed, all the measures were next repeated with a prism of a special glass known to be transparent to the invisible rays. (It was first attempted to use the linear thermopile; but the head was insufficient, and the linear bolometer was substituted.) With this, as many as thirteen ordinates were measured (representing the proportionate heat at as many points), their respective indices of refraction being determined by the known refracting angle of the prism and the observed deviations on the circle of the spectro-bolometer.

[1] Read before the National academy of sciences at its Washington meeting, April, 1883.

In a late communication to the academy, I gave the results of a recent research upon the connection between indices of refraction and wave-lengths, which enable us to deduce the normal spectrum (invisible as well as visible) from the prismatic one. It appeared to me when I was engaged in the first investigation, which to all but students of the subject must seem abstruse, that its results were of a kind which could never have much other than a theoretical interest: but it happens that this their first application is of a utilitarian character; for, having thus converted the prismatic values into corresponding ones on the wave-length scale, I was able to represent the conclusions from both by the normal maps which I now have here, and which exhibit the results of the analysis of the radiant heat which has come through the chimney. Let us remember that this radiant energy differs wholly in its qualities in different parts, and that the quality is shown by the wave-length number on the horizontal scale; the amount, by the height of the curve at that point. Near the part with wave-length .5 it gives the eye the sensation we call blueness, and near .7 it appears as a dull red, bringing very little light; at the point .9 or .10 it makes on the most sensitive eye no impression whatever, but has the power of passing freely through the glass chimney; near .3 the glass, so transparent to the light, is almost wholly opaque to the energy: so that each part has some quality peculiar to itself. By far the most important of these qualities, for our present purpose, is that of giving light. If we then analyze the radiant energy which comes through the chimney, the result is shown in our lower curve. The energy, which is what the gas supplies at the cost of the production from the coal, is for our present purpose regarded as saved or wasted, according as it is visible (light) or invisible (dark heat). The energy first becomes measurable in the blue, where there is very little of it, but where all there is, is effective as light; it increases steadily to the extreme red ordinarily visible where there is a great deal of it, but of a quality which is only interpreted by the eye as a dull reddish glow of little value for lighting; and then goes on increasing where it passes into complete invisibility, and still continues to increase as (for the present purpose) *pure waste*, till its maximum is reached at a wave-length of 1.5 or 1.6, – something like three times the length of the visible spectrum below the lowest visible ray. The energy at any point being proportional to the height, the entire radiant heat is proportional to the area of the curve. If we draw it on such a scale that this whole area equals 100, we can see the percentage expended in any kind of radiation at a glance. The small, nearly triangular area to the left of the line at .7, for instance, represents all the radiant energy useful as light; and this area being by measurement 2.4, while that of the whole curve is 100, we see that 2.4% are employed as light, and the remainder, 97.6%, are wasted. But this refers to the *radiant* heat alone, and takes no account of that expended in heating the air by convection currents. I have heard this latter estimated at three-quarters of the whole, but have not myself measured it. Admitting that this is approximately correct, however, it follows, that, since only one-quarter is radiated, it is 2.4% of this one-quarter only, which is light, and that finally less than 1% of the whole is used, and more than 99% wasted.

It is instructive to take an amount of solar energy exactly equal to that we have just analyzed in the gas-burner, and notice how totally different it is in kind. The upper curve shows the distribution of such a small sample of the sun's radiation as shall be exactly equal to that from the argand burner. Of 1,000 parts of sun energy, 340 appear as light, and 660 as dark heat, if we take the dividing-line between light and darkness at the same point (wave-length 0.0007 mm.) in each curve. If we look at the quality of the light, the difference is enhanced. The sun-curve attains its greatest height in the yellow, which here means that the energy is not only most efficient in making us see (that is, is most available as light), but that of this light the energy is again most effective in a part to which the eye is most sensitive, while, of the small amount of energy employed by the gas in making us see at all, most (as shown by the height of the curve in the dark red) is spend in rays to which the eye is not sensitive, and which give the gas its well-known inferiority in quality (of color) to sunlight, even where the quantity is the same.

Similar curves obtained for the electric light would be interesting, but I have not undertaken them.

We are accustomed to indicator and other diagrams in the use of the steam-engine, showing us how our energy is being generated; but it is singular that so little has been done in the present direction in showing us with what economy it is being employed. I think interest attaches to these curves from a purely scientific stand-point, and they were made with no ulterior purpose. Yet in looking at them I can but be so impressed with their utilitarian applications that I will ask leave to make a remark in conclusion with reference to this.

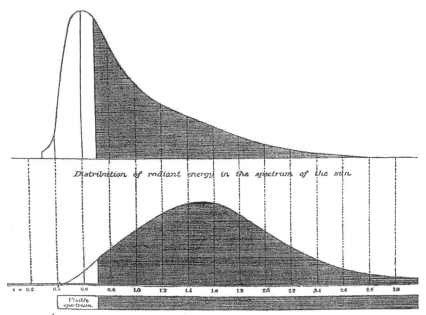

Distribution of radiant energy in the spectrum of an argand gas-burner.

The gas-plant of this country is said to be some $30,000,000; and (except so far as it is used in heating) it appears from what has just been said, that it is mostly wasted as compared with the results possibly attainable, and in the sense that it does not realize one one-hundredth of what an ideally perfect lighting-agent might get from the coal now used. Though this ideal light will never be fully realized, it is undoubtedly possible to do what we see actually done in sunlight; and thus whoever can, without altering the quantity, effect this change in the quality of the radiation from gas, will add millions to the national wealth.

S. P. Langley, "The New Astronomy. I. Spots on the Sun" (excerpt), *The Century* (1885)

THE visitor to Salisbury Plain sees around him a lonely waste, utterly barren except for a few recently planted trees, and otherwise as desolate as it could have been when Hengist and Horsa landed in Britain, for its monotony is still unbroken except by the funeral mounds of ancient chiefs, which dot it to its horizon, and contrast strangely with the crowded life and fertile soil which everywhere surrounds its borders. In the midst of this loneliness rise the rude, enormous monoliths of Stonehenge, circles of gray stones which seem as old as time, and were there, as we now are told, the temple of a people which had already passed away, and whose worship was forgotten, when our Saxon forefathers first saw the place.

In the center of the inner circle is a stone which is believed once to have been the alter, while beyond the outmost ring, quite away to the north-east upon the open plain, still stands a solitary stone, set up there evidently with some special object by the same unknown builders. Seen under ordinary circumstances, it is difficult to divine its connection with the others; but we are told that once in each year, upon the morning of the longest day, the level shadow of this distant, isolated stone is projected at sunrise to the very center of the ancient sanctuary, and falls just upon the alter. The primitive man who devised this was both astronomer and priest, for he not only adored the risen god whose first beams brought him light and warmth, but could mark its place; and, though utterly ignorant of its nature, had evidently learned enough of its motions to embody his simple astronomical knowledge in a record so exact and so enduring that, though his very memory has gone, common men are still interested in it; for, as I learned when viewing the scene, people are accustomed to come from all the surrounding country, and pass in this desolate spot the short night preceding the longest day of the year, to see the shadow touch the altar at the moment of sunrise.

Most great national observatories, like Greenwich or Washington, are the perfected development of that kind of astronomy of which the builders of Stonehenge represent the infancy. Those primitive men could know where the sun would rise on a certain day, and make their observation of its place, as we see, very well, without knowing anything of its physical nature. At Greenwich the moon has been observed with scarcely an intermission for one hundred and fifty years, but we should mistake greatly did we suppose that it was for the purpose of seeing what it was made of, or of making discoveries in it. This immense mass of Greenwich observations is for quite another purpose – for the very practical purpose of forming the lunar tables, which, by means of the moon's place among the stars, will tell the navigator in distant oceans where he is, and conduct the fleets of England safely home.

In the observatory at Washington one may see a wonderfully exact instrument, in which circles of brass have replaced circles of stone, all so bolted between massive piers, that the sun can be observed by it but once daily, as it crosses the meridian. This instrument is the completed attainment along that long line of progress in one direction, of which the solitary stone at Stonehenge marks the initial step – the attainment, that is, purely of precision of measurement; for the astronomer to-day can still use his circles for the special purpose of fixing the sun's place in the heavens, without any more knowledge of that body's chemical constitution than had the man who built Stonehenge.

Yet the object of both is, in fact, the same. It is true that the functions of astronomer and priest have become divided in the advance of our modern civilization, which has committed the special cultivation of the religious aspect of these problems to a distinct profession; while the modern observer has possibly exchanged the emotions of awe and wonder for a more exact knowledge of the equinox than was possessed by his primitive brother, who both observed and adored. Still, both aim at the common end, not of learning what the sun is made of, but of where it will be at a certain moment; for the prime object of astronomy, until very lately indeed, has still been to say *where* any heavenly body is, and not *what* it is. It is this precision of measurement, then, which has always – and justly – been a paramount object of this oldest of the sciences, not only as a good in itself, but as leading to great ends; and it is this which the poet of Urania has chosen rightly to note as its characteristic, when he says:

> "That little Vernier, on whose slender lines
> The midnight taper trembles as it shines,
> Tells through the mist where dazzled Mercury burns,
> And marks the point where Uranus returns."

But within a comparatively few years a new branch of astronomy has arisen, which studies sun, moon, and stars for what they are in themselves, and in relation to ourselves. Its study of the sun, beginning with its external features (and full of novelty and interest, even, as regards those), led to the further inquiry as to what it was made of, and then to finding the unexpected relations which it bore to the earth and our own daily lives on it, the conclusion being that, in a physical sense, it made us and re-creates us, as it were, daily, and that the knowledge of the intimate ties which unite man with it brings results of the most practical and important kind, which a generation ago were unguessed at.

This new branch of inquiry is sometimes called Celestial Physics, sometimes Solar Physics, and is sometimes more rarely referred to as the New Astronomy. I will call it here by this title, and try to tell the reader something about it which may interest him, beginning with the sun.

The whole of what we have to say about the sun and stars presupposes a knowledge of their size and distance, and we may take it for granted that the reader has at some time or another heard such statements as that the moon's distance is two hundred and forty thousand miles, and the sun's ninety-three million (and very probably has forgotten them again as of no practical concern). He will not be offered here the kind of statistics which he would expect in a college text-book; but we must linger a moment on the threshold of our subject – the nature of these bodies – to insist on the real meaning of such figures as those just quoted. We are accustomed to look on the sun and moon as far off together is the sky; and though we know the sun is greater, we are apt to think of them vaguely as things of a common order of largeness, away among the stars. It would be safe to say that, though nine out of ten intelligent readers have learned that the sun is larger than the moon, and, in fact, larger than the earth itself, most of them do not at all realize that the difference is so enormous that if we could hollow out the sun's globe and place the earth in the center, there would still be so much room that the moon might go on moving in her present orbit at two hundred and forty thousand miles from the earth, – *all within the globe of the sun itself,* – and have plenty of room to spare.

As to the distance of ninety-three million miles, a cannon-ball would travel it in about fifteen years. It may help us to remember that at the speed attained by the Limited Express on our railroads a train which had left the sun for the earth when the *Mayflower* sailed from Delfhaven with the Pilgrim Fathers, and which ran at that rate day and night, would in 1884 still be a journey of some years away from its terrestrial station. The fare, at the customary rates, it may be remarked, would be rather over two million five hundred thousand dollars, so that it is clear that we should need both money and leisure for the journey.

Perhaps the most striking illustration of the sun's distance is given by expressing it in terms of what the physiologists would call velocity of nerve transmission. It has been found that sensation is not absolutely instantaneous, but that it occupies a very minute time in traveling along the nerves; so that if a child puts its finger into the candle, there is a certain almost inconceivably small space of time, say the one-hundredth of a second, before he feels the heat. In case, then, a child's arm were long enough to touch the sun, it can be calculated from this known rate of transmission that the infant would have to live to be a man of over a hundred before it know that its fingers were burned.

Trying with the help of these still inadequate images, we may get some idea of the real size and distance of the sun. I could wish not to have to dwell so long upon figures, that seem, however, indispensable; but we are not done with these, and are ready to turn to the telescope and see what the sun itself look like.

The sun, as we shall learn later, is a star, and not a particularly large star. It is, as has been said, "only a private in the host of heaven," but it is one of that host; it is one of those glittering points to which we have been brought near. Let us keep in mind, then, from the first, what we shall see confirmed later, that there is an essentially similar constitution in them all, and not forget that when we study the sun, as we now begin to do, we are studying the stars also.

HENRY AUGUSTUS ROWLAND (1848–1901)

Physicist.
Born in Honesdale, Pennsylvania. Died in Baltimore, Maryland.

ROWLAND was interested in experimentation and researching the nature of things from his youth. His family expectations were for a classical education and the ministry, but he would have none of it and was able to convince his family to allow him to attend Rensselaer Technological Institute. He graduated in engineering in 1870 and attempted a non-academic job but bored quickly and returned to Rensselaer to teach. In his free time, Rowland conducted research on electricity and magnetism. When the American Journal of Science showed no interest in publishing his work on magnetic permeability, he sent the paper on to James Clerk Maxwell, the famed Scottish physicist and mathematician, who understood better what Rowland was doing and had the paper published in an English journal. (The paper would eventually prove important for calculating designs for dynamos and transformers.) Still, no one in America particularly noticed. So when Johns Hopkins University was being organized, the founders asked the advice of some European scientists on who an outstanding physicist would be and were surprised that the Europeans thought so highly of Rowland. He was hired as the first chair of physics at Johns Hopkins University and remained there until his death. Rowland would continue to do seminal work in electricity and magnetism for the rest of his career, but it is perhaps one particular engineering accomplishment that remains most remembered for the effect it had on spectroscopy. Rowland engineered a screw of nearly flawless pitch for the ruling engine which greatly decreased the false, noisy background of the readings as well as a spherically curved plate for the diffraction gratings to replace the flat, less accurate plates. These changes started a whole new era of spectroscopy. Rowland was also an ardent proponent of basic research and helped define the role of the academy for the twentieth century.

Henry A. Rowland, "Screw," *Encyclopedia Britannica*, vol. 21, 9th edition (1875–1889)

THE screw is the simplest instrument for converting a uniform motion of rotation into a uniform motion of translation (see "Mechanics," vol. xv, p. 754). Metal screws requiring no special accuracy are generally cut by taps and dies. A tap is a cylindrical piece of steel having a screw on its exterior with sharp cutting edges; by forcing this with a revolving motion into a hole of the proper size, a screw is cut on its interior forming what is known as a nut or female screw. The die is a nut with sharp cutting edges used to screw upon the outside of round pieces of metal and thus produce male screws. More accurate screws are cut in a lathe by causing the carriage carrying the tool to move uniformly forward, thus a continuous spiral line is cut on the uniformly revolving cylinder fixed between the lathe centres. The cutting tool may be an ordinary form of lathe tool or a revolving saw-like disk (see "Machine Tools," vol. xv, p. 153).

Errors of Screws. – For scientific purposes the screw must be so regular that it moves forward in its nut exactly the same distance for each given angular rotation around its axis. As the mountings of a screw introduce many errors, the final and exact test of its accuracy can only be made when it is finished and set up for use. A large screw can, however, be roughly examined in the following manner: (1) See whether the surface of the threads has a perfect polish. The more it departs from this, and approaches the rough, torn surface as cut by the lathe tool, the worse it is. A perfect screw has a perfect polish. (2) Mount upon it between the centres of a lathe and the slip a short nut which fits perfectly. If the nut moves from end to end with equal friction, the screw is uniform in diameter. If the nut is long, unequal resistance may be due to either an error of run or a bend in the screw. (3) Fix a microscope on the lathe carriage and focus its single cross-hair on the edge of the screw and parallel to its axis. If the screw runs true at every point, its axis is straight. (4) Observe whether the short nut runs from end to end of the screw without a wabbling motion when the screw is turned and the nut kept from revolving. If it wabbles the screw is said to be drunk. One can see this error better by fixing a long pointer to the nut, or by attaching to it a mirror and observing an image in it with a telescope. The following experiment will also detect this error: (5) Put upon the screw two well-fitting and rather short nuts, which are kept from revolving by arms bearing against a straight edge parallel to the axis of the screw. Let one nut carry an arm which supports a microscope focused on a line ruled on the other nut. Screw this combination to different parts of the screw. If during one revolution the microscope remains in focus, the screw is not drunk; and if the cross-hairs bisect the lines in every position, there is no error of run.

Making Accurate Screws. – To produce a screw of a foot or even a yard long with errors not exceeding 1/1000th of an inch is not difficult. Prof. Wm A. Rogers, of Harvard Observatory, has invented a process in which the tool of the lathe while cutting the screw is moved so as to counteract the errors of the lathe screw. The screw is then partly ground to get rid of local errors. But, where the highest accuracy is needed, we must resort in the case of screws, as in all other cases, to grinding. A long, solid nut, tightly fitting the screw in one position, cannot be moved freely to another position unless the screw is very accurate. If grinding material is applied and the nut is constantly tightened, it will grind out all errors of run, drunkenness, crookedness, and irregularity of size. The condition is that the nut must be long, rigid and capable of being tightened as the grinding proceeds; also the screw must be ground longer than it will finally be needed so that the imperfect ends may be removed.

The following process will produce a screw suitable for ruling gratings for optical purposes. Suppose it is our purpose to produce a screw which is finally to be 9 inches long, not including bearings, and 1 1/8 in. in diameter. Select a bar of soft Bessemer steel, which has not the hard spots usually found in cast steel, and about 1 3/8 inches in diameter and 30 lone. Put it between lathe centres and turn it down to one inch diameter everywhere, except about 12 inches in the centre, where it is left a little over 1 1/8 inches in diameter for cutting the screw. Now cut the screw with a triangular thread a little sharper than 60°. Above all, avoid a fine screw, using about 20 threads to the inch.

The grinding nut, about 11 inches long, has now to be made. Fig. 1 represents a section of the nut, which is made of brass, or better, of Bessemer steel. It consists of four segments,– a, a, which can be drawn about the screw by two collars, b, b, and the screw c. Wedges between the segments prevent too great pressure on the screw. The final clamping is effected by the rings and screws, d, d, which enclose the flanges, e, of the segments. The screw is now placed in a lathe and surrounded by water whose temperature can be kept constant to 1° C., and the nut placed on it. In order that the weight of the nut may not make the ends too small, it must either be counterbalanced by weights hung from a rope passing over pulleys in the ceiling, or the screw must be vertical during the whole process. Emery and oil seem to be the only available grinding materials, though a softer silica powder might be used towards the end of the operation to clean off the emery and prevent future wear. Now grind the screw in the nut, making the nut pass backwards and forwards over the screw, its whole range being nearly 20 inches at first. Turn the nut end for end every ten minutes and continue for two weeks finally making the range of the nut only about 10 inches, using finer washed emery and moving the lathe slower to avoid heating. Finish with a fine silica powder or rouge. During the process, if the thread becomes too blunt, recut the nut by a *short* tap so as not to change the pitch at any point. This must, of course, not be done less than five days before the finish. Now cut to the proper length; centre again in the lathe under a microscope, and turn the bearings. A screw so ground

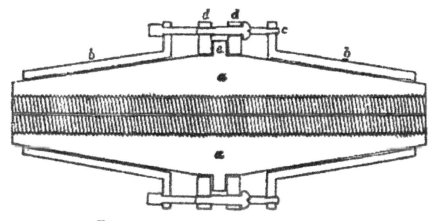

FIG. 1.—Section of Grinding Nut.

has less errors than from any other system of mounting. The periodic error especially will be too small to be discovered, though the mountings and graduation and centering of the head will introduce it; it must therefore finally be corrected.

Mounting of Screws. – The mounting must be devised most carefully, and is, indeed, more difficult to make without error than the screw itself. The principle which should be adopted is that no workmanship is perfect; the design must make up for its imperfections. Thus the screw can never be made to run true on its bearings, and hence the device of resting one end of the carriage on the nut must be rejected. Also all rigid connection between the nut and the carriage must be avoided, as the screw can never be adjusted parallel to the ways on which the carriage rests. For many purposes, such as ruling optical gratings, the carriage must move accurately forward in a straight line as far as the horizontal plane is concerned, while a little curvature in the vertical plane produces very little effect. These conditions can be satisfied by making the ways V-shaped and grinding with a grinder somewhat shorter than the ways. By contrast reversals and by lengthening or shortening the strokes, they will finally become nearly perfect. The vertical curvature can be sufficiently tested by a short carriage carrying a delicate spirit level. Another and very efficient form of ways in V-shaped with a flat top and nearly vertical sides. The carriage rests on the flat top and is held by springs against one of the nearly vertical sides. To determine with accuracy whether the ways are straight, fix a flat piece of glass on the carriage and rule a line on it by moving it under a diamond; reverse and rule another line near the first, and measure the distance apart at the centre and at the two ends by a micrometer. If the centre measurement is equal to the mean of the two end ones, the line is straight. This is better than the method with a mirror mounted on the carriage and a telescope. The screw itself must rest in bearings, and the end motion be prevented by a point bearing against its flat end, which is protected by hardened steel or a flat diamond. Collar bearings introduce periodic errors. The secret of success is so to design the nut and its connections as to eliminate all adjustments of the screw and indeed all imperfect workmanship. The connection must also be such as to give means of correcting any residual periodic errors or errors of run which may be introduced in the mountings or by the wear of the machine.

The nut is shown in Fig 2. It is made in two halves, of wrought iron filled with boxwood or lignum vitae plugs, on which the screw is cut. To each half a long piece of sheet steel is fixed which bears against a guiding edge, to be described presently. The two halves are held to the screw by springs, so that each moves forward almost independently of the other. To join the nut to the carriage, a ring is attached to the latter, whose plane is vertical and which can turn round a vertical axis. Hence each half does its share independently of the other in moving the carriage forward. Any want of parallelism between the screws and the ways or eccentricity in the screw mountings thus scarcely affects the forward motion of the carriage. The guide against which the steel pieces of the nut rest can be made of such form as to correct any small error of run due to wear of the screw. Also, by causing it to move backwards and forwards periodically, the periodic error of the head and mountings can be corrected.

In making gratings for optical purposes the periodic error must be very perfectly eliminated, since the periodic displacement of the lines only one-millionth of an inch

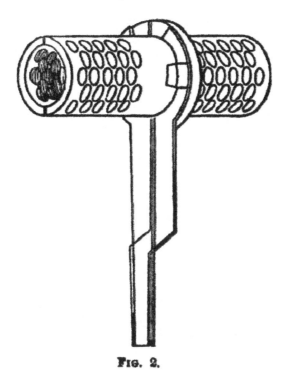

Fig. 2.

from their mean position will produce "ghosts" in the spectrum.[1] Indeed, this is the most sensitive method of detecting the existence of this error, and it is practically impossible to mount the most perfect of screws without introducing it. A very practical method of determining this error is to rule a short grating with very long lines on a piece of common thin plate glass; cut it in two with a diamond and superimpose the two halves with the rulings together and displaced sideways over each other one-half the pitch of the screw. On now looking at the plates in a proper light so as to have the spectral colors show through it, dark lines will appear, which are wavy if there is a periodic error and straight if there is none. By measuring the comparative amplitude of the waves and the distance apart of the two lines, the amount of the periodic error can be determined. The phase of the periodic error is best found by a series of trials after setting the corrector at the proper amplitude as determined above.

A machine properly made as above and kept at a constant temperature should be able to make a scale of 6 inches in length, with errors at no point exceeding 1/100000th of an inch. When, however, a grating of that length is attempted at the rate of 14,000 lines to the inch, four days and nights are required, and the result is seldom perfect, possibly on account of the wear of the machine or changes of temperature. Gratings, however, less than 3 inches long are easy to make.

---

[1] In a machine made by the present writer for ruling gratings the periodic error is entirely due to the graduation and centering of the head. The uncorrected periodic error from this cause displaces the lines 1/300000th of an inch, which is sufficient to entirely ruin all gratings made without correcting it.

ALBERT ABRAHAM MICHELSON (1852–1931)

Physicist.
Born in Strelno, Prussia. Died in Pasadena, California.

Michelson and his parents immigrated to America when he was two, and he grew up in San Francisco. After graduating from the U.S. Naval Academy, he completed a two year cruise, returned to the academy to teach, worked with Simon Newcomb in Washington, and then traveled to Europe for advanced study in physics. After leaving the Navy, he was professor of physics at Case School of Applied Science, Cleveland, Ohio, then at Clark University, Worchester, Massachusetts, and then was appointed professor of physics at the newly founded University of Chicago. He left for a time to serve in WWI but returned to Chicago until 1929 when he left to work at Mount Wilson Observatory in California. Michelson excelled in many aspects of physics but was particularly brilliant in optics, receiving the Nobel Prize for physics in 1907 (the first American to receive the Nobel Prize in the sciences). He is best known for his work on measuring the speed of light and the Michelson-Morley experiment. In 1881, while studying in Germany, Michelson wanted to perform experiments on the velocity of light and so invented the interferometer – a setup of light source, mirrors, and detectors that split and then superpositioned light waves into a new pattern – and attempted to investigate the effect of the Earth's motion on the velocity of light through the luminiferous aether. At the time of Michelson's experiments it was assumed that light waves were most like sound waves and so would need to travel through some medium in order to exist, just as sound needed air. The proposed medium for light was called the luminiferous aether. Aether was a concept that went back to antiquity and was believed to be the substance that filled all of space. Michelson was trying to discover how the motion of the earth through the aether would affect the speed of light (along the same lines as differences in air movement affect the speed of sound). The initial experiments Michelson performed in 1881 gave no result and so Michelson assumed his method was at fault. Then in 1885–87 he worked with Edward W. Morley to perfect the interferometer and the design of the experiment and ran the tests again. Again they failed to find any effect of aether on light. However, now Michelson and Morley trusted their findings and realized that "no affect" was truly significant. It was the brilliant sort of failure that scientific experimentation was all about, providing the first strong evidence against the theory of a luminiferous aether and preparing the way for Einstein and his theory of special relativity. Michelson continued to investigate light the rest of this life, producing such classics as *Velocity of Light* (1902) and *Studies in Optics* (1927). Later in life he became more interested in astronomy and, using light interference and increasingly sensitive versions of his interferometer, was able to measure the diameter of the star Betelgeuse.

EDWARD WILLIAM MORLEY (1838–1923)

Physicist, chemist.
Born in Newark, New Jersey. Died in Hartford, Connecticut.

MORLEY graduated from Williams College in 1860, received a degree from Andover Theological Seminary in 1863, and worked for the Sanitary Commission during the Civil War. After the war, he chose science over the pulpit and became chair of natural history and chemistry at Western Reserve College, Cleveland, Ohio. In addition to his work with Michelson, Morley worked extensively on studies of oxygen and the relative atomic weights of hydrogen and oxygen. While Morley finished second in balloting for the Nobel Prize in chemistry in 1902, he did receive the Rumford Medal from the Royal Society in 1907.

Albert A. Michelson and Edward W. Morley, "On the Relative Motion of the Earth and the Luminiferous Ether," *American Journal of Science* (1887)

THE discovery of the aberration of light was soon followed by an explanation according to the emission theory. The effect was attributed to a simple composition of the velocity of light with the velocity of the earth in its orbit. The difficulties in this apparently sufficient explanation were overlooked until after an explanation on the undulatory theory of light was proposed. This new explanation was at first almost as simple as the former. But it failed to account for the fact proved by experiment that the aberration was unchanged when observations were made with a telescope filled with water. For if the tangent of the angle of aberration is the ratio of the velocity of the earth to the velocity of light, then, since the latter velocity in water is three-fourths its velocity in a vacuum, the aberration observed with a water telescope should be four-thirds of its true value.[1]

On the undulatory theory, according to Fresnel, first, the ether is supposed to be at rest except in the interior of transparent media, in which secondly, it is supposed to move with a velocity less than the velocity of the medium in the ratio $(n^2 - 1)/n^2$, where n is the index of refraction. These two hypotheses give a complete and

---

[1] It may be noticed that most writers admit the sufficiency of the explanation according to the emission theory of light; while in fact the difficulty is even greater than according to the undulatory theory. For on the emission theory the velocity of light must be greater in the water telescope, and therefore the angle of aberration should be less; hence, in order to reduce it to its true value, we must make the absurd hypothesis that the motion of the water in the telescope carries the ray of light in the opposite direction!

satisfactory explanation of aberration. The second hypothesis, notwithstanding its seeming improbability, must be considered as fully proved, first, by the celebrated experiment of Fizeau,[2] and secondly, by the ample confirmation of our own work.[3] The experimental trial of the first hypothesis forms the subject of the present paper.

If the earth were a transparent body, it might perhaps be conceded, in view of the experiments just cited, that the inter-molecular ether was at rest in space, notwithstanding the motion of the earth in its orbit; but we have no right to extend the conclusion from these experiments to opaque bodies. But there can hardly be question that the ether can and does pass through metals. Lorentz cites the illustration of a metallic barometer tube. When the tube is inclined the ether in the space above the mercury is certainly forced out, for it is incompressible.[4] But again we have no right to assume that it makes its escape with perfect freedom, and if there be any resistance, however slight, we certainly could not assume an opaque body such as the whole earth to offer free passage through its entire mass. But as Lorentz aptly remarks: "quoi qui'l en soit, on fera bien, à mon avis, de ne pas se laisser guider, dans une question aussi importante, par des considérations sur le degré de probabilité ou de simplicité de l'une ou de l'autre hypothèse, mais de s'addresser a l'experience pour apprendre à connaitre l'état, de repos ou de mouvement, dans lequel se trouve l'ether à la surface terrestre."[5]

In April, 1881, a method was proposed and carried out for testing the question experimentally.[6]

In deducing the formula for the quantity to be measured, the effect of the motion of the earth through the ether on the path of the ray at right angles to this motion was overlooked.[7] The discussion of this oversight and of the entire experiment forms the subject of a very searching analysis by H. A. Lorentz, who finds that this effect can by no means be disregarded. In consequence, the quantity to be measured had in fact but one-half the value supposed, and as it was already barely beyond the limits of errors of experiment, the conclusion drawn from the result of the experiment might well be questioned; since, however, the main portion of the theory remains unquestioned, it was decided to repeat the experiment with such modifications as would insure a theoretical result much too large to be masked by experimental errors. The theory of the method may be briefly stated as follows:

Let *sa*, fig. 1, be a ray of light which is partly reflected in *ab*, and partly transmitted in *ac*, being returned by the mirrors *b* and *c*, along *ba* and *ca*. *ba* is partly transmitted along *ad*,

---

[2] Comptes Rendus, xxxiii, 349, 1851; Pogg. Ann. Ergänzungsband, iii. 457, 1853; Ann. Chim. Phys., III, lvii, 385, 1859.

[3] Influence of Motion of the Medium on the Velocity of Light. This Journal, III, xxxi, 377, 1886.

[4] It may be objected that it may escape by the space between the mercury and the walls; but this could be prevented by amalgamating the walls.

[5] Archives Néerlandaises, xxi, 2me livr.

[6] *The Relative Motion of the Earth and the Luminiferous Ether* by Albert A. Michelson, this Jour., III xxii, 120.

[7] It may be mentioned here that the error was pointed out-to the author of the former paper by M. A. Potier of Paris, in the winter of 1881.

[8] De l'Influence du Mouvement de la Terre sur les Phen. Lum. Archives Néerlandaises, xxi, 2me livr., 1886.

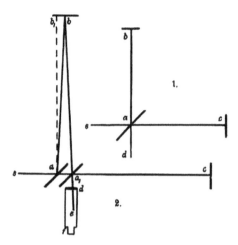

and *ca* is partly reflected along *ad*. If then the paths *ab* and *ac* are equal, the two rays interfere along *ad*. Suppose now, the ether being at rest, that the whole apparatus moves in the direction sc, with the velocity of the earth in its orbit, the directions and distances traversed by the rays will be altered thus:— The ray *sa* is reflected along *ab*, fig. 2; the angle *bab*, being equal to the aberration =*a*, is returned along *ba*/, (*aba*/ =2*a*), and goes to the focus of the telescope, whose direction is unaltered. The transmitted ray goes along *ac*, is returned along *ca*/, and is reflected at *a*/, making *ca*/*e* equal 90—*a*, and therefore still coinciding with the first ray. It may be remarked that the rays *ba*/ and *ca*/, do not now meet exactly in the same point *a*/, though the difference is of the second order; this does not affect the validity of the reasoning. Let it now be required to find the difference in the two paths *aba*/, and *aca*/.

> Let V = velocity of light.
> v = velocity of the earth in its orbit,
> D = distance *ab* or *ac*, (fig. 1).
> T = time light occupies to pass from *a* to *c*.
> $T/$ = time light occupies to return from *c* to *a*/, (fig. 2.)

 Then $| T = D/(V - v), T/ = D/(V + v)$. The whole time of going and coming is $2D [V^2/(V^2 - v^2)] = 2D [1 + (v^2/V^2)]$, and the distance traveled in this time is $2D [V^2/(V^2 - v^2)] = 2D [1 + (v^2/V^2)]$, neglecting terms of the fourth order. The length of the other path is evidently $2D\sqrt{[1 + (v^2/V^2)]}$ or to the same degree of accuracy, $2D [1 + (v^2/2V^2)]$. The difference is therefore $D(v^2/V^2)$. If now the whole apparatus be turned through 90°, the difference will be in the opposite direction, hence the displacement of the interference fringes should be $2D(v^2/V^2)$. Considering only the velocity of the earth in its orbit, this would be $2D \times 10^{-8}$. If, as was the case in the first experiment, $D = 2 \times 10^6$ waves of yellow light, the displacement to be expected would be 0.04 of the distance between the interference fringes.

In the first experiment one of the principal difficulties encountered was that of revolving the apparatus without producing distortion; and another was its extreme sensitiveness to vibration. This was so great that it was impossible to see the interference fringes except at brief intervals when working in the city, even at two o'clock in the morning. Finally, as before remarked, the quantity to be observed, namely, a displacement of something less than a twentieth of the distance between the interference fringes may have been too small to be detected when masked by experimental errors.

The first named difficulties were entirely overcome by mounting the apparatus on a massive stone floating on mercury; and the second by increasing, by repeated reflection, the path of the light to about ten times its former value.

The apparatus is represented in perspective in fig. 3, in plan in fig. 4, and in vertical section in fig. 6. The stone *a* (fig. 5) is about 1.5 meter square and 0.3 meter thick. It rests on an annular wooden float *bb*, 1.5 meter outside diameter, 0.7 meter inside diameter, and 0.25 meter thick. The float rests on mercury contained in the cast-iron trough *cc*, 1.6 centimeter thick, and of such dimensions as to leave a clearance of about one centimeter around the float. A pin *d*, guided by arms *gggg*, fits into a socket *e* attached to the float. The pin may be pushed into the socket or be withdrawn, by a lever pivoted at *f*. This pin keeps the float concentric with the trough, but does not bear any part of the weight of the stone. The annular iron trough rests on a bed of cement on a low brick pier built in the form of a hollow octagon.

At each corner of the stone were placed four mirrors *dd ee* (fig. 4). Near the center of the stone was a plane-parallel glass *b*. These were so disposed that light from an argand burner *a*, passing through a lens, fell on *b* so as to be in part reflected to *d/* ; the two pencils followed the paths indicated in the figure, *bdedbf* and *bd/e/ db/f* respectively, and were observed by the telescope *f*. Both *f* and *a* revolved with the stone. The mirrors were of speculum metal carefully worked to optically plane surfaces five centimeters in diameter, and the glasses *b* and *e* were plane-parallel

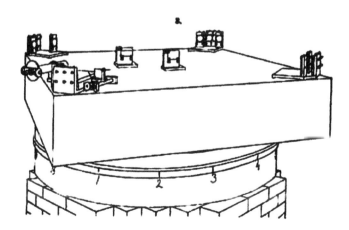

and of the same thickness. 1.25 centimeter; their surfaces measured 5.0 by 7.5 centimeters. The second of these was placed in the path of one of the pencils to compensate for the passage of the other through the same thickness of glass. The whole of the optical portion of the apparatus was kept covered with a wooden cover to prevent air currents and rapid changes of temperature.

The adjustment was effected as follows: The mirrors having been adjusted by screws in the castings which held the mirrors, against which they were pressed by springs, till light from both pencils could be seen in the telescope, the lengths of the two paths were measured by a light wooden rod reaching diagonally from mirror to mirror, the distance being read from a small steel scale to tenths of millimeters. The difference in the lengths of the two paths was then annulled by moving the mirror *e*, This mirror had three adjustments; it had an adjustment in altitude and one in azimuth, like all the other mirrors, but finer; it also had an adjustment in the direction of the incident ray, sliding forward or backward, but keeping very accurately parallel to its former plane. The three adjustments of this mirror could be made with the wooden cover in position.

The paths being now approximately equal, the two images of the source of light or of some well-defined object placed in front of the condensing lens, were made to coincide, the telescope was now adjusted for distinct vision of the expected interference bands, and sodium light was substituted for white light, when the interference bands appeared. These were now made as clear as possible by adjusting the mirror *e*; then white light

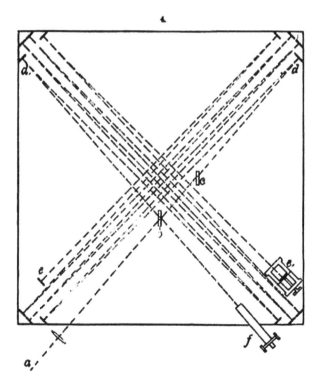

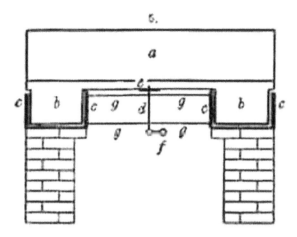

was restored, the screw altering the length of path was very slowly moved (one turn of a screw of one hundred threads to the inch altering the path nearly 1000 wavelengths) till the colored interference fringes reappeared in white light. These were now given a convenient width and position, and the apparatus was ready for observation.

The observations were conducted as follows: Around the cast-iron trough were sixteen equidistant marks. The apparatus was revolved very slowly (one turn in six minutes) and after a few minutes the cross wire of the micrometer was set on the clearest of the interference fringes at the instant of passing one of the marks. The motion was so slow that this could be done readily and accurately. The reading of the screw-head on the micrometer was noted, and a very slight and gradual impulse was given to keep up the motion of the stone; on passing the second mark, the same process was repeated, and this was continued till the apparatus had completed six revolutions. It was found that by keeping the apparatus in slow uniform motion, the results were much more uniform and consistent than when the stone was brought to rest for every observation; for the effects of strains could be noted for at least half a minute after the stone came to rest, and during this time effects of change of temperature came into action.

The following tables give the means of the six readings; the first, for observations made near noon, the second, those near six o'clock in the evening. The readings are divisions of the screw-heads. The width of the fringes varied from 40 to 60 divisions, the mean value being near 50, so that one division means 0.02 wave-length. The rotation in the observations at noon was contrary to, and in the evening observations, with, that of the hands of a watch.

The results of the observations are expressed graphically in fig. 6. The upper is the curve for the observations at noon, and the lower that for the evening observations. The dotted curves represent *one-eighth* of the theoretical displacements. It seems fair to conclude from the figure that if there is any displacement due to the relative motion of the earth and the luminiferous ether, this cannot be much greater than 0.01 of the distance between the fringes.

Noon Observations.

	16.	1.	2.	3.	4.	5.	6.	7.	8.	9.	10.	11.	12.	13.	14.	15.	16.
July 8	44.7	44.0	43.5	39.7	35.2	34.7	34.3	32.5	28.2	26.2	23.8	23.2	20.3	18.7	17.5	16.8	13.7
July 9	57.4	57.3	58.2	59.2	58.7	60.2	60.8	62.0	61.5	63.3	65.8	67.3	69.7	70.7	73.0	70.2	72.2
July 11	27.3	23.5	22.0	19.3	19.2	19.3	18.7	18.8	16.2	14.3	13.3	12.8	13.3	12.3	10.2	7.3	6.5
Mean	43.1	41.6	41.2	39.4	37.7	38.1	37.9	37.8	35.3	34.6	34.3	34.4	34.4	33.9	33.6	31.4	30.8
Mean in w.l.	.862	.832	.824	.788	.754	.762	.758	.756	.706	.692	.686	.688	.688	.678	.672	.628	.616
	.706	.692	.686	.688	.688	.678	.672	.628	.616								
Final mean	.784	.762	.755	.738	.721	.720	.715	.692	.661								

P. M. Observations.

	16.	1.	2.	3.	4.	5.	6.	7.	8.	9.	10.	11.	12.	13.	14.	15.	16.
July 8	61.2	63.3	63.3	68.2	67.7	69.3	70.3	69.8	69.0	71.3	71.3	70.5	71.2	71.2	70.5	72.5	75.7
July 9	26.0	26.0	28.2	29.2	31.5	32.0	31.3	31.7	33.0	35.8	36.5	37.3	38.8	41.0	42.7	43.7	44.0
July 12	66.8	66.5	66.0	64.3	62.2	61.0	61.3	59.7	58.2	55.7	53.7	54.7	55.0	58.2	58.5	57.0	56.0
Mean	51.3	51.9	52.5	53.9	53.8	54.1	54.3	53.7	53.4	54.3	53.8	54.2	55.0	56.8	57.2	57.7	58.6
Mean in w.l.	1.026	1.038	1.050	1.078	1.076	1.082	1.086	1.074	1.068	1.086	1.076	1.084	1.100	1.136	1.144	1.154	1.172
	1.068	1.086	1.076	1.084	1.100	1.136	1.144	1.154	1.172								
Final mean	1.047	1.062	1.063	1.081	1.088	1.109	1.115	1.114	1.120								

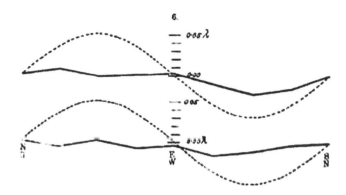

Considering the motion of the earth in its orbit only, this displacement should be $2D(v^2/V^2) = 2D \times 10^{-8}$. The distance D was about eleven meters, or 2×10^7 wavelengths of yellow light; hence the displacement to be expected was 0.4 fringe. The actual displacement was certainly less than the twentieth part of this, and probably less than the fortieth part. But since the displacement is proportional to the square of the velocity, the relative velocity of the earth and the ether is probably less than one sixth the earth's orbital velocity, and certainly less than one-fourth.

In what precedes, only the orbital motion of the earth is considered. If this is combined with the motion of the solar system, concerning which but little is known with certainty, the result would have to be modified; and it is just possible that the resultant velocity at the time of the observations was small though the chances are much against it. The experiment will therefore be repeated at intervals of three months, and thus all uncertainty will be avoided.

It appears, from all that precedes, reasonably certain that if there be any relative motion between the earth and the luminiferous ether, it must be small; quite small enough entirely to refute Fresnel's explanation of aberration. Stokes has given a theory of aberration which assumes the ether at the earth's surface to be at rest with regard to the latter, and only requires in addition that the relative velocity have a potential; but Lorentz shows that these conditions are incompatible. Lorentz then proposes a modification which combines some ideas of Stokes and Fresnel, and assumes the existence of a potential, together with Fresnel's coefficient. If now it were legitimate to conclude from the present work that the ether is at rest with regard to the earth's surface, according to Lorentz there could not be a velocity potential, and his own theory also fails.

Supplement

It is obvious from what has gone before that it would be hopeless to attempt to solve the question of the motion of the solar system by observations of optical phenomena at the surface of the earth. But it is not impossible that at even moderate distances above the level of the sea, at the top of an isolated mountain peak, for instance, the relative motion might be perceptible in an apparatus like that used in these experiments. Perhaps if the experiment should ever be tried in these circumstances, the cover should be of glass, or should be removed.

It may be worth while to notice another method for multiplying the square of the aberration sufficiently to bring it within the range of observation, which has presented itself during the preparation of this paper. This is founded on the fact that reflection from surfaces in motion varies from the ordinary laws of reflection.

Let *ab* (fig. 1) be a plane wave falling on the mirror *mn* at an incidence of 45°. If the mirror is at rest, the wave front after reflection will be *ac*.

Now suppose the mirror to move in a direction which makes an angle *a* with its normal, with a velocity ω. Let V be the velocity of light in the ether supposed stationary, and let *cd* be the increase in the distance the light has to travel to reach *d*. In this time the mirror will have moved a distance $cd/\sqrt{2}\cos\alpha$. We have $cd/ad = \omega\sqrt{2} \cos \alpha/V$ which put $=$ r, and $ac/ad = 1 -$ r.

In order to find the new wave front, draw the arc *fg* with *b* as a center and *ad* as radius; the tangent to this arc from *d* will be the new wave front, and the normal to the tangent from *b* will he the new direction. This will differ from the direction *ba* by the angle θ which it is required to find. From the equality of the triangles *adb* and *edb* it follows that $\theta = 2\varphi$, $ab = ac$, $\tan adb = \tan[45° - (\theta/2)] = [1 - \tan (\theta/2)]/[1 + \tan (\theta/2)] = ac/ad = 1 -$ r, or neglecting terms of the order r^2, $\theta = r + (r^2/2) = \sqrt{2}\omega \cos \alpha/V + (\omega^2/V^2) \cos^2 \alpha$.

Now let the light fall on a parallel mirror facing the first, we should then have $\theta/ = (-\sqrt{2}\omega \cos\alpha/V) + (\omega^2/V^2)\cos^2\alpha$ and the total deviation would be $\theta + \theta / = 2\rho^2\cos^2\alpha$ where ρ is the angle of aberration, if only the orbital motion of the earth is considered. The maximum displacement obtained by revolving the whole apparatus through 90° would be $\Delta = 2\rho^2 = 2 = 0.004''$. With fifty Buch couples the displacement would be 0•2″. But astronomical observations in circumstances far less favorable than those in which these may be taken have been made to hundredths of a second; so that this new method bids fair to be at least as sensitive as the former.

The arrangement of apparatus might be as in fig. 2; *s* in the focus of the lens *a*, in a slit; *bb cc* are two glass mirrors optically plane and so silvered as to allow say one-twentieth of the-light to pass through, and reflecting say ninety per cent. The intensity of the light falling on the observing telescope *df* would be about one-millionth of the original intensity, so that if sunlight or the electric arc were used it could still be readily seen. The mirrors *bb/* and *cc/*, would differ from parallelism sufficiently to separate the successive images. Finally, the apparatus need not be mounted so as to revolve, as the earth's rotation would be sufficient.

If it were possible to measure with sufficient accuracy the velocity of light without returning the ray to its starting point, the problem of measuring the first power of the relative velocity of the earth with respect to the ether would be solved. This may not be as hopeless as might appear at first sight, since the difficulties are entirely mechanical and may possibly be surmounted in the course of time.

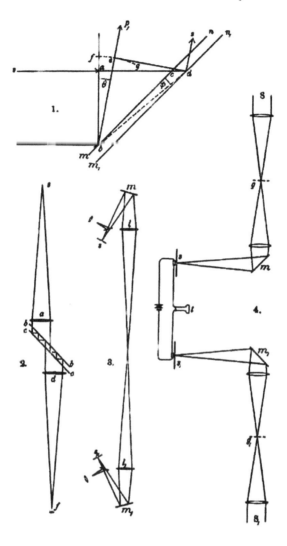

For example, suppose (fig. 3) m and $m_{/}$, two mirrors revolving with equal velocity in opposite directions. It is evident that light from s will form a stationary image at $s_{/}$ and similarly light from $s_{/}$ will form a stationary image at s. If now the velocity of the mirrors be increased sufficiently, their phases still being exactly the same, both images will be deflected from s and $s_{/}$, in inverse proportion to the velocities of light in the two directions; or, if the two deflections are made equal, and the difference of phase of the mirrors be simultaneously measured, this will evidently be proportional to the difference of velocity in the two directions. The only real difficulty lies in this measurement. The following is perhaps a possible solution : $gg_{/}$ (fig. 4) are two gratings on which sunlight is concentrated. These are placed so that after falling on the revolving mirrors m and $m_{/}$, the light forms images of the gratings at s and $s_{/}$, two very sensitive selenium cells in circuit with a battery and a telephone. If everything be symmetrical, the sound in the telephone will be a maximum. If now one of the slits *s* be displaced through half the distance between the image of the grating bars, there will be silence. Suppose now that the two deflections having been made exactly equal, the slit is adjusted for silence. Then if the experiment be repeated when the earth's rotation has turned the whole apparatus through 180°, and the deflections are again made equal, there will no longer be silence, and the angular distance through which *s* must be moved to restore silence will measure the required difference in phase.

There remain three other methods, all astronomical, for attacking the problem of the motion of the solar system through space.

1. The telescopic observation of the proper motions of the stars. This has given us a highly probably determination of the direction of this motion, but only a guess as to its amount.

2. The spectroscopic observation of the motion of stars in the line of sight. This could furnish data for the relative motions only, though it seems likely that by the immense improvements in the photography of stellar spectra, the information thus obtained will be far more accurate than any other.

3. Finally there remains the determination of the velocity of light by observations of the eclipses of Jupiter's satellites. If the improved photometric methods practiced at the Harvard observatory make it possible to observe these with sufficient accuracy, the difference in the results found for the velocity of light when Jupiter is nearest to and farthest from the line of motion will give, not merely the motion of the solar system with reference to the stars, but with reference to the luminiferous ether itself.

STEPHEN A. FORBES (1844–1930)

Entomologist, ecologist.
Born in Silver Creek, Illinois. Died in Urbana, Illinois.

RAISED on an Illinois farm, Forbes was something of a Civil War hero, being captured and held as prisoner for four months at the age of 18, reaching the rank of captain by the age of 20. After the war he studied medicine until his interests turned to natural history. During his career he held a number of key positions at Illinois institutions (successively and simultaneously), including state entomologist, professor of zoology and then of entomology at the University of Illinois, chief of the Illinois Natural History Survey, and founder of the state Laboratory of Natural History. He published widely, over 500 papers, particularly on insects and fish. His work was unique at the time for its merger of extensive fieldwork, innovative quantitative methods, and conceptual insights. He is still read for his studies on food chain relationships and his recognition of community ecology and predator–prey interactions. His most famous work is undoubtedly the 1887 paper excerpted here, "The Lake as a Microcosm," which foreshadowed the development of both the community and ecosystem concept. He is considered the founder of aquatic ecology and one of the principals in the development of the field of ecology generally. Strikingly modern in his outlook, Forbes championed the idea that ecological knowledge was fundamental to human well-being and did not discriminate between basic and applied ecological science.

Stephen A. Forbes, "The Lake as a Microcosm" (excerpt), *Bulletin of the Peoria Scientific Association* (1887)

A lake is to the naturalist a chapter out of the history of a primeval time, for the conditions of life there are primitive, the forms of life are, as a whole, relatively low and ancient, and the system of organic interactions by which they influence and control each other has remained substantially unchanged from a remote geological period.

The animals of such a body of water are, as a whole, remarkably isolated – closely related among themselves in all their interests, but so far independent of the land about them that if every terrestrial animal were suddenly annihilated, it would doubtless be long before the general multitude of the inhabitants of the lake would feel the effects of this event in any important way. One finds in a single body of water a far more complete and independent equilibrium of organic life and activity than on any equal body of land. It is an islet of older, lower life in the midst of the higher more recent life of the surrounding region. It forms a little world within itself – a microcosm within which all the elemental forces are at work and the play of life goes on in full, but on so small a scale as to bring it easily within the mental grasp.

Nowhere can one see more clearly illustrated what may be called the *sensibility* of such an organic complex, expressed by the fact that whatever affects any species belonging to it, must speedily have its influence of some sort upon the whole assemblage. He will thus be made to see the impossibility of studying any form completely, out of relation to the other forms; the necessity for taking a comprehensive survey of the whole as a condition to a satisfactory understanding of any part. If one wishes to become acquainted with the black bass, for example, he will learn but little if he limits himself to that species. He must evidently study also the species upon which it depends for its existence, and the various conditions upon which *these* depend. He must likewise study the species with which it comes in competition, and the entire system of conditions affecting their prosperity, and by the time he has studied all these sufficiently he will find that he has run through the whole complicated mechanism of the aquatic life of the locality, both animal and vegetable, of which his species forms but a single element.

It is under the influence of these general ideas that I propose to examine briefly to-night the lacustrine life of Illinois, drawing my data from collections and observations made during recent years by myself and my assistants of the State Laboratory of Natural History.

The lakes of Illinois are of two kinds, fluviatile and water-shed. The fluviatile lakes, which are much more numerous and important, are appendages of the river systems of the State, being situated in the river bottoms and connected with the adjacent streams by periodical overflows. Their fauna is therefore substantially that of the rivers themselves, and the two should, of course, be studied together.

They are probably in all cases either parts of former river channels, which have been cut off and abandoned by the current as the river changed its course, or else are tracts of the high-water beds of streams over which, for one reason or another, the periodical deposit of sediment has gone on less rapidly than over the surrounding area, and which have thus come to form depressions in the surface which retain the waters of overflow longer than the higher lands adjacent. Most of the numerous "horse-shoe lakes" belong to the first of these varieties, and the "bluff lakes" situated along the borders of the bottoms, are many of them examples of the second.

These fluviatile lakes are most important breeding grounds and reservoirs of life – especially as they are protected from the filth and poison of towns and manufactories by which the running waters of the state are yearly more deeply defiled.

The amount and variety of animal life contained in them as well as in the streams related to them, is extremely variable, depending chiefly on the frequency, extent, and duration of the overflows. This is, in fact, the characteristic and peculiar feature of life in these waters. There is perhaps no better illustration of the methods by which the flexible system of organic life adapts itself, without injury, to widely and rapidly fluctuating conditions. Whenever the waters of the river remain for a long time far beyond their banks, the breeding grounds of fishes and other animals are immensely extended, and their food supplies increased to a corresponding degree. The slow or stagnant backwaters of such an overflow afford the best situations possible for the development of myriads of Entomostraca, which furnish, in turn, abundant food for young fishes of all descriptions. There thus results a sudden outpouring of life – an extraordinary multiplication of nearly every species, most prompt and rapid, generally speaking, in such as have the highest reproductive rate, that is to say, in those which produce the largest average number of eggs and young for each adult.

The first to feel this tremendous impulse are the Protophytes and Protozoa, upon which most of the Entomostraca and certain minute insect larvæ depend for food. This sudden development of their food resources causes, of course, a corresponding increase in the numbers of the latter classes, and, through them, of all sorts of fishes. The first fishes to feel the force of this tidal wave of life, are the rapidly-breeding, non-predaceous kinds; and the last, the game fishes, which derive from the others their principal food supplies. Evidently each of these classes must act as a check upon the one preceding it. The development of animalcules is arrested, and soon sent back below its highest point by the consequent development of Entomostraca; the latter, again, are met, checked, and reduced in number by the innumerable shoals of fishes with which the water speedily swarms; and the lower fishes, springing up at first in excessive ratio, are soon driven back to a lower limit by the following excessive increase of the higher carnivorous kinds. In this way a general adjustment of numbers to the new conditions would finally be reached spontaneously; but long before any such settled balance can be established, often of course before the full effect of this upward influence has been exhibited, a new cause of disturbance intervenes in the *disappearance of the overflow*. As the waters retire the lakes are again defined. The teeming life which they contain is restricted within daily narrower bounds, and a fearful slaughter follows. The lower and more defenceless animals are penned up more and more closely with their predaceous enemies, and these thrive for a time to an extraordinary degree. To trace the further consequences of this oscillation would take me too far. Enough has been said to illustrate the general idea that the life of waters subject to periodical expansions of considerable duration, is peculiarly unstable and fluctuating, that each species swings, pendulum-like, but irregularly, between a highest and a lowest point, and that this fluctuation affects the different classes successively, in the order of their dependence upon each other for food.

Where a water-shed is a nearly level plateau with slight irregularities of the surface, many of these will probably be imperfectly drained, and the accumulating waters will form either marshes or lakes, according to the depth of the depressions. Highland marshes of this character are seen in Ford, Livingston, and adjacent counties,[1] between

---

[1] All now drained and brought under cultivation.

the headwaters of the Illinois and Wabash systems; and an area of water-shed lakes occurs in Lake and McHenry counties, in northeastern Illinois.

The latter region is everywhere broken by low, irregular ridges of glacial drift, with no rock but boulders anywhere in sight. The intervening hollows are of every variety, from mere sink-holes, either dry or occupied by ponds, to expanses of several square miles, forming marshes or lakes.

This is, in fact, the southern end of a broad lake belt which borders Lakes Michigan and Superior on the west and south, extending through eastern and northern Wisconsin and northeastern Minnesota, and occupying the plateau which separates the headwaters of the St. Lawrence from those of the Mississippi. These lakes are of glacial origin, some filling beds excavated in the solid rock, and others collecting the surface waters in hollows of the drift. The latter class, to which all the Illinois lakes belong, may lie either parallel to the line of glacial action, occupying valleys between adjacent lateral moraines, or transverse to that line, and bounded by terminal moraines. Those of our own State all drain at present into the Illinois, through the Des Plaines and Fox; but, as the terraces around their borders indicate a former water level considerably higher than the present one, it is likely that some of them once emptied eastward into Lake Michigan. Several of these lakes are clear and beautiful sheets of water, with sandy or gravelly beaches, and shores bold and broken enough to relieve them from monotony. Sportsmen long ago discovered their advantages; and club-houses and places of summer resort are rapidly springing up on the borders of the most attractive and easily accessible. They offer also an unusually rich field to the naturalist; and their zoölogy and botany should be better known.

The conditions of aquatic life are here in marked contrast to those afforded by the fluviatile lakes already mentioned. Connected with each other or with adjacent streams only by slender rivulets; varying but little in level with the change of the season, and scarcely at all from year to year; they are characterized by an isolation, independence, and uniformity which can be found nowhere else within our limits.

Among these Illinois lakes I did considerable work during October of two successive years, using the sounding line, deep sea thermometer, towing net, dredge, and trawl in six lakes of northern Illinois, and in Geneva Lake, Wisconsin, just across the line. Upon one of these Illinois lakes I spent a week in October, and an assistant, Prof. Garman, now of the University, spent two more, making as thorough a physical and zoölogical survey of this lake as was possible at that season of the year.

★   ★   ★

When one sees acres of the shallower water black with water-fowl, and so clogged with weeds that a boat can scarcely be pushed through the mass; when, lifting a handful of the latter he finds them covered with shells and alive with small crustaceans; and then, dragging a towing net for a few minutes, finds it lined with myriads of diatoms and other microscopic Algae, and with multitudes of Entomostraca, he is likely to infer that these waters are everywhere swarming with life, from top to bottom, and from shore to shore. If, however, he will haul a dredge for an hour or so in the deepest water he can find, he will invariably discover an area singularly barren of both plant and animal life, yielding scarcely anything but a small bivalve mollusk, a few low worms, and red larvæ of gnats. These inhabit a black, deep, and almost impalpable mud or ooze, too soft and unstable to afford foot hold to plants, even if the lake is shallow enough to admit a sufficient quantity of light to its bottom to support vegetation. It is doubtless to this character of the bottom that the barrenness of the interior parts of these lakes is due; and this again is caused by the selective influence of gravity upon the mud and detritus washed down by rains. The heaviest and coarsest of this material necessarily settles nearest the margin, and only the finest silt reaches the remotest parts of the lake, which, filling most slowly, remain, of course, the deepest. The largest lakes, are not, therefore, as a rule by any means the most prolific of life, but this shades inward rapidly from the shore, and becomes at no great distance almost as simple and scanty as that of a desert.

Among the weeds and lily-pads upon the shallows and around the margin, the Potamogeton, Myriophyllum Ceratophyllum, Anacharis and Chara, and the common Nelumbium – among these the fishes chiefly swim or lurk, by far the commonest being the barbaric bream or "pumpkin seed" of northern Illinois, splendid with its greens and scarlet and purple and orange. Little less abundant is the common perch (*Perca lutea*), in the larger lakes, – in the largest outnumbering the bream, itself. The whole sunfish family, to which the latter belongs, is, in fact, the dominant group in these lakes. Of the one hundred and thirty-two fishes of Illinois only thirty-seven are found in these waters – about twenty-eight per cent – while eight out of our seventeen sunfishes have been taken there. Next, perhaps, one searching the pebbly beaches, or scanning the weedy tracts, will be struck by the small number of minnows or cyprinoids which catch the eye, or come out, in the net. Of our thirty-three Illinois cyprinoids, only six occur there – about eighteen per cent – and only three of these are common. These are in part replaced by shoals of the beautiful little silversides (*Labidesthes sicculus*) a spiny-finned fish, bright, slender, active, and voracious – as well supplied with teeth as a perch, and far better equipped for self-defense than the soft-bodied, and toothless cyprinoids. Next, we note that of our twelve catfishes only two have been taken in these lakes – one the common bullhead (*Ictalurus nebulosus*) which occurs everywhere, and the other an insignificant stone cat, not as long as one's thumb. The suckers, also, are much less abundant in this region, the buffalo fishes not appearing at all in our collections. Their family is represented by the worthless carp, by two redhorse, by the carp sucker and the common sucker (*Catostomus commersonii*), and one other species. Even the hickory shad – an ichthyological weed in the Illinois – we have not found in these lakes at all. The sheepshead, so common here, is also conspicuous there by its absence. The yellow bass, not rare in this river, we should not expect in these lakes, because it is rather a southern species; but why the white bass, abundant here, in Lake Michigan, and in the

Wisconsin lakes, should be wholly absent from the lakes of the Illinois plateau, I am unable to imagine. If it occurs there at all, it must be rare, as I could neither find nor hear of it.

A characteristic, abundant, and attractive little fish is the log perch (*Percina caprodes*) – the largest of the darters, slender, active, barred like a zebra, spending much of its time in chase of Entomostraca among the water plants, or prying curiously about among the stones for minute insect larvæ. Six darters in all, out of the eighteen from the state, are on our list from these lakes. The two black bass are the popular game fishes – the large-mouthed species being much the most abundant. The pickerels, gars, and dogfish are there about as here, but the shovel fish does not occur.

Of the peculiar fish fauna of Lake Michigan – the burbot, white fish, trout, lake herring or cisco, etc., not one species occurs in these smaller lakes and all attempts to transfer any of them have failed completely. The cisco is a notable fish of Geneva Lake, Wisconsin, but does not reach Illinois except in lake Michigan. It is useless to attempt to introduce it, because the deeper areas of the interior lakes are too limited to give it sufficient range of cool water in midsummer.

In short, the fishes of these lakes are substantially those of their region – excluding the Lake Michigan series (for which the lakes are too small and warm) and those peculiar to creeks and rivers. Possibly the relative scarcity of catfishes is due to the comparative clearness and cleanness of the waters. I see no good reason why minnows should be so few, unless it be the abundance of pike and Chicago sportsmen.

Concerning the molluscan fauna, I will only say that it is poor in bivalves – as far as our observations go – and rich in univalves. Our collections have been but partly determined, but they give us three species of Valvata, seven of Planorbis, four Amnicolas, a Melantho, two Physas, six limnæas and an Ancylus among the Gasteropoda, and two Unios, an Anodonta, a Sphærium and a Pisidium among the Lamelli branchiates. *Pisidium variabile* is by far the most abundant mollusk in the oozy bottom in the deeper parts of the lakes; and crawling over the weeds are multitudes of small Amnicolas and Valvatas.

The entomology of these waters I can merely touch upon, mentioning only the most important and abundant insect larvæ. Hiding under stones and driftwood, well aware, no doubt, what enticing morsels they are to a great variety of fishes, we find a number of species of Ephemerid larvæ whose specific determination we have not yet attempted. Among the weeds are the usual larvæ of dragon flies – Agrionina and Libellulidæ, familiar to every one; swimming in open water the predaceous larvæ of Corethra; wriggling through the water or buried in the mud the larvæ of Chironomus – the shallow water species white, and those from the deeper ooze of the central parts of the lakes, blood red and larger. Among Chara on the sandy bottoms are a great number and variety of interesting case worms – larvæ of Phryganeidæ – most of them inhabiting tubes of a slender conical form made of a viscid secretion exuded from the mouth and strengthened and thickened by grains of sand – fine or coarse. One of these cases, nearly naked, but usually thinly covered with diatoms, is especially worthy of note, as it has been reported nowhere in this country except in our collections, and, was, indeed, recently described from Brazil as new. Its generic name is Lageno-psyche, but its species undetermined. These larvæ are also much eaten by fishes.

Among the worms we have of course a number of species of leeches and of planarians – in the mud minute Anguillulidæ, like vinegar eels, and a slender Lumbriculus, which makes a tubular mud burrow for itself in the deepest water, and also the curious *Nais probiscidia* – notable for its capacity of multiplication by transverse division.

The crustacean fauna of these lakes is more varied than that of any other group. About forty species were noted in all. Crawfishes were not especially abundant, and all captured belonged to a single species – *Cambarus virilis*. Two amphipods occurred frequently in our collections; one, less common here but very abundant farther south – *Crangonyx gracilis* – and one, *Allorchestes dentata*, probably the commonest animal in these waters, crawling everywhere in myriads over the submerged water plants. An occasional *Gammarus fasciatus* was also taken in the dredge. A few isopod Crustacea occur, belonging to *Mancasellus tenax*, Harger – a species not previously found in the state.

I have reserved for the last the Entomostraca – minute crustaceans of a surprising number and variety, and of a beauty often truly exquisite. They belong wholly, in our waters, to the three orders, Copepoda, Cladocera, and Ostracoda – the first predaceous upon still smaller organisms and upon each other, and the two others chiefly vegetarian. Twenty-one species of Cladocera have been recognized in our collections, these representing sixteen genera. It is an interesting fact that twelve of these species are found also in the fresh waters of Europe. Five cyprids have been recognized, two of them common to Europe, and also an abundant Diaptomus, a variety of a European species. Several Cyclops were collected which have not yet been determined.

These Entomostraca swarm in microscopic myriads among the weeds along the shore, some swimming freely, and others creeping in the mud or climbing over the leaves of plants. Some prefer the open water, in which they throng locally like flocks of birds, coming to the surface preferably by night, or on dark days, and sinking to the bottom usually to avoid the sunshine. These pelagic forms, as they are called, are often exquisitely transparent, and hence almost invisible in their native element – a charming device of Nature to protect them against their enemies in the open lake, where there is no chance of shelter or escape. Then with an ingenuity in which one may almost detect the flavor of a sarcastic humor, Nature has turned upon these favored children and endowed their most deadly enemies with a like transparency, so that wherever the towing net brings to light a host of these crystalline Cladocera, there it discovers, also, swimming, invisible, among them, a lovely pair of robbers and beasts of prey – the delicate Leptodora and the Corethra larva.

These slight, transparent, pelagic forms are much more numerous in Lake Michigan than in any of the smaller lakes, and peculiar forms occur there commonly, which are rare in the larger lakes of Illinois and entirely wanting in the smallest.

The vertical range of the animals of Geneva Lake showed clearly that the barrenness of the interiors of these small bodies of water was not due to the greater depth – or at least not to that alone. While there were a few species of crustaceans and case-worms which occurred there abundantly near shore, but rarely, or not at all, at depths greater than four fathoms, and may hence be called littoral species, there was, on the whole, little diminution either in quantity or variety of animal life, until about fifteen fathoms had been reached. Dredgings at four and five fathoms were nearly or quite as fruitful as any made. On the other hand, the barrenness of the bottom at twenty to twenty-three fathoms was very remarkable. The total products of four hauls of the dredge and one of the trawl at that depth, aggregating fully a mile and a half of continuous dragging, would easily go into a two-dram vial, and represented only nine species of animals –not counting dead shells and fragments which had probably floated in from shallower waters. The greater part of this little collection was composed of specimens of Lumbriculus and larvæ of Chironomus. There were a few Corethra larvæ, a single Gammarus, three small leeches, and some sixteen mollusks, all but four of which belonged to Pisidium. The others were two Sphæriums, *Valvata 3*–carinata, and a *V. sincera*. None of the species taken here were peculiar, but all were of the kinds found in the smaller lakes, and all occurred also in shallower water. It is evident that these interior regions of the lakes must be as destitute of fishes as they are of plants and lower animals.

While none of the deep-water animals of the Great Lakes were found in Geneva Lake, other evidences of zoölogical affinity were detected. The towing net yielded almost precisely the assemblage of species of Entomostraca found in Lake Michigan, including many specimens of *Limnocolonus macrurus*, Sars.; and peculiar long, smooth leeches, common in Lake Michigan, but not occurring in the small Illinois lakes, were also found in Geneva. Many *Valvata 3-carinata* lacked the middle carina, as in Long Lake and other *isolated* lakes of this region.

Comparing the Daphnias of Lake Michigan with those of Geneva Lake, Wis. (nine miles long and twenty-three fathoms in depth), those of Long Lake, Ill. (one and a half miles long and six fathoms deep), and those of other still smaller lakes of that region, and the swamps and smaller ponds as well, we shall be struck by the inferior development of the Entomostraca of the larger bodies of water, in numbers, in size and robustness, and in reproductive power. Their smaller numbers and size are doubtless due to the relative scarcity of food. The system of aquatic animal life rests essentially upon the vegetable world, although perhaps less strictly than does the terrestrial system; and in a large and deep lake vegetation is much less abundant than in a narrower and shallower one, not only relatively to the amount of water but also to the area of the bottom. From this deficiency of plant life results a deficiency of food for Entomostraca, whether of Algæ, of Protozoa or of higher forms, and hence, of course, a smaller number of the Entomostraca themselves, with more slender bodies suitable for more rapid locomotion and wider range.

The difference of reproductive energy, as shown by the much smaller egg-masses borne by the species of the larger lakes depends upon the vastly greater destruction to which the paludinal crustacea are subjected. Many of the latter occupy waters liable to be exhausted by drought, with a consequent enormous waste of Entomostracan life. The opportunity for reproduction is here greatly limited – in some situations to early spring alone – and the chances for destruction of the summer eggs in the dry and often dusty soil are so numerous that only the most prolific species can maintain themselves under such conditions.

Further, the marshes and shallower lakes are the favorite breeding grounds of fishes, which migrate to them in spawning time, if possible, and it is from the Entomostraca found here that most young fishes get their earliest food supplies – a danger from which power the deepwater species are measurably free. Not only is a high reproductive therefore rendered unnecessary among the latter by their freedom from many dangers to which the shallow-water species are exposed, but in view of the relatively small amount of food available for them, a high rate of multiplication would be a positive injury, and could result only in wholesale starvation.

All these lakes of Illinois and Wisconsin, together with the much larger Lake Mendota at Madison (in which also I have done much work with dredge, trawl, and seine), differ in one notable particular both from Lake Michigan and from the larger lakes of Europe. In the latter, the bottoms in the deeper parts yield a peculiar assemblage of animal forms, which range but rarely into the littoral region, while in our inland lakes no such deep water fauna occurs, with the exception of the cisco and the large red Chironomus larva. At Grand Traverse Bay, in Lake Michigan, I found at a depth of one hundred fathoms a very odd fish of the sculpin family (*Triglopsis thompsoni*, Gir.), which, until I collected it, had been known only from the stomachs of fishes; and there also was an abundant crustacean, Mysis – the "opossum shrimp," as it is sometimes called – the principal food of these deep lake sculpins. Two remarkable amphipod crustaceans also belong in a peculiar way to this deep water. In the European lakes the same Mysis occurs in the deepest part, with several other forms not represented in our collections,–two of these being blind crustaceans related to those which in this country occur in caves and wells.

Comparing the other features of our lake fauna with that of Europe, we find a surprising number of Entomostraca identical; but this is a general phenomenon, as many of the more abundant Cladocera and Copepoda of our small wayside pools are either European species, or differ from them so slightly that it is doubtful if they ought to be called distinct.

★ ★ ★

Perhaps no phenomenon of life in such a situation is more remarkable than the steady balance of organic nature, which holds each species within the limits of a uniform average number, year after year, although every one is always doing its best to break across its boundaries, on every side. The reproductive rate is usually enormous and the struggle for existence is correspondingly severe. Every animal within these bounds has its enemies, and Nature seems to have taxed her skill and ingenuity to the utmost to furnish these enemies with contrivances for the destruction of their prey in myriads. For every defensive device with which she has armed an animal, she has invented a still more effective apparatus of destruction, and bestowed it upon some foe, thus striving with unending pertinacity, to outwit herself, and yet life does not perish in the lake, nor even oscillate to any considerable degree; but on the contrary the little community secluded here is as prosperous as if its state were one of profound and perpetual peace. Although every species has to fight its way, inch by inch, from the egg to maturity, yet no species is exterminated, but each is maintained at a regular average number which we shall find good reason to believe is the greatest for which there is, year after year, a sufficient supply of food.

I will bring this paper to a close, already too long postponed, by endeavoring to show how this beneficent order is maintained in the midst of a conflict seemingly so lawless.

It is a self-evident proposition that a species cannot maintain itself continuously, year after year, unless its birth-rate at least equals its death-rate. If it is preyed upon by another species, it must produce regularly an excess of individuals for destruction, or else it must certainly dwindle and disappear. On the other hand, the dependent species evidently must not appropriate, on an average, any more than the surplus and excess of individuals upon which it preys, for if it does so, it will regularly diminish its own food supply, and thus indirectly, but surely, exterminate itself. The interests of both parties will therefore be best served by an adjustment of their respective rates of multiplication, such that the species devoured shall furnish an excess of numbers to supply the wants of the devourer, and that the latter shall confine its appropriations to the excess thus furnished. We thus see that there is really a close *community of interest* between these two seemingly deadly foes.

And next we note that this common interest is promoted by the process of natural selection; for it is the great office of this process to eliminate the unfit. If two species standing to each other in the relation of hunter and prey are or become badly adjusted in respect to their rates of increase, so that the one preyed upon is kept very far below the normal number which might find food, even if they do not presently obliterate each other, the pair are placed at a disadvantage in the battle for life, and must suffer accordingly. Just as certainly as the thrifty business man who lives within his income will finally dispossess his shiftless competitor who can never pay his debts, the well-adjusted aquatic animal will, in time crowd out his poorly-adjusted competitors for food and for the various goods of life. Consequently we may believe that in the long run and as a general rule, those species which have survived, are those which have reached a fairly close adjustment in this particular.[2]

Two ideas are thus seen to be sufficient to explain the order evolved from this seeming chaos; the first that of a general community of interests among all classes

---

[2] For a fuller statement of this argument, see Bul. Ill. State Lab. Nat. Hist. Vol. 1, No. 3, pages 5 to 10.

of organic beings, and the second that of the beneficent power of natural selection which compels such adjustments of the rates of destruction and of multiplication of the various species as shall best promote this common interest.

Have these facts and ideas derived from a study of our aquatic microcosm any general application on a higher plane? We have here an example of the triumphant beneficence of the laws of life applied to conditions seemingly the most unfavorable possible for any mutually helpful adjustment. In this lake, where competitions are fierce and continuous beyond any parallel in the worst periods of human history, where they take hold not on the goods of life, merely, but always upon life itself; where mercy and charity and sympathy and magnanimity and all the virtues are utterly unknown; where robbery and murder and the deadly tyranny of strength over weakness are the unvarying rule; where what we call wrong-doing is always triumphant, and what we call goodness would be immediately fatal to its possessor – even here, out of these hard conditions, an order has been evolved which is the best conceivable without a total change in the conditions themselves; an equilibrium has been reached and is steadily maintained that actually accomplishes for all the parties involved the greatest good which the circumstances will at all permit. In a system where life is the universal good, but the destruction of life the well nigh universal occupation, an order has spontaneously risen which constantly tends to maintain life at the highest limit – a limit far higher, in fact, with respect to both quality and quantity, than would be possible in the absence of this destructive conflict. Is there not, in this reflection, solid ground for a belief in the final beneficence of the laws of organic nature? If the system of life is such that a harmonious balance of conflicting interests has been reached where every element is either hostile or indifferent to every other, may we not trust much to the outcome where, as in human affairs, the spontaneous adjustments of nature are aided by intelligent effort, by sympathy, and by self-sacrifice?

CLINTON HART MERRIAM (1855–1942)

Zoologist, biogeographer, naturalist.
Born in New York. Died in Berkeley, California.

Merriam's father was a successful businessman and U.S. Congressman. Through his father he met Spencer Baird and worked as a naturalist on the Hayden Geological Survey in the summer of 1872. He studied biology and anatomy at Yale's Sheffield Scientific School before going on to complete his medical degree at Columbia's College of Physicians and Surgeons in 1879. He practiced for a few years but turned his full attention back to biology and natural history around 1883. In 1886, Merriam began work for the United States Department of Agriculture in the Entomology Division. C. Hart Merriam had a tendency to pursue his own scientific interests and ignore the economic and agricultural aspects of scientific study, which were part of the federal agency's intended focus. In a different time, Merriam's pure science focus might have been overlooked but after the 1870s there were increasing calls for reform from various progressive movements with strong agricultural and land interests. Granges, alliances, populists were all political forces that traveled West with the growing population. This sometimes created tension between Merriam and the other USDA scientists, officials, and agricultural advocates. Fortunately for all involved, an earlier alliance with wealthy railroad magnate Edward H. Harriman on scientific work in Alaska led to Harriman's widow, Mary, setting up a trust fund to underwrite Merriam's research activities. He left the USDA and, working as a research associate with the Smithsonian Institution, followed his own interests, which shifted about this time to ethnographic studies of the Indian tribes of the West. In addition to his ethnographic work, he is probably best remembered for his "life zones" concept. The concept greatly influenced the developing science of ecology and while today factors other than just latitude and elevation are recognized in the distribution of biota – including aspect, fire history, soil type, and biotic interactions – Merriam's zones are still a major starting point.

C. Hart Merriam, "Laws of Temperature Control of the Geographic Distribution of Terrestrial Animals and Plants," *National Geographic Magazine* (1894)[1]

The tendency of animals and plants to multiply beyond the means of subsistence and to spread over all available areas is well understood. What naturalists wish to know is

---

[1] The present abstract of the principal results of an investigation carried on under the Department of Agriculture is here published by permission of the Honorable J. Sterling Morton, Secretary of Agriculture. The temperature data have been furnished by the United States Weather Bureau, a branch of

not how species are dispersed, but how they are checked in their efforts to overrun the earth. Geographic barriers are rare, except in the case of oceans, and since even these were formerly bridged at the north, another cause must be sought. This has been found in the group of phenomena commonly hidden under the word climate, and nearly a century ago it was shown by Humboldt that temperature is the most important of these climatic factors.

In the northern hemisphere animals and plants are distributed in circumpolar belts or zones, the boundaries of which follow lines of equal temperature rather than parallels of latitude. They conform in a general way, therefore, with the elevation of the land, sweeping northward over the lowlands and southward over the mountains. Between the pole and the equator there are three primary belts – Boreal, Austral and Tropical – each of which may be subdivided into minor belts and areas. In the United States the Boreal and Austral regions have each been split into three secondary transcontinental zones. The Boreal are known as the *Arctic, Hudsonian* and *Canadian*; the Austral as the *Transition, Upper Austral* and *Lower Austral*. The subordinate faunas and floras need not be here considered.

The area of overlapping of Boreal and Austral types is confined in most parts of the country to the narrow Transition zone, but along the Pacific coast it reaches all the way from southern California to Puget sound. This Pacific coast strip has always proved a stumbling-block to students of geographic distribution of life in America, but has now become the means of verifying the fundamental laws governing this distribution, as shown later.

But while the boundaries of the several zones rarely coincide with absolute mechanical barriers, being fixed in the main by temperature, difference of opinion prevails as to the period during which the temperature exerts its restraining influence, and no formula for the expression of the temperature control has been heretofore discovered. None of the temperature data computed and platted on maps as isotherms are available in locating the exact boundaries of the zones, because these isotherms invariably show the temperature of arbitrary periods, such as months, seasons and years – periods whose beginning and ending have reference to a particular time of year rather than a particular degree or quantity of heat. Thus the temperature for July, which is by far the most important of those commonly shown in isotherms, bears an inconstant relation to the hottest part of the year. In certain localities the four hottest weeks may fall within the month of July, but in other localities they cover the period from the middle of June to the middle of July; in others from the middle of July to the middle of August, and in others still from the early part of August to early September. Similarly, the isotherms showing the mean annual temperature fail to conform to the boundaries of the life zones, although in the far south they may be nearly coincident. The mean summer temperature is obviously inapplicable because of the varying length of the season in different localities.

the Department of Agriculture. A preliminary announcement of results was made by the author before the Philosophical Society of Washington May 26, 1894.

Several years ago I endeavored to show that the distribution of terrestrial animals and plants is governed by the temperature of the period of growth and reproductive activity, not by the temperature of the whole year; but how to measure the temperatures concerned was not then worked out. The period of growth and reproductive activity is of variable duration; according to latitude, altitude and local conditions of each particular locality. In the tropics and a few other areas it extends over nearly the whole year, while within the Arctic circle and on the summits of high mountains it is of less than two months' duration.[2] It is evident, therefore, that while in the tropics there may be a close agreement between the mean annual temperature and the life zones, in the north the widest discrepancy exists between them.

At one time I believed that the mean temperature of the actual period of reproductive activity in each locality was the factor needed,[3] but such means are almost impossible to obtain, and subsequent study has convinced me that the real temperature control may be better expressed by other data.

For more than a century physiological botanists have maintained that the various events in the life of plants, as leafing, flowering and maturing of fruit, take place when the plant has been exposed to a definite quantity of heat, which quantity is the sure total of the daily temperatures above a minimum assumed to be necessary for functional activity. The minimum used by Boussingault and early botanists generally was the freezing point (0° C. or 32° F.), but Marie-Davy and other recent writers believe that 6° C. or 43° F.[4] more correctly indicates the temperature of the awakening of plant life in spring. In either case the substance of the theory is that *the same stage of vegetation is attained in any year when the sum of the mean daily temperatures reaches the same value*, which value or total is essentially the same for the same plant in all localities. This implies that the period necessary for the accomplishment of a definite physiological act, blossoming for instance, may be short or long, according to local climatic peculiarities, but the total quantity of heat must be the same. The total amount of heat necessary to advance a plant to a given stage came to be known as the physiological constant of that stage. Linsser believed this law to be fallacious and maintained that the *physiological constant* of any particular stage of vegetation was *not the sum total* of heat acquired that time, but the *ratio* or proportion of this sum to the sum total for the entire season. Thus Linsser's physiological constant is the *ratio* of the sum of the mean daily temperatures at the time when any particular stage of vegetation is attained to the sum total for the year. This formula was based on the belief that plants of the same species living in different places arrive at the same phase of development by utilizing the same proportion of the total heat which they receive in the course of a season.

---

[2] See N. Am. Fauna, No. 3, September, 1890, pp. 26, 27, 29–32; also Presidential Address, Biological Soc. Wash., vol. vii, April, 1892, pp. 45, 46.

[3] I began work on this line about fifteen years ago and continued at intervals for ten years before convinced of its impracticability.

[4] The exact equivalent of 6° C. is 42°.8 F.

Students of geographic distribution may dismiss this phase of the inquiry as not pertinent to the problem in hand, for we are concerned with the physiological constant *of the species itself*, not of any stage or period in its life history. But what is the physiological constant of a species, and how can it be measured? If it is true that the same stage of vegetation is attained in different years when the sum of the mean daily temperatures reaches the same value, it is obvious that *the physiological constant of a species must be the total quantity of heat or sum of positive temperatures required by that species to complete its cycle of development and reproduction.* The difficulty in computing such sums is in fixing the end of the period during which temperature exerts its influence upon the organism. In the case of plants this can be done by direct observation of a particular individual or crop, in connection with careful thermometric readings covering the whole period of vegetative activity, and data of this sort have been actually recorded by certain European phenologists, but I am not aware that an attempt has been made to correlate the facts thus obtained with the boundaries of the life zones. Since, however, all forms of life are affected by temperature and it is manifestly impracticable to ascertain by direct observation the total quantity of heat necessary to enable the various species of mammals, birds and reptiles to complete the annual cycle of reproduction, and since the areas inhabited by definite assemblages of animals and plants have been found to be essentially coincident, it is evident that a more generalized formula is necessary. If the computation can be transferred from the *species* to the *zone* it inhabits–if a *zone constant* can be substituted for a *species constant*–the problem will be well nigh solved. This I have attempted to do. In conformity with the usage of botanists, a minimum temperature of 6° C. (43° F.) has been assumed as marking the inception of the period of physiological activity in plants and of reproductive activity in animals. The effective temperatures or degrees of normal mean daily heat in excess of this minimum have been added together for each station, beginning when the normal mean daily temperature rises higher than 6° C. in spring and continuing until it falls to the same point at the end of the season. The sums thus obtained have been platted on a large scale map of the United States,[5] and isotherms have been run which are found to conform in a most gratifying manner to the northern boundaries of the several life zones, as may be seen on comparing a reduced copy of this map (see plate 12) with a map of the life zones (see plate 14). The latter, it may be observed, is identical; save a few corrections in minor details, with the third edition of my Bio-geographic map of North America (prepared a year ago and published in the Annual Report of the Secretary of Agriculture for 1893).[6] While the available data are not so numerous as might be desired, the stations in many instances being too far apart, still enough are at hand to justify the belief that *animals and plants are restricted in northward distribution by the total quantity of heat during the season of growth and reproduction.*[7]

---

[5] Gannett's "Nine-sheet contour map," published by the U.S. Geological Survey.

[6] The only changes worth mentioning are the introduction of the Tropical along the lower Colorado valley, the extension of the Tropical across the peninsula of Florida, and the extension of the Transition along the Pacific coast strip.

[7] In the case of certain sensitive species another factor enters into the problem, namely, *killing frosts*, for a few species are excluded by the occurrence of frosts from areas having a sufficient total quantity of heat for their needs.

The isotherm indicating a sum total of 5,500° C. (10,000° F.) coincides with the northern limit of distribution of Transition zone species, agreeing in the main with the dividing line between the two primary life regions of the northern hemisphere – Austral and Boreal. But in areas where extensive overlapping of Austral and Boreal types occurs, as along the Pacific coast from southern California northward to Puget sound, it will be observed that the isotherm in question points, as elsewhere, to the northern limit of Austral types and bears no relation whatever to the southward limit of Boreal types. It is evident, therefore, that the southward range of Boreal species, and perhaps of others also, is regulated by some cause other than the total quantity of heat. This cause was believed to be the mean temperature of the hottest part of the year,[8] for it is reasonable to suppose that Boreal species in ranging southward will encounter, sooner or later, a degree of heat they are unable to endure. The difficulty is in ascertaining the *length of the period* whose mean temperature acts as a barrier. It must be short enough to be included within the hottest part of the summer in high northern latitudes, and would naturally increase in length from the north southward. For experimental purposes, and without attempting unnecessary refinement, the mean normal temperature of the six hottest consecutive weeks of summer was arbitrarily chosen and platted on a large contour map of the United States, as in the case of the total quantity of heat. On comparing a reduced copy of this map (plate 13) with the zone map (plate 14) it appears that the isotherms conform to the southern boundaries of the Boreal Transition and Upper Austral life zones, and that the isotherm of 18° C. (61°.4 F.) agrees almost precisely with the southern boundary of the Boreal region. The coincidence is indeed so close as to justify the belief that *animals and plants are restricted in southward distribution by the mean temperature of a brief period covering the hottest part of the year.*

If the isotherm of 18° C. (64°.4 F.) for the six hottest consecutive weeks (see plate 13) is compared with that of 5,500° C. (10,000° F.), showing the sum of positive temperatures (see plate 12), it will be observed that the two are coincident in the main except in a few localities. The principal discrepancy is along the Pacific coast from Puget Sound to southern California. In this strip maps 12 and 13 not only fail to agree, but are fundamentally different, showing that no constant relation exists between the mean temperature of the six hottest consecutive weeks and the total of heat for the season. The mean temperature of the hottest part of the year from about latitude 35° northward along the coast is truly boreal, being as low as the mean of the corresponding period in northern Maine and other points well within the Boreal zone. The mean of the six consecutive hottest weeks at several points on the coast of California is as follows: At Eureka, on Humboldt bay, 13°.5 C. (56° F.);[9] at San Francisco, 15°.5 C. (60° F.); at Monterey and Ventura, 17°.5 C. (63° 5 F).[10] Strange as it may seem, San Francisco has a lower normal mean temperature

---

[8] This was indicated by mean summer temperatures platted from time to time during the past fifteen years, but the length of the period was never satisfactorily ascertained.

[9] In the following mean temperatures, fractions smaller than one-half a degree are ignored.

[10] Santa Barbara, between Monterey and Ventura, has a slightly higher mean (67° F), which is explained by its situation on a low, narrow coastal plain facing the south, with a range of mountains immediately on the north.

during the hottest part of the year than Eastport, Maine, the mean at Eastport being 16° C. (61° F.). On the other hand, the sum of positive temperatures (the normal mean daily temperatures above 6° C.) at San Francisco is more than 10,000° Fahrenheit higher than at Eastport, being 11,290° C. (20,360° F.) at the former and only 5.470° C. (9,880° F.) at the latter locality. At no point in the Pacific coast strip is the sum of the positive temperatures known to fall below 7,330° C. (13,600° F.), and it reaches 8,200° C. (14,800° F.) at Tatoosh island, off cape Flattery, the extreme northwestern point of the United States. Even at cape Flattery, therefore, the total of heat for the season is 260° C. (500° F.) greater than at Eastport, Maine, though the latter is the more southern locality and has the higher mean summer temperature.

The data at hand for the Pacific coast strip are amply sufficient to demonstrate two important facts: (1) that the temperature of the summer season, the hottest part of the year, is phenomenally low for the latitude and altitude – so low, indeed, as to enable Boreal types to push south to latitude 35°; (2) that the total quantity of heat (the sum of the positive temperatures) for the entire season is phenomenally high for the latitude – so high, indeed, as to enable Austral types to push north to Puget Sound. The total of heat is even greater at Puget Sound than at Philadelphia, Pittsburg, Cleveland, Indianapolis, Keokuk, or Omaha, though five hundred miles north of the latitude of these places. In other words, the mean temperature of the hottest part of the year is sufficiently low for Boreal species, while the total quantity of heat is sufficiently great for Austral species.

It is evident, therefore, that the principal climatic factors that permit Boreal and Austral types to live together along the Pacific coast are *a low summer temperature combined with a high sum total of heat*. The temperature is remarkably equable throughout the year; it never rises high for any length of time, and killing frosts are rare.

The study of the accompanying maps was the means of leading me, *first*, to the explanation of the anomalous distribution of species on the Pacific coast, where for a distance of more than a thousand miles a curious intermingling of northern and southern forms occurs; and, second, to what I now conceive to be the true theory of the temperature control of the geographic distribution of species.

The fundamental laws here developed, phrased for the northern hemisphere, may be briefly formulated as follows:

(1)   *The northward distribution of animals and plants is determined by the total quantity of heat – the sum of the effective temperatures.*
(2)   *The southward distribution of Boreal, Transition zone, and Upper Austral species is determined by the mean temperature of the hottest part of the year.*

Zone Temperatures

Boreal Zones. – The distinctive temperatures of the three Boreal zones (Arctic, Hudsonian and Canadian) are not positively known, but the southern limit of the Boreal as a whole is marked by the isotherm of 18° C. (64°.4 F.) for the six hottest consecutive weeks of summer. It seems probable, from the few data, available, that the limiting temperatures of the southern boundaries of the Hudsonian and Arctic zones are respectively 14° C. (57°.2 F.) and 10° C. (50° F.) for the same period.

Transition Zone species require a total quantity of heat of at least 5,500° C. (10,000° F.), but cannot endure a summer temperature the mean of which for the six hottest consecutive weeks exceeds 22° C. (71°.6 F.). The northern boundary of the Transition zone, therefore, is marked by the isotherm showing a sum of normal positive temperatures of 5,500° C. (10,000° F.), while its southern boundary is coincident with the isotherm of 22° C. (71°.6 F.) for the six hottest consecutive weeks.

Upper Austral species require a total quantity of heat of at least 6,400° C. (11,500° F.), but apparently cannot endure a summer temperature the mean of which for the six hottest consecutive weeks exceeds 26° C. (78°.8 F.). The northern boundary of the Upper Austral zone, therefore, is marked by the isotherm showing a sum of normal positive temperatures of 6,400° C. (11,500° F.), while its southern boundary agrees very closely with the isotherm of 26° C. (78°.8 F.) for the six hottest consecutive weeks.

Lower Austral species require a total quantity of heat of at least 10,000° C. (18,000° F.). The northern boundary of the Lower Austral zone, therefore, is marked by the isotherm showing a sum of normal positive temperatures of 10,000° C. (18,000° F.). A formula expressing the temperature-control of its southern boundary has not yet been found.

Tropical species require a total quantity of heat of at least 14,400° C. (26,000° F.); and, since the Tropical Life region is a broad equatorial belt, it is probable that both its northern and southern boundaries are marked by the isotherm showing a sum of normal positive temperatures of 14,400° C. (26,000° F.).

An interesting fact respecting the relative values of the zones is brought out by the isotherms showing the total quantity of heat necessary for each. It appears that the Transition and Upper Austral zones are not of equal value, but that together they are the exact equivalent of the Lower Austral zone.

Secondary Causes Affecting Distribution

It is not the purpose of the present essay to discuss the secondary causes affecting distribution. At the same time it seems desirable to contrast for a moment the influence of humidity, which is by far the most potent of the secondary causes, with that of temperature, which has been shown to be the primary controlling cause. Humidity governs details of distribution of numerous species of plants, reptiles and birds, and of a few species of mammals, within the several temperature zones. Thus the palmetto, the green chameleon, the chuck-wills widow and the rice field mouse inhabit humid parts of the Lower Austral zone (the Austroriparian area), while the mesquite, the leopard lizard, the sickle-billed thrashers and the four-toed kangaroo rats find their homes in arid parts of the same zone (the Lower Sonoran area).

That humidity is less potent than temperature as a controlling factor in distribution may be shown in several ways. The numerical evidence I have given on a previous

Regions	Zones	GOVERNING TEMPERATURES.			
		Northern limit. Sum of normal mean daily temperatures above 6° C. (43° F.).		Southern limit. Normal mean temperature of six hottest consecutive weeks.	
		C.	F.	C.	F.
Boreal	Arctic..............	....	....	10°*	50°*
	Hudsonian.........	....	....	14°*	57°.2*
	Canadian..........	....	....	18°	64°.4
Austral	Transition[1].........	5,500°	10,000°[4]	22°	71°.6
	Upper Austral[2]......	6,400°	11,500°	26°	78°.8
	Lower Austral[3]......	10,000°	18,000°	...	...
Tropical	....	14,500°	26,000°	...	...

* Estimated from insufficient data.

[1] The Transition zone comprises three principal subdivisions: an eastern or Alleghenian humid area, a western arid area, and a Pacific coast humid area.

[2] The Upper Austral zone comprises two principal subdivisions: an eastern or Carolinian area and a western or Upper Sonoran area.

[3] The Lower Austral zone comprises two principal subdivisions: an eastern or Austroriparian area and a western or Lower Sonoran area.

[4] The Fahrenheit equivalents of Centigrade sum temperatures are stated in round numbers to avoid small figures of equivocal value.

occasion.[11] Equally convincing is the circumstance that many genera restricted to particular conditions of temperature range completely across the continent, inhabiting alike the humid and arid sub-divisions of their respective zones; but no genus restricted to particular conditions of humidity ranges north and south across the several temperature zones.

Humidity and other secondary causes determine the presence or absence of particular species in particular localities *within* their appropriate zones, but temperature predetermines the possibilities of distribution; it fixes the limits beyond which species cannot pass; it defines broad transcontinental belts within which certain forms may thrive if other conditions permit, but outside of which they cannot exist, be the other conditions never so favorable.

Explanation of Maps

The temperature maps show the isotherms that conform to the boundaries of the life zones and the data on which they are based. The spots show the actual positions of the temperature stations.

[11] Presidential Address, Ibid., pp. 47–49.

NAT. GEOG. MAG. VOL VI. 1894 PL. 12

DISTRIBUTION OF THE TOTAL QUANTITY OF HEAT DURING SEASON OF GROWTH AND REPRODUCTIVE ACTIVITY.
[SUM OF DAILY MEAN TEMPERATURES ABOVE 6° C.]

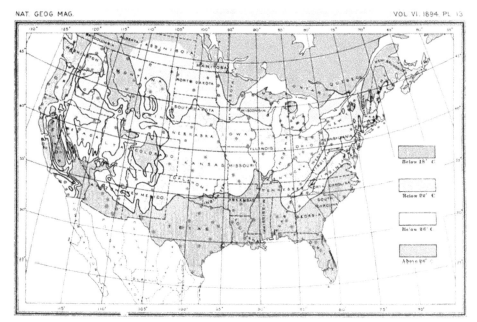

NAT. GEOG. MAG. VOL VI. 1894 PL. 13

MEAN TEMPERATURE OF HOTTEST SIX CONSECUTIVE WEEKS OF YEAR.

NAT. GEOG. MAG.

VOL. VI. 1894. PL. 14

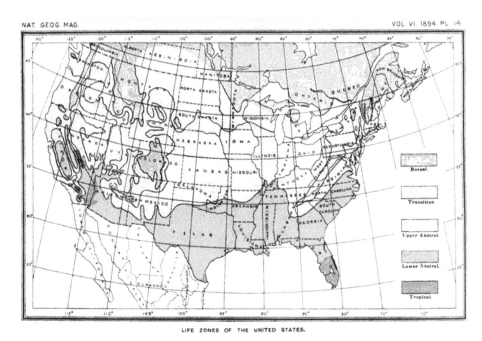

LIFE ZONES OF THE UNITED STATES.

In map 12, showing the distribution of the sum of effective temperatures, the isotherms conform with the northern boundaries of the life zones (as shown on map 14) as follows: The isotherm of 14,500° C. conforms with the northern boundary of the Tropical; of 10,000° C. with that of the Lower Austral; 6,400° C. with that of the Upper Austral, and 5,500° C. with that of the Transition.

In map 13, showing the normal mean temperature of the six hottest consecutive weeks, the isotherms conform with the southern boundaries of the life zones (as shown on map 14) as follows: The isotherm of 18° C. with the southern boundary of the Boreal; of 22° C. with the southern boundary of the Transition, and 26° C. with the southern boundary of the Upper Austral.

HENRY CHANDLER COWLES (1869–1939)

Botanist, ecologist.
Born in Kensington, Connecticut. Died in Chicago, Illinois.

Cowles received his undergraduate degree from Oberlin College and his Ph.D. from the University of Chicago in 1898. The excerpt offered here is from his seminal doctoral work on vegetative succession on Lake Michigan sand dunes. Cowles would remain at Chicago the rest of his career, as professor from 1911 and department chair from 1925. He retired in 1934. Cowles was one of the pioneers of dynamic plant ecology and his dissertation work was fundamental in the field of ecological succession. Cowles's geological background was vital in helping him assess the dynamic interaction between plant communities and their geological substructure. Cowles focused subsequent study on the varied habitats – forests, dunes, prairies – of Lake Michigan and Northern Illinois. He was a strong advocate of field study as a teaching method, was instrumental in the creation of forest reserves in Illinois, and co-founded the Ecological Society of America.

Henry Chandler Cowles, "The Ecological Relations
of the Vegetation on the Sand Dunes of Lake Michigan"
(excerpt), *Botanical Gazette* (1899)

"Part I. Geographical Relations of the Dune Floras"

I. Introduction

The province of ecology is to consider the mutual relations between plants and their environment. Such a study is to structural botany what dynamical geology is to structural geology. Just as modern geologists interpret the structure of the rocks by seeking to find how and under what conditions similar rocks are formed today, so ecologists seek to study those plant structures which are changing at the present time, and thus to throw light on the origin of plant structures themselves.

Again, ecology is comparable to physiography. The surface of the earth is composed of a myriad of topographic forms, not at all distinct, but passing into one another by a series of almost perfect gradations; the physiographer studies landscapes in their making, and writes on the origin and relationships of topographic forms. The ecologist employs the methods of physiography, regarding the flora of a pond or swamp or hillside not as a changeless landscape feature, but rather as a panorama, never twice alike. The ecologist, then, must study the order of succession of the plant societies in the development of a region, and he must endeavor to discover the laws which govern the panoramic changes. Ecology, therefore, is a study in dynamics. For its most ready application, plants should be found whose tissues and organs are actually changing at the present time in response to varying conditions. Plant formations should be found which are rapidly passing into other types by reason of a changing environment.

These requirements are met *par excellence* in a region of sand dunes. Perhaps no topographic form is more unstable than a dune. Because of this instability plant societies, plant organs, and plant tissues are obliged to adapt themselves to a new mode of life within years rather than centuries, the penalty for lack of adaptation being certain death. The sand dunes furnish a favorable region for the pursuit of ecological investigations because of the comparative absence of the perplexing problems arising from previous vegetation. Any plant society is the joint product of present and past environmental conditions, and perhaps the latter are much more potent than most ecologists have thought. As will be shown in another place, even the sand dune floras are often highly modified by preexisting conditions, but on the whole the physical forces of the present shape the floras as we find them. The advancing dune buries the old plant societies of a region, and with their death there pass away the influences which contributed so largely to their making. In place of the rich soil which had been accumulating by centuries of plant and animal decay, and in place of the complex reciprocal relations between the plants, as worked out by a struggle of centuries, the advance of a dune makes all things new. By burying the past, the dune offers to plant life a world for conquest, subject almost entirely to existing physical conditions. The primary motive, then, which prompted this present study was the feeling that nowhere else could many of the living problems of ecology be solved more clearly; that nowhere else could ecological principles be subjected to a more rigid test.

This particular investigation was also prompted by the fact that the previous ecological studies of sand-dune floras have been carried on chiefly in European countries, and almost exclusively along marine coasts. There has been considerable difference of opinion as to the influence of salty soils and atmospheres upon the vegetation. It would seem that a comparison of dunes along an inland fresh water lake with those along the sea should yield instructive results.

★   ★   ★

III. The ecological factors

The distribution of the plants in the various dune associations is governed by physical and biotic agencies which will be considered somewhat in detail in another place. At this point it seems advisable to give a general survey of these factors, especially in so far as they affect the distribution of plant societies in the region as a whole.

Light and heat

Nearly all of the dune societies are characterized by a high degree of exposure to *light*. Particularly is this true of the beach and the active dunes. The intensity of direct illumination is greatly increased by reflection; the glare of the white sand is almost intolerable on a bright summer day. The *temperature* relation is even more marked in its influence upon plant life. Because of the absence of vegetation and the general exposure of sand dunes the temperature is higher in summer and lower in winter than in most localities. The great divergence between the temperature extremes is still further increased by the low specific heat of sand. On sandy slopes protected from cold winds the vegetation renews its activity very early in the spring, because of the strong sunlight and the ease with which the surface layers of sand are heated. Willow shoots half-buried in the sand frequently develop fully a week in advance of similar shoots a few centimeters above the surface. Similarly in the autumn the activity of plant life ceases early largely because of the rapid cooling of the superficial layers of sand, as well as because of direct exposure to the cold.

Wind

The wind in one of the most potent of all factors in determining the character of the dune vegetation. The winds constantly gather force as they sweep across the lake, and when they reach the shore quantities of sand are frequently picked up and carried on. The force with which this sand is hurled against all obstacles in its path may be realized if one stoops down and faces it. The carving of the dead and living trees which are exposed to these natural sand-blasts is another evidence of their power. Fleshy fungi have been found growing on the windward side of logs and stumps completely petrified, as it were, by sand-blast action; sand grains are imbedded in the soft plant body and as it grows the imbedding is continued, so that finally the structure appears like a mass of sand cemented firmly together by the fungus. The bark of the common osier dogwood is red on the leeward side, but white to the windward because the colored outer layers of the bark have been wholly worn away. On the windward side of basswood limbs the softer portions are carved away while the tougher fibers remains as a reticulated network. On the leeward side of these same limbs, the outer bark is intact and even covered more or less with lichens.

The indirect action of the wind produces effects that are considerably more far-reaching than any other factor, for it is the wind which is primarily responsible for sand dunes and hence their floras. But more directly than this, the wind plays a prominent part in modifying the plant societies of the dunes. The wind is the chief destroyer of plant societies. Its methods of destruction are twofold. Single trees or entire groups of plants frequently have the soil blown away from under them, leaving the roots exposed high above the surface; as will be shown later this process is sometimes continued until entire forests are undermined, the debris being strewn about in great abundance.

Again, swamps, forests, and even low hills may be buried by the onward advance of a dune impelled by the winds; in place of a diversified landscape there results from this an all but barren waste of sand.

Soil

The soil of the dunes is chiefly quartz sand, since quartz is so resistant to the processes of disintegration. The quartz particles are commonly so light colored that the sand as a whole appears whitish; closer examination reveals many grains that are not white, especially those that are colored by iron oxide. With the quartz there are conspicuous grains of black sand, largely hornblende and magnetite. These black grains often accumulate in streaks, persistent for considerable depths and apparently sifted by the wind; large quartz grains are mingled with these grains of magnetite and hornblende so that it would seem as if grains of higher specific gravity are sifted out together with those of greater absolute weight. The sand of the dunes is remarkably uniform in the size of the particles as compared with beach sand; this feature is due to the selective action of the wind, since the latter agent is unable to pick up and carry for any distance the gravel or large sand particles of the beach.

As is well known, soil made up chiefly of quartz sand has certain marked peculiarities which strongly influence the vegetation. The particles are relatively very large; hence the soil is extremely porous and almost devoid of cohesion between the grains. These features are of especial importance in their effect upon the water and heat relations as shown elsewhere. As a rule, sandy soils are poor in nutrient food materials, nor do they rapidly develop a rich humus soil because of the rapid oxidation of the organic matter.

Water

A factor of great importance, here as everywhere, is the water relation. Nothing need to be said of atmospheric moisture, since that is sufficient to develop a rich vegetation if properly conserved, as is shown by the luxuriance of neighboring floras. Because of the peculiar physical properties of quartz sand, precipitated water quickly percolates to the water level and becomes unavailable to plants with short roots. The water capacity of sand is also slight, not is there such pronounced capillarity as is characteristic of many other soils. Again, the evaporation from a sandy surface is commonly quite rapid. All of these features combine to furnish a scanty supply of water to the tenants of sandy soil. The rapid cooling of sand on summer nights may, however, result in a considerable condensation of dew, and thus, in a small way, compensate for the other disadvantages.

The ecological factors thus far mentioned act together harmoniously and produce a striking composite effect upon the vegetation. A flora which is subjected to periods of drought is called a xerophytic flora and its component species have commonly worked out various xerophytic structural adaptations of one sort or another. Again, a flora which is subjected to extreme cold, especially when accompanied by severe winds, takes on various structural adaptations similar to those that are characteristic of alpine and arctic floras. The dune flora is a composite flora, showing both xerophytic and arctic structures. In those situations which are most exposed to cold winds, one finds the best illustrations of the arctic type, while the desert or xerophilous type is shown in its purest expression on protected inland sandy hills. The discussion of the various arctic and desert structures and their relations to each other will be deferred to the second part of this paper.

Other factors

Certain other factors are of minor importance in determining the character of the dune flora. *Forest fires* occur occasionally, and, as will be shown later, they may considerably shorten the lifetime of a coniferous plant society.

Near cities the vegetation is unfavorably influenced by *smoke* and other products issuing from chimneys. In the neighborhood of the oil refineries at Whiting, Ind., the pine trees especially have been injured or destroyed. A careful study would probably show many plant species that have suffered a similar fate.

The *topography* is often a factor of considerable importance. Dune areas are conspicuous for their diversified topography. This factor determines to a great extent the water relation which has been previously considered, the hills and slopes being of course much drier than the depressions. The topography indirectly affects the soil, since it is mainly in the depressions that humus can rapidly accumulate. The direction of slope is a matter of importance, as will be shown in discussing the oak dunes; the greater exposure of the southern slopes to the sun results in a drier soil and a more xerophytic flora on that side.

Animals do not appear to exert any dominating influences on the dune floras. The dispersal of pollen and fruits by their agency is common here as elsewhere; so, too, the changes that animal activities produce in the soil. Near the cities the influence of man is seen, although such influences are slight unless the sand is removed bodily for railroad grading and other purposes.

The influence of *plants*, which so often becomes the dominant factor, is relatively inconspicuous on the dunes. The most important function which dune plants perform for other plants is in the contribution of organic food materials to the soil. The oxidation or removal of decaying vegetation is so complete on the newer dunes that the accumulation of humus is not important. On the more established dunes the mold becomes deeper and deeper, and, after the lapse of centuries, the sandy soil beneath may become buried so deeply that a mesophytic flora is able to establish itself where once there lived the tenants of an active dune. The advance of a wandering dune often results in the burial of a large amount of organic matter; when this matter becomes unburied years afterward it may again furnish a soil for plants. Many fossil soil lines have thus been uncovered on the Sleeping Bear dunes at Glen Haven, Mich.

IV. The plant societies

A plant society is defined as a group of plants living together in a common habitat and subjected to similar life conditions. The term is taken to be the English equivalent of Warming's *Plantesamfund*, translated into the German as *Pflanzenverein*. The term formation, as used by Drude and others, is more comprehensive, in so far as it is not synonymous. It may be well to consider the individual habitat groups in a given locality as plant societies, while all of these groups taken together comprise a formation of that type, thus giving to the word formation a value similar to its familiar geological application. For example, one might refer to particular sedge swamp societies near Chicago, or, on the other hand, to the sedge swamp formation as a whole; by this application formation becomes a term of generic value, plant society of specific value.

Plant societies may be still further subdivided into patches or zones; the former more or less irregular, the latter more or less radially symmetrical. Patches are to be found in any plant society, where one or another constituent becomes locally dominant; zones are conspicuously developed on the beach and in sphagnous swamps. The term patch or zone has a value like that of variety in taxonomy. Authors disagree, here as everywhere, upon the content and values of the terms employed; this disagreement is but an expression of the fact that there are few if any sharp lines in nature. The above, or any other terminology, is largely arbitrary and adopted only as a matter of convenience.

★   ★   ★

i. *The lower beach*

The lower beach has been defined as the zone of land washed by the waves of summer storms. It might almost be defined as that portion of the beach which is devoid of vegetation. Perhaps there is not flora in the temperate zone quite so sparse as that of the lower beach, unless we except bare rocks and alkaline deserts. A survey of the life conditions in this zone reveals at once the reason for the scanty vegetation. Land life is excluded because of the frequency and violence of storms; the waves tear away the sand in one spot only to deposit it in another. Even though a seed had the temerity to germinate, the young plant would soon be destroyed by the breakers. Nor is there great likelihood that seeds will find a lodgment in this unstable location. As will be seen later the seeds ripened by tenants of the middle beach are almost entirely scattered away from the lake instead of toward it. The action of both wind and wave tends to carry seeds away from the lower beach. Again, few seeds could endure the alternate extremes of cold and head, wetting and drying so characteristic of this zone.

Water life is excluded because of the extreme xerophytic conditions which commonly prevail on the lower beach. While algae may propagate themselves in the shallow pools or even in the wet sand during a prolonged season of wet weather, a cessation of activity if not death itself soon follows the advent of dry weather. During a period of rainy weather in the autumn of 1897 green patches were observed in wet sand a few meters from the mouth of a creek near Porter, Ind. Microscopic observation showed that the green coloration was due to the presence of millions of motile Chlamydomonas forms. These unicellular biciliate algae were in process of active locomotion in the water held by capillarity between the grains of sand. In all probability these forms migrated to the beach from the waters of the creek during a period of wet weather. It is possible that they might pass into resting stages and live through a season of drought, were it not for the wind which gathers much of its dune material from the lower beach.

Thus the lower beach is a barren zone between two zones of life. Below it there exist algae and other hydrophytic forms which flourish which flourish in the fury of the breakers; above it there exists the flora of the middle beach, a flora adapted to the most intense xerophytic conditions. At no particular time, perhaps, are the conditions too severe for some type of life; vegetation is excluded because of the alternation of opposite extremes.

ii. *The middle beach*

The middle beach is situated between the upper limits of the summer and winter waves, comparatively dry in summer but washed by the high storms of winter. It may also be defined as the zone of succulent annuals. The upper limit of this beach is commonly marked by a line of driftwood and débris. The instability of the beach conditions is often shown by the presence of a number of such lines, marking wave limits for different seasons. A very heavy storm will carry the débris line far up on the upper beach, to all intents and purposes carrying the middle beach just so much farther inland, as the flora of the next season testifies. Another season may be without the visitation of heavy storms and the middle beach will encroach upon the territory of the lower beach. The limits of the middle beach are altered more permanently by changes in the lower beach. In many places the lower beach is growing outwards, reclaiming land from the lake, while at other points the lake encroaches upon the land. Speaking broadly, the middle beach advances or recedes *pari passu* with the advance or recession of the lower beach. To some extent the debris line register these changes, as their notable departure from persistent parallelism may indicate; however, there is a considerable lack of parallelism in the débris lines of a single season, owing to variations in the direction of the wind and other factors.

The life conditions in this zone are exceedingly severe, and result in a flora of the most pronounced xerophytic characters. The fury of the winter storms as they wash over the middle beach, tearing up here and depositing there, excludes almost entirely the possibility of survival through the period. In other words, biennials and perennials are practically excluded from maturing flowers and fruits, although their vegetative structures may flourish for a single season. In the summer the xerophilous conditions are extreme. Nowhere in the dune region are the winds more severe than here; the middle beach is close enough to the lake to feel all the force of its winds and yet far enough away for the wind to pick up sand from the lower beach and bring to bear upon the floras the intense severity of the sand-blast. No flora is more exposed to the extreme desiccating influences of the summer sun than that which grows upon the bare and open beach. Even though the roots can readily penetrate to the water level, the great exposure of the aerial organs to wind and sun results in the working out of that most perfect of all xerophytic organs, the succulent leaf. Just as succulent plants inhabit deserts where no other high grade plants can grow, so, too, they are able to withstand the severe conditions of the beach.

Along the entire eastern shore of the lake, the dominant plant of the middle beach is *Cakile Americana*. At many points this succulent crucifer is almost the only plant species found in this zone, and it is always the chief character species. Two other fleshy annuals are very common tenants of the middle beach: *Corispermum hyssopifolium* and one of the spurges, *Euphorbia polygonifolia*. It is a matter of interest to observe that two of these three character plants of the middle beach, Cakile and Euphorbia, are also characteristic inhabitants of the beach on the Atlantic coast. The significance of the presence of these and other marine forms along the shore of Lake Michigan will be discussed in another place. The above plants are rarely distributed uniformly over the middle beach. The favorite place for growth is along the lines of debris previously referred to; along these lines a greater number of seeds find lodgment than elsewhere, because the waves wash them up from lower levels and the protection of the driftwood prevents the winds from carrying them on farther. Then, too the driftwood may furnish some protection to the growing plants, especially protection from winds which might otherwise uproot them. Cakile and Euphorbia reach their culmination on the beach, and when found farther inland it is chiefly on the upper beach or on windward slopes of active dunes. Corispermum, on the other hand, appears to be rather more characteristic of the active dunes than of the beach. Cakile is much the hardiest of the three species, venturing farther out toward the lower beach than either of the other two. Of the three, Cakile is the most succulent and perhaps thus best adapted to the extreme xerophilous conditions to which beach plants are subjected. Euphorbia, however, has a copious supply of latex and its prostrate habit would seem to offer some advantages for existence on the beach. Cakile and Corispermum are readily dispersed by the wind, the latter by means of its winged seeds, while the former is a sort of tumbleweed; broken Cakile plants are common sights all over the dunes in the autumn and winter. Corispermum and Euphorbia become less and less common toward the north; at Charlevoix and Petoskey, Cakile is almost the only plant growing on the middle beach, and even this latter species is less common than farther to the south. Thus is seems as though the life conditions on the middle beach are more severe northward than southward, as indeed might be expected.

JOSIAH WILLARD GIBBS (1839–1903)

Physicist, chemist, mathematician.
Born in New Haven, Connecticut. Died in New Haven, Connecticut.

Gibbs was simply the greatest American scientist of the nineteenth century. He was so uninterested in self-promotion, however, that his brilliance went unrecognized for a long time. No doubt the theoretical nature of his work did not make its importance immediately apparent, that and perhaps the fact that Gibbs tended to publish in the *Transactions of the Connecticut Academy of Sciences*, a journal edited by his librarian brother-in-law and rarely read in Connecticut let alone Europe. His father was a Yale professor in sacred literature and so Gibbs was naturally a Yale graduate as well, excelling in Latin and Mathematics. He remained at Yale to work on his doctorate, the first in engineering in the United States. He went to Europe with his sisters for a grand tour of Paris, Berlin and Heidelberg from 1866–1869. Gibbs spent much of his time there studying, attending lectures, and meeting with other scientist. It seems by all accounts to have been a pivotal experience. He returned to Yale as a tutor but was appointed professor of mathematical physics by 1871. He still had not published any work. He did finally begin to publish in 1873, at the age of 34. Between 1876 and 1878 Gibbs wrote the parts of what would collectively be titled "On the Equilibrium of Heterogeneous Substances," one of his greatest efforts and a key part of the theoretical foundation for chemical thermodynamics (including his important phase rule). In the years that followed he would produce seminal work on vector analysis (1880–84), the electromagnetic theory of light (1882–89), and statistical mechanics (1889). He would be well-known in his life time throughout Europe, yet unknown at home except by his colleagues and students. People who knew him well always speak of a kind, soft-spoken, retiring genius. He never married, never really left Yale again after his long sojourn to Europe, but lived with his sister in their father's house, walking to school every day, and making the most of his quiet, profound gifts.

J. Willard Gibbs, "On the Equilibrium of Heterogeneous Substances, abstract,"[1] *American Journal of Science and Arts* (1878)

It is an inference naturally suggested by the general increase of entropy which accompanies the changes occurring in any isolated material system that when the entropy of the system has reached a maximum, the system will be in a state of equilibrium. Although this principle has by no means escaped the attention of physicists, its importance does

[1] *Transactions of the Connecticut Academy of Arts and Sciences*, vol. iii, pp. 108–248 and 343–524.

not appear to have been duly appreciated. Little has been done to develop the principle as a foundation for the general theory of thermodynamic equilibrium.

The principle may be formulated as follows, constituting a criterion of equilibrium:

I. *For the equilibrium of any isolated system it is necessary and sufficient that in all possible variations of the state of the system which do not alter its energy, the variation of its entropy shall either vanish or be negative.*

The following form, which is easily shown to be equivalent to the preceding, is often more convenient in application:

II. *For the equilibrium of any isolated system it is necessary and sufficient that in all possible variations of the state of the system which do not alter its entropy, the variation of its energy shall either vanish or be positive.*

If we denote the energy and entropy of the system by ε and η respectively, the criterion of equilibrium may be expressed by either of the formulae

$$(\delta\eta)_\varepsilon \leq 0, \tag{1}$$

$$(\delta\varepsilon)_\eta \geq 0. \tag{2}$$

Again, if we assume that the temperature of the system is uniform, and denote its absolute temperature by t, and set

$$\Psi = \varepsilon - t\eta, \tag{3}$$

the remaining conditions of equilibrium may be expressed by the formula

$$(\delta\Psi)_t \geq 0, \tag{4}$$

the suffixed letter, as in the preceding cases, indicating that the quantity which it represents is constant. This condition, in connection with that of uniform temperature, may be shown to be equivalent to (1) or (2). The difference of the values of Ψ for two different states of the system which have the same temperature represents the work which would be expended in bringing the system from one state to the other by a reversible process and without change of temperature.

If the system is incapable of thermal changes, like the systems considered in theoretical mechanics, we may regard the entropy as having the constant value zero. Conditions (2) and (4) may then be written

$$\delta\varepsilon \geq 0, \ \delta\Psi \geq 0,$$

and are obviously identical in signification, since in this case $\Psi = \varepsilon$.

Conditions (2) and (4), as criteria of equilibrium, may therefore both be regarded as extensions of the criterion employed in ordinary statics to the more general case of a thermodynamic system. In fact, each of the quantities $-\varepsilon$ and $-\Psi$ (relating to a system without sensible motion) may be regarded as a kind of force-function for the system, — the former as the force-function *for constant entropy* (*i.e.*, when only such states of the system are considered as have the same entropy) and the latter as the force-function *for constant temperature* (*i.e.*, when only such states of the system are considered as have the same uniform temperature).

In the deduction of the particular conditions of equilibrium for any system, the general formula (4) has an evident advantage over (1) or (2) with respect to the brevity of the processes of reduction, since the limitation of constant temperature applies to every part of the system taken separately, and diminishes by one the number of independent variations in the state of these parts which we have to consider. Moreover, the transition from the systems considered in ordinary mechanics to thermodynamic systems is most naturally made by this formula, since it has always been customary to apply the principles of theoretical mechanics to real systems on the supposition (more or less distinctly conceived and expressed) that the temperature of the system remains constant, the mechanical properties of a thermodynamic system maintained at a constant temperature being such as might be imagined to belong to a purely mechanical system, and admitting of representation by a force-function, as follows directly from the fundamental laws of thermodynamics.

Notwithstanding these considerations, the author has preferred in general to use condition (2) as the criterion of equilibrium, believing that it would be useful to exhibit the conditions of equilibrium of thermodynamic systems in connection with those quantities which are most simple and most general in their definitions, and which appear most important in the general theory of such systems. The slightly different form in which the subject would develop itself, if condition (4) had been chosen as a point of departure instead of (2), is occasionally indicated.

Equilibrium of masses in contact. – The first problem to which the criterion is applied is the determination of the conditions of equilibrium for different masses in contact, when uninfluenced by gravity, electricity, distortion of the solid masses, or capillary tensions. The statement of the result is facilitated by the following definition.

If to any homogeneous mass in a state of hydrostatic stress we suppose an infinitesimal quantity of any substance to be added, the mass remaining homogeneous and its entropy and volume remaining unchanged, the increase of the energy of the mass divided by the quantity of the substance added is the *potential* for that substance in the mass considered.

In addition to equality of temperature and pressure in the masses in contact, it is necessary for equilibrium that the potential for every substance which is an independently variable component of any of the different masses shall have the same value in all of which it is such a component, so far as they are in contact with one another. But if a substance, without being an actual component of a certain mass in the given state of the system, is capable of being absorbed by it, it is sufficient if the value of the potential for that substance in that mass is not less than in any contiguous mass of which the substance is an actual component. We may regard these conditions as sufficient for equilibrium with respect to infinitesimal variations in the composition and thermodynamic state of the different masses in contact. There are certain other conditions which relate to the possible formation of masses entirely different in composition or state from any initially existing. These conditions are best regarded as determining the stability of the system, and will be mentioned under that head.

Anything which restricts the free movement of the component substances, or of the masses as such, may diminish the number of conditions which are necessary for equilibrium.

Equilibrium of osmotic forces. – If we suppose two fluid masses to be separated by a diaphragm which is permeable to some of the component substances and not to others, of the conditions of equilibrium which have just been mentioned, those will still subsist which relate to temperature and the potentials for the substances to which the diaphragm is permeable, but those relating to the potentials for the substances to which the diaphragm is impermeable will no longer be necessary. Whether the pressure must be the same in the two fluids will depend upon the rigidity of the diaphragm. Even when the diaphragm is permeable to all the components without restriction, equality of pressure in the two fluids is not always necessary for equilibrium.

Effect of gravity. – In a system subject to the action of gravity, the potential for each substance, instead of having a uniform value throughout the system, so far as the substance actually occurs as an independently variable component, will decrease uniformly with increasing height, the difference of its values at different levels being equal to the difference of level multiplied by the force of gravity.

Fundamental equations. – Let ε, η, v, t and p denote respectively the energy, entropy, volume, (absolute) temperature, and pressure of a homogeneous mass, which may be either fluid or solid, provided that it is subject only to hydrostatic pressures, and let m_1, m_2, ... m_n denote the quantities of its independently variable components, and $\mu_1, \mu_2, ... \mu_n$ the potentials for these components. It is easily shown that ε is a function of η, v, $m_1, m_2, ... m_n$, and that the complete value of $d\varepsilon$ is given by the equation

$$d\varepsilon = td\eta - pdv + \mu_1 dm_1 + \mu_2 dm_2 ... + \mu_n dm_n. \tag{5}$$

Now ε if is known in terms of μ, v, m_1, ... m_n, we can obtain by differentiation t, p, μ_1, ... μ_n in terms of the same variables. This will make $n + 3$ independent known relations between the $2n + 5$ variables, ε, η, v, m_1, m_2 ... m_n, t, p, μ_1, μ_2, ... μ_n. These are all that exist, for of these variables, $n + 2$ are evidently independent. Now upon these relations depend a very large class of the properties of the compound considered,—we may say in general, all its thermal, mechanical, and chemical properties, so far as *active tendencies* are concerned, in cases in which the form of the mass does not require consideration. A single equation from which all these relations may be deduced may be called a fundamental equation. An equation between ε, η, v, m_1, m_2, ... m_n is a fundamental equation. But there are other equations which possess the same property.

If we suppose the quantity Ψ to be determined for such a mass as we are considering by equation (3), we may obtain by differentiation and comparison with (5)

$$d\Psi = -\eta dt - pdv + \mu_1 dm_1 + \mu_2 dm_2 \ldots + \mu_n dm_n. \tag{6}$$

If, then, Ψ is known as a function of t, v, m_1, m_2, ... m_n, we can find η, p, μ_1, μ_2, ... μ_n in terms of the same variables. If we then substitute for Ψ in our original equation its value taken from equation (3) we shall have again $n + 3$ independent relations between the same $2n + 5$ variables as before.

Let

$$\zeta = \varepsilon - t\eta + pv, \tag{7}$$

then, by (5),

$$d\zeta = -\eta dt + vdp + \mu_1 dm_1 + \mu_2 dm_2 \ldots + \mu_n dm_n. \tag{8}$$

If, then, ζ is known as a function of t, p, m_1, m_2, ... m_n, we can find η, v, μ_1, μ_2, ... μ_n in terms of the same variables. By eliminating, we may obtain again $n + 3$ independent relations between the same $2n + 5$ variables as at first.[2]

If we integrate (5), (6) and (8), supposing the quantity of the compound substance considered to vary from zero to any finite value, its nature and state remaining unchanged, we obtain

$$\varepsilon = t\eta - pv + \mu_1 m_1 + \mu_2 m_2 \ldots + \mu_n m_n, \tag{9}$$
$$\Psi = -pv + \mu_1 m_1 + \mu_2 m_2 \ldots + \mu_n m_n, \tag{10}$$
$$\zeta = \mu_1 m_1 + \mu_2 m_2 \ldots + \mu_n m_n. \tag{11}$$

If we differentiate (9) in the most general manner, and compare the result with (5), we obtain

$$-vdp + \eta dt + m_1 d\mu_1 + m_2 d\mu_2 \ldots + m_n d\mu_n = 0, \tag{12}$$

or

[2] The properties of the quantities $-\Psi$ and $-\zeta$ regarded as functions of the temperature and volume, and temperature and pressure, respectively, the composition of the body being regarded as invariable, have been discussed by M. Massieu in a memoir entitled " Sur les fonctions caracteristiques des divers fluides et sur la thearie des vapeurs" (*Mem. Sarants Strang.*, t xxii.) A brief sketch of his method in a form slightly different from that ultimately adopted is given in *Comptes Rendus.* t. lxix, (1869) pp. 858 and 1057, and a report on his memoir by M. Bertrand in (*Comptes Rendus*, t. lxxi, p. 257.) M. Massieu appears to have been the first to solve the problem of representing all the properties of a body of invariable composition which are concerned in reversible processes by means of a single function.

$$dp = \frac{\eta}{v} dt + \frac{m_1}{v} d\mu_1 + \frac{m_2}{v} d\mu_2 \ldots + \frac{m_n}{v} d\mu_n = 0. \tag{13}$$

Hence, there is a relation between the $n + 2$ quantities t, p, μ_1, μ_2, $\ldots$ μ_n, which, if known, will enable us to find in terms of these quantities all the ratios of the $n + 2$ quantities η, v, m_1, m_2, $\ldots$ m_n. With (9), this will make $n + 3$ independent relations between the same $2n + 5$ variables as at first.

Any equation, therefore, between the quantities

	ε,	η,	v,	m_1,	$m_2, \ldots m_n$,
or	Ψ,	t,	v,	m_1,	$m_2, \ldots m_n$,
or	ζ,	t,	v,	m_1,	$m_2, \ldots m_n$,
or		t,	v,	μ_1,	$\mu_2, \ldots \mu_n$,

is a fundamental equation, and any such is entirely equivalent to any other.

Coexistent phases. – In considering the different homogeneous bodies which can be formed out of any set of component substances, it is convenient to have a term which shall refer solely to the composition and thermodynamic state of any such body without regard to its size or form. The word *phase* has been chosen for this purpose. Such bodies as differ in composition or state are called different phases of the matter considered, all bodies which differ only in size and form being regarded as different examples of the same phase. Phases which can exist together, the dividing surfaces being plane, in an equilibrium which does not depend upon passive resistances to change, are called *coexistent*.

The number of independent variations of which a system of coexistent phases is capable is $n + 2 - r$, where r denotes the number of phases, and n the number of independently variable components in the whole system. For the system of phases is completely specified by the temperature, the pressure, and the n potentials, and between these $n + 2$ quantities there are r independent relations (one for each phase), which characterize the system of phases.

When the number of phases exceeds the number of components by unity, the system is capable of a single variation of phase. The pressure and all the potentials may be regarded as functions of the temperature. The determination of these functions depends upon the elimination of the proper quantities from the fundamental equations in p, t, μ_1, μ_2, etc., for the several members of the system. But without a knowledge of these fundamental equations, the values of the differential co-efficients such as dp/dt may be expressed in terms of the entropies and volumes of the different bodies and the quantities of their several components. For this end we have only to eliminate the differentials of the potentials from the different equations of the form (12) relating to the different bodies. In the simplest case, when there is but one component, we obtain the well-known formula $dp/dt = (\eta' - \eta'')/(v' - v'') = Q/(t(v' - v''))$, in v', v'', η', η'', which denote the volumes and entropies of a given quantity of the substance in the two phases, and Q the heat which it absorbs in passing from one phase to the other.

It is easily shown that if the temperature of two coexistent phases of two components is maintained constant, the pressure is in general a maximum or minimum when the composition of the phases is identical. In like manner, if the pressure of the phases is maintained constant, the temperature is in general a maximum or minimum when the composition of the phases is identical. The series of simultaneous values of t and p for which the composition of two coexistent phases is identical separates those simultaneous values of t and p for which no coexistent phases are possible from those for which there are two pairs of coexistent phases.

If the temperature of three coexistent phases of three components is maintained constant, the pressure is in general a maximum or minimum when the composition of one of the phases is such as can be produced by combining the other two. If the pressure is maintained constant, the temperature is in general a maximum or minimum when the same condition in regard to the composition of the phases is fulfilled.

Stability of fluids. – A criterion of the stability of a homogeneous fluid, or of a system of coexistent fluid phases, is afforded by the expression

$$\varepsilon - t'\eta + p'v - \mu_1'm_1 - \mu_2'm_2 \dots - \mu_n'm_n, \tag{14}$$

in which the values of the accented letters are to be determined by the phase or system of phases of which the stability is in question, and the values of the unaccented letters by any other phase of the same components, the possible formation of which is in question. We may call the former constants, and the latter variables. Now if the value of the expression, thus determined, is always positive for any possible values of the variables, the phase or system of phases will be stable with respect to the formation of any new phases of its components. But if the expression is capable of a negative value, the phase or system is at least *practically* unstable. By this is meant that, although, strictly speaking, an infinitely small disturbance or change may not be sufficient to destroy the equilibrium, yet a very small change in the initial state will be sufficient to do so. The presence of a small portion of matter in a phase for which the above expression has a negative value will in general be sufficient to produce this result. In the case of a system of phases, it is of course supposed that their contiguity is such that the formation of the new phase does not involve any transportation of matter through finite distances.

The preceding criterion affords a convenient point of departure in the discussion of the stability of homogeneous fluids. Of the other forms in which the criterion may be expressed, the following is perhaps the most useful: –

If the pressure of a fluid is greater than that of any other phase of its independent variable components which has the same temperature and potentials, the fluid is stable with respect to the formation of any other phase of these components; but if its pressure is not as great as that of some such phase, it will be practically unstable.

Stability of fluids with respect to continuous changes of phase. – In considering the changes which may take place in any mass, we have often to distinguish between infinitesimal changes in existing phases, and the formation of entirely new phases. A phase of a fluid may be stable with respect to the former kind of change, and unstable with respect to the latter. In this case, it may be capable of continued existence in virtue of properties which prevent the commencement of discontinuous changes. But a phase which is unstable with respect to continuous changes is evidently incapable of permanent existence on a large scale except in consequence of passive resistances to change. To obtain the conditions of stability with respect to continuous changes, we have only to limit the application of the variables in (14) to phases adjacent to the given phase. We obtain results of the following nature.

The stability of any phase with respect to continuous changes depends upon the same conditions with respect to the second and higher differential coefficients of the density of energy regarded as a function of the density of entropy and the densities of the several components, which would make the density of energy a minimum, if the necessary conditions with respect to the first differential coefficients were fulfilled.

Again, it is necessary and sufficient for the stability with respect to continuous changes of all the phases within any given limits, that within those limits the same conditions should be fulfilled with respect to the second and higher differential coefficients of the pressure regarded as a function of the temperature and the several potentials, which would make the pressure a minimum, if the necessary conditions with respect to the first differential coefficients were fulfilled.

The equation of the limits of stability with respect to continuous changes may be written

$$\left(\frac{d\mu_n}{d\gamma_n}\right)_{t,\mu_1,\dots\mu_{n-1}} = 0, \quad \text{or} \quad \left(\frac{d^2 p}{d\mu_n^2}\right)_{t,\mu_1,\dots\mu_{n-1}} = \infty \tag{15}$$

where γ_n denotes the density of the component specified or $m_n \div v$. It is in general immaterial to what component the suffix $_n$ is regarded as relating.

Critical phases. – The variations of two coexistent phases are sometimes limited by the vanishing of the difference between them. Phases at which this occurs are called *critical phases.* A critical phase, like any other, is capable of $n + 1$ independent variations, n denoting the number of independently variable components. But when subject to the condition of remaining a critical phase, it is capable of only $n - 1$ independent variations. There are therefore two independent equations which characterize critical phases. These may be written

$$\left(\frac{d\mu_n}{d\gamma_n}\right)_{t,\mu_1,\dots\mu_{n-1}} = 0, \quad \left(\frac{d^2 \mu_n}{d\gamma_n^2}\right)_{t,\mu,\dots\mu} = 0. \tag{16}$$

It will be observed that the first of these equations is identical with the equation of the limit of stability with respect to continuous changes. In fact, stable critical phases are situated at that limit. They are also situated at the limit of stability with respect to discontinuous changes. These limits are in general distinct, but touch each other at critical phases.

Geometrical illustrations. – In an earlier paper,[3] the author has described a method of representing the thermodynamic properties of substances of invariable composition by means of surfaces. The volume, entropy, and energy of a constant quantity of the substance are represented by rectangular coordinates. This method corresponds to the first kind of fundamental equation described above. Any other kind of fundamental equation for a substance of invariable composition will suggest an analogous geometrical method. In the present paper, the method in which the coordinates represent temperature, pressure, and the potential, is briefly considered. But when the composition of the body is variable, the fundamental equation cannot be completely represented by any surface or finite number of surfaces. In the case of three components, if we regard the temperature and pressure as constant, as well as the total quantity of matter, the relations between ζ, m_1, m_2, m_3 may be represented by a surface in which the distances of a point from the three sides of a triangular prism represent the quantities m_1, m_2, m_3, and the distance of the point from the base of the prism represents the quantity ζ. In the case of two components, analogous relations may be represented by a plane curve. Such methods are especially useful for illustrating the combinations and separations of the components, and the changes in states of aggregation, which take place when the substances are exposed in varying proportions to the temperature and pressure considered.

Fundamental equations of ideal gases and gas-mixtures. – From the physical properties which we attribute to ideal gases, it is easy to deduce their fundamental equations. The fundamental equation in ε, η, v, and m for an ideal gas is

$$c \, \log \frac{\varepsilon - Em}{cm} = \frac{\eta}{m} - H + a \log \frac{m}{v} \, ; \tag{17}$$

that in ψ, t, v, and m is

$$\psi = Em + mt \left(c - H - c \log t + a \log \frac{m}{v} \right); \tag{18}$$

that in p, t, and μ is

$$p = ae^{\frac{H-c-a}{a}} \, t^{\frac{c+a}{a}} \, e^{\frac{\mu-E}{at}} \, , \tag{19}$$

where e denotes the base of the Naperian system of logarithms. As for the other constants, c denotes the specific heat of the gas at constant volume, a denotes the constant value of $pv \div mt$, E and H depend upon the zeros of energy and entropy. The two last equations may be abbreviated by the use of different constants. The properties of fundamental equations mentioned above may easily be verified in each case by differentiation.

The law of Dalton respecting a mixture of different gases affords a point of departure for the discussion of such mixtures and the establishment of their fundamental equations. It is found convenient to give the law the following form:

The pressure in a mixture of different gases is equal to the sum of the pressures of the different gases as existing each by itself at the same temperature and with the same value of its potential.

A mixture of ideal gases which satisfies this law is called an *ideal gas-mixture*. Its fundamental equation in p, t, μ_1, μ_2, etc. is evidently of the form

$$p = \Sigma_1 \left(a_1 e^{\frac{H_1 - c_1 - a_1}{a_1}} \ t^{\frac{c_1 + a_1}{a_1}} \ e^{\frac{\mu_1 - E_1}{a_1 t}} \right) \tag{20}$$

where Σ_1 denotes summation with respect to the different components of the mixture. From this may be deduced other fundamental equations for ideal gas-mixtures. That in ψ, t, v, m_1, m_2, etc. is

$$\psi = \Sigma_1 \left(E_1 m_1 + m_1 t \left(c_1 - H_1 - c_1 \ \log t + a_1 \ \log \frac{m_1}{v} \right) \right). \tag{21}$$

Phases of dissipated energy of ideal gas-mixtures. – When the proximate components of a gas-mixture are so related that some of them can be formed out of others, although not necessarily in the gas-mixture itself at the temperatures considered, there are certain phases of the gas-mixture which deserve especial attention. These are the *phases of dissipated energy*, i.e., those phases in which the energy of the mass has the least value consistent with its entropy and volume. An atmosphere of such a phase could not furnish a source of mechanical power to any machine or chemical engine working within it, as other phases of the same matter might do. Nor can such phases be affected by any catalytic agent. A *perfect catalytic agent* would reduce any other phase of the gas-mixture to a phase of dissipated energy. The condition which will make the energy a minimum is that the potentials for the proximate components shall satisfy an equation similar to that which expresses the relation between the units of weight of these components. For example, if the components were hydrogen, oxygen and water, since one gram of hydrogen with eight grams of oxygen are chemically equivalent to nine grams of water, the potentials for these substances in a phase of dissipated energy must satisfy the relation

$$\mu_H + 8\mu_O = 9\mu_W.$$

Gas-mixtures with convertible component. – The theory of the phases of dissipated energy of an ideal gas-mixture derives an especial interest from its possible application to the case of those gas-mixtures in which the chemical composition and resolution of the components can take place in the gas-mixture itself, and actually does take place, so that the quantities of the proximate components are entirely determined by the quantities of a smaller number of ultimate components, with the temperature and pressure. These may be called *gas-mixtures with convertible components*. If the general laws of *ideal* gas-mixtures apply in any such case, it may easily be shown that the phases of dissipated energy are the only phases which can exist. We can form a fundamental equation which shall relate solely to these phases. For this end, we first form the equation in p, t, μ_1, μ_2 etc. for the gas-mixture, regarding its proximate components as *not convertible*. This equation will contain a potential for every proximate component of the gas-mixture. We then eliminate one (or more) of these potentials by means of the relations which exist between them in virtue of the convertibility of the components to which they relate, leaving the potentials which relate to those substances which naturally express the ultimate composition of the gas-mixture.

The validity of the results thus obtained depends upon the applicability of the laws of ideal gas-mixtures to cases in which chemical action takes place. Some of these laws are generally regarded as capable of such application, others are not so regarded. But it may be shown that in the very important case in which the components of a gas are convertible at certain temperatures, and not at others, the theory proposed may be established without other assumptions than such as are generally admitted.

It is, however, only by experiments upon gas-mixtures with convertible components, that the validity of any theory concerning them can be satisfactorily established.

The vapor of the peroxide of nitrogen appears to be a mixture of two different vapors, of one of which the molecular formula is double that of the other. If we suppose that the vapor conforms to the laws of an ideal gas-mixture in a state of dissipated energy, we may obtain an equation between the temperature, pressure, and density of the vapor, which exhibits a somewhat striking agreement with the results of experiment.

Equilibrium of stressed solid. – The second paper[4] commences with a discussion of the conditions of internal and external equilibrium for solids in contact with fluids with regard to all possible states of strain of the solids. These conditions are deduced by analytical processes from the general condition of equilibrium (2). The condition of equilibrium which relates to the dissolving of the solid at a surface where it meets a fluid may be expressed by the equation

$$\mu_1 = \frac{\varepsilon - t\eta + pv}{m},\tag{22}$$

where ε, η, v, and m_1 denote respectively the energy, entropy, volume, and mass of the solid, if it is homogeneous in nature and state of strain,—otherwise, of any small portion which may be treated as thus homogeneous, $-\mu_1$ the potential in the fluid for the substance of which the solid consists, p the pressure in the fluid and therefore one of the principal pressures in the solid, and t the temperature. It will be observed that when the

---

[4] See footnote, p. 184.

pressure in the solid is isotropic, the second member of this equation will represent the potential in the solid for the substance of which it consists [see (9)], and the condition reduces to the equality of the potential in the two masses, just as if it were a case of two fluids. But if the stresses in the solid are not isotropic, the value of the second member of the equation is not entirely determined by the nature and state of the solid, but has in general three different values (for the same solid at the same temperature, and in the same state of strain) corresponding to the three principal pressures in the solid. If a solid in the form of a right parallelopiped is subject to different pressures on its three pairs of opposite sides by fluids in which it is soluble, it is in general necessary for equilibrium that the composition of the fluids shall be different.

The *fundamental equations* which have been described above are limited, in their application to solids, to the case in which the stresses in the solid are isotropic. An example of a more general form of fundamental equation for a solid, is afforded by an equation between the energy and entropy of a given quantity of the solid, and the quantities which express its state of strain, or by an equation between ψ [see (3)] as determined for a given quantity of the solid, the temperature, and the quantities which express the state of strain.

Capillarity. – The solution of the problems which precede may be regarded as a first approximation, in which the peculiar state of thermodynamic equilibrium about the surfaces of discontinuity is neglected. To take account of the condition of things at these surfaces, the following method is used. Let us suppose that two homogeneous fluid masses are separated by a surface of discontinuity, *i.e.*, by a very thin non-homogeneous film. Now we may imagine a state of things in which each of the homogeneous masses extends without variation of the densities of its several components, or of the densities of energy and entropy, quite up to a geometrical surface (to be called the dividing surface) at which the masses meet. We may suppose this surface to be sensibly coincident with the physical surface of discontinuity. Now if we compare the actual state of things with the supposed state, there will be in the former in the vicinity of the surface a certain (positive or negative) excess of energy, of entropy, and of each of the component substances. These quantities are denoted by ε^S, η^S, m_1^S, m_2^S, etc. and are treated as belonging to the surface. The ^{S} is used simply as a distinguishing mark, and must not be taken for an algebraic exponent.

It is shown that the conditions of equilibrium already obtained relating to the temperature and the potentials of the homogeneous masses, are not affected by the surfaces of discontinuity, and that the complete value of $\delta\varepsilon^S$ is given by the equation

$$\delta\varepsilon^S = t\delta\eta^S + \sigma\delta s + \mu_1\delta m_1^S + \mu_2\delta m_2^S + \text{etc.,} \qquad (23)$$

in which s denotes the area of the surface considered, t the temperature, μ_1, μ_2, etc., the potentials for the various components in the adjacent masses. It may be, however, that some of the components are found only at the surface of discontinuity, in which case the letter μ with the suffix relating to such a substance denotes, as the equation shows, the rate of increase of energy at the surface per unit of the substance added, when the entropy, the area of the surface, and the quantities of the other components are unchanged. The quantity σ may regard as defined by the equation itself, or by the following, which is obtained by integration:

$$\varepsilon^S = t\eta^S + \sigma s + \mu_1 m_1^S + \mu_2 m_2^S + \text{etc.} \tag{24}$$

There are terms relating to variations of the curvatures of the surface which might be added, but it is shown that we can give the dividing surface such a position as to make these terms vanish, and it is found convenient to regard its position as thus determined. It is always sensibly coincident with the physical surface of discontinuity. (Yet in treating of plane surfaces, this supposition in regard to the position of the dividing surface is unnecessary, and it is sometimes convenient to suppose that its position is determined by other considerations.)

With the aid of (23), the remaining condition of equilibrium for contiguous homogeneous masses is found, viz:

$$\sigma(c_1 + c_2) = p' - p'', \tag{25}$$

where p', p'' denote the pressures in the two masses, and c_1, c_2 the principal curvatures of the surface. Since this equation has the same form as if a tension equal to σ resided at the surface, the quantity σ is called (as is usual) the *superficial tension*, and the dividing surface in the particular position above mentioned is called the *surface of tension*.

By differentiation of (24) and comparison with (23), we obtain

$$d\sigma = -\eta_s dt - \Gamma_1 d\mu_1 - \Gamma_2 d\mu_2 - \text{etc.,} \tag{26}$$

where η_s, Γ_1, Γ_2, etc. are written for η^S/s, m_1^S/s, m_2^S/s etc., and denote the superficial densities of entropy and of the various substances. We may regard σ as a function of t, μ_1, μ_2, etc., from which if known η_s, Γ_1, Γ_2, etc. may be determined in terms of the same variables. An equation between σ, t, μ_1, μ_2, etc. may therefore be called a *fundamental equation for the surface of discontinuity*. The same may be said of an equation between ε^S, η^S, S, m_1^S, m_2^S, etc.

It is necessary for the stability of a surface of discontinuity that its tension shall be as small as that of any other surface which can exist between the same homogeneous masses with the same temperature and potentials. Beside this condition, which relates to the nature of the surface of discontinuity, there are other conditions of stability, which relate to the possible motion of such surfaces. One of these is that the tension shall be positive. The others are of a less simple nature, depending upon the extent and form of the surface of discontinuity, and in general upon the whole system of which it is a part. The most simple case of a system with a surface of discontinuity is that of two coexistent phases separated by a spherical surface, the outer mass being of indefinite extent. When the interior mass and the surface of discontinuity are formed entirely of substances which are components of the surrounding mass, the equilibrium is always unstable; in other cases, the equilibrium may be stable. Thus, the equilibrium of a drop of water in an atmosphere of vapor is unstable, but may be made stable by the addition of a little salt. The analytical conditions which determine the stability or instability of the system are easily found, when the temperature and potentials of the system are regarded as known, as well as the fundamental equations for the interior mass and the surface of discontinuity.

The study of surfaces of discontinuity throws considerable light upon the subject of the stability of such phases of fluids as have a less pressure than other phases of the same components with the same temperature and potentials. Let the pressure of the phase of which the stability is in question be denoted by p' and that of the other phase of the same temperature and potentials by p''. A spherical mass of the second phase and of a radius determined by the equation

$$2\sigma = (p'' - p') r, \tag{27}$$

would be in equilibrium with a surrounding mass of the first phase. This equilibrium, as we have just seen, is instable, when the surrounding mass is indefinitely extended. A spherical mass a little larger would tend to increase indefinitely. The work required to form such a spherical mass, by a reversible process, in the interior of an infinite mass of the other phase, is given by the equation

$$W = \sigma S - (p'' - p') v''. \tag{28}$$

The term σS represents the work spent in forming the surface, and the term $(p'' - p') v''$ the work gained in forming the interior mass. The second of these quantities is always equal to two-thirds of the first. The value of W is therefore positive, and the phase is in strictness stable, the quantity W affording a kind of measure of its stability. We may easily express the value of W in a form which does not involve any geometrical magnitudes, viz.,

$$W = \frac{16\pi\sigma^3}{3(p'' - p')^2}, \tag{29}$$

where p'', p' and σ may be regarded as functions of the temperature and potentials. It will be seen that the stability, thus measured, is infinite for an infinitesimal difference of pressures, but decreases very rapidly as the difference of pressures increases. These conclusions are all, however, practically limited to the case in which the value of r, as determined by equation (27), is of sensible magnitude.

With respect to the somewhat similar problem of the stability of the surface of contact of two phases with respect to the formation of a new phase, the following results are obtained. Let the phases (supposed to have the same temperature and potentials) be denoted by A, B, and C; their pressures by p_A, p_B and p_C; and the tensions of the three possible surfaces by σ_{AB}, σ_{BC}, σ_{AC}. If p_C is less than

$$\frac{\sigma_{BC} p_A + \sigma_{AC} p_B}{\sigma_{BC} + \sigma_{AC}},$$

there will be no tendency toward the formation of the new phase at the surface between A and B. If the temperature or potentials are now varied until p_C is equal to the above expression, there are two cases to be distinguished. The tension σ_{AB} will be either equal to or less. (A greater value could only relate to an unstable and therefore unusual surface.) If $\sigma_{AB} = \sigma_{AC} + \sigma_{BC}$, a farther variation of the temperature or potentials, making p_C greater than the above expression, would cause the phase C to be formed at the surface between A and B. But if $\sigma_{AB} < \sigma_{AC} + \sigma_{BC}$, the surface between A and B would remain stable, but with rapidly diminishing stability, after p_C has passed the limit mentioned.

The conditions of stability for a line where several surfaces of discontinuity meet, with respect to the possible formation of a new surface, are capable of a very simple expression. If the surfaces A-B, B-C, C-D, D-A, separating the masses A, B, C, D, meet along a line, it is necessary for equilibrium that their tensions and directions at any point of the line should be such that a quadrilateral α, β, γ, δ may be formed with sides representing in direction and length the normals and tensions of the successive surfaces. For the stability of the system with reference to the possible formation of surfaces between A and C, or between B and D, it is farther necessary that the tensions σ_{AC} and σ_{BD} should be greater than the diagonals $\alpha\gamma$ and $\beta\delta$ respectively. The conditions of stability are entirely analogous in the case of a greater number of surfaces. For the conditions of stability relating to the formation of a new phase at a line in which three surfaces of discontinuity meet, or at a point where four different phases meet, the reader is referred to the original paper.

Liquid films. – When a fluid exists in the form of a very thin film between other fluids, the great inequality of its extension in different directions will give rise to certain peculiar properties, even when its thickness is sufficient for its interior to have the properties of matter in mass. The most important case is where the film is liquid and the contiguous fluids are gaseous. If we imagine the film to be divided into elements of the same order of magnitude as its thickness, each element extending through the film from side to side, it is evident that far less time will in general be required for the attainment of approximate equilibrium between the different parts of any such element and the contiguous gases than for the attainment of equilibrium between all the different elements of the film.

There will accordingly be a time, commencing shortly after the formation of the film, in which its separate elements may be regarded as satisfying the conditions of internal equilibrium, and of equilibrium with the contiguous gases, while they may not satisfy all the conditions of equilibrium with each other. It is when the changes due to this want of complete equilibrium take place so slowly that the film appears to be at rest, except so far as it accommodates itself to any change in the external conditions to which it is subjected, that the characteristic properties of the film are most striking and most sharply defined. It is from this point of view that these bodies are discussed. They are regarded as satisfying a certain well-defined class of conditions of equilibrium, but as not satisfying at all certain other conditions which would be necessary for complete equilibrium, in consequence of which they are subject to gradual changes, which ultimately determine their rupture.

The elasticity of a film (*i.e.*, the increase of its tension when extended,) is easily accounted for. It follows from the general relations given above that, when a film has more than one component, those components which diminish the tension will be found in greater proportion on the surfaces. When the film is extended, there will not be enough of these substances to keep up the same volume- and surface-densities as before, and the deficiency will cause a certain increase of tension. It does not follow that a thinner film has always a greater tension than a thicker formed of the same liquid. When the phases within the films as well as without are the same, and the surfaces of the films are also the same, there will be no difference of tension. Nor will the tension of the same film be altered, if a part of the interior drains away in the course of time, without affecting the surfaces. If the thickness of the film is reduced by evaporation, its tension may be either increased or diminished, according to the relative volatility of its different components.

Let us now suppose that the thickness of the film is reduced until the limit is reached at which the interior ceases to have the properties of matter in mass. The elasticity of the film, which determines its stability with respect to extension and contraction, does not vanish at this limit. But a certain kind of instability will generally arise, in virtue of which inequalities in the thickness of the film will tend to increase through currents in the interior of the film. This probably leads to the destruction of the film, in the case of most liquids. In a film of soap-water, the kind of instability described seems to be manifested in the breaking out of the black spots. But the sudden diminution in thickness which takes place in parts of the film is arrested by some unknown cause, possibly by viscous or gelatinous properties, so that the rupture of the film does not necessarily follow.

Electromotive force.–The conditions of equilibrium may be modified by electromotive force. Of such cases a galvanic or electrolytic cell may be regarded as the type. With respect to the potentials for the ions and the electrical potential the following relation may be noticed:

When all the conditions of equilibrium are fulfilled in a galvanic or electrolytic cell, the electromotive force is equal to the difference in (the values of the potential for any ion at the surfaces of the electrodes multiplied by the electro-chemical equivalent of that ion, the greater potential of an anion being at the same electrode as the greater electrical potential, and the reverse being true of a cation.

The relation which exists between the electromotive force of a *perfect electro-chemical apparatus* (*i.e.*, a galvanic or electrolytic cell which satisfies the condition of reversibility), and the changes in the cell which accompany the passage of electricity, may be expressed by the equation

$$d\varepsilon = \left(V' - V''\right)de + td\eta + dW_{\mathrm{G}} + dW_{\mathrm{P}}, \tag{30}$$

in which $d\varepsilon$ denotes the increment of the intrinsic energy in the apparatus, $d\eta$ the increment of entropy, de the quantity of electricity which passes through it, V' and V'' the electrical potentials in pieces of the same kind of metal connected with the anode and cathode respectively, dW_{G} the work done by gravity, and dW_{P} the work done by the pressures which act on the external surface of the apparatus. The term dW_{G} may generally be neglected. The same is true of dW_{P}, when gases are not concerned. If no heat is supplied or withdrawn the term $td\eta$ will vanish. But in the calculation of electromotive forces, which is the most important application of the equation, it is convenient and customary to suppose that the temperature is maintained constant. Now this term $td\eta$ which represents the heat absorbed by the cell, is frequently neglected in the consideration of cells of which the temperature is supposed to remain constant. In other words, it is frequently assumed that neither heat or cold is produced by the passage of an electrical current through a perfect electro-chemical apparatus (except that heat which may be indefinitely diminished by increasing the time in which a given quantity of electricity passes), unless it be by processes of a secondary nature, which are not immediately or necessarily connected with the process of electrolysis.

That this assumption is incorrect is shown by the electromotive force of a gas battery charged with hydrogen and nitrogen, by the currents caused by differences in the concentration of the electrolyte, by electrodes of zinc and mercury in a solution of sulphate of zinc, by *a priori* considerations based on the phenomena exhibited in the direct combination of the elements of water or of hydrochloric acid, by the absorption of heat which M. Favre has in many cases observed in a galvanic or electrolytic cell, and by the fact that the solid or liquid state of an electrode (at its temperature of fusion) does not affect the electromotive force.

BIBLIOGRAPHY

Abir-Am, Pnina G. and Dorinda Outram, eds. *Uneasy Careers and Intimate Lives: Women in Science, 1789–1979*. New Brunswick: Rutgers University Press, 1987.

Bailey, L. H. "Neo-Lamarckism and Neo-Darwinism." *American Naturalist* 28, no. 332 (Aug. 1894): 661–78.

Baker, Ray Palmer. "Rensselaer Polytechnic Institute and the Beginnings of Science in the United States." *The Scientific Monthly* 19, no. 4 (Oct. 1924): 337–56.

Bates, Ralph. *Scientific Societies in the United States*. 2nd ed. New York: Columbia University Press, 1958.

Baym, Nina. *American Women of Letters and the Nineteenth-Century Sciences: Style of Affiliation*. New Brunswick, NJ: Rutgers University Press, 2001.

Bernal, John D. *Science and Industry in the Nineteenth Century*. London: Routledge and Kegan Paul, 1953.

Blumenthal, Henry. *American and French Culture, 1800–1900: Interchanges in Art, Science, Literature, and Society*. Baton Rouge: Louisiana State University, 1975.

Botkin, Daniel B. *Our Natural History: The Lessons of Lewis and Clark*. New York: Berkley Publishing, 1995.

Boorstin, Daniel J. *The Lost World of Thomas Jefferson*. New York: Henry Holt and Co., 1948.

Bozeman, Theodore Dwight. "Science and Nineteenth-Century American Culture: A Note on George H. Daniels' *Science in the Age of Jackson*." *Isis* 63, no. 3 (Sept. 1972): 397–402.

———. *Protestants in an Age of Science: The Baconian Ideal and Antebellum American Religious Thought*. Chapel Hill: University of North Carolina Press, 1977.

Brent, Joseph. *Charles Sanders Peirce: A Life*. Bloomington: Indiana University Press, 1993.

Brown, Chandos Michael. *Benjamin Silliman: A Life in the Young Republic*. Princeton, NJ: Princeton University Press, 1990.

Browther, J. G. *Famous American Men of Science*. Freeport, NY: Books for Libraries Press, 1937.

Bruce, Robert. *The Launching of Modern American Science, 1846–1876*. New York: Alfred A. Knopf, 1987.

Cannon, Susan Faye. "Humboldtian Science." In *Science in Culture: The Early Victorian Period*. New York: Science History Publications, 1978.

Carey, Charles W., Jr. *African Americans in Science: An Encyclopedia of People and Progress*. Vol. 1. Santa Barbara, CA: ABC-CLIO Inc., 2008.

Conklin, Edwin G. "The Early History of the American Naturalist." *American Naturalist* 78, no. 774 (1944): 29–37.

Cohen, I. Bernard. "Some Reflections on the State of Science in America during the Nineteenth Century." *Proceedings of the National Academy of Sciences* 45, no. 5 (15 May 1959): 666–77.

———. "Science in America: The Nineteenth Century." In *Paths of American Thought*, ed. Arthur Schlesinger, Jr. and Morton White. Boston: Houghton Mifflin, 1963.

Coolidge, Julian Lowell. *The Development of Harvard University since the Inauguration of President Eliot, 1869–1929*, ed. Samuel Eliot Morison. Cambridge, MA: Harvard University Press, 1930.

Cravens, Hamilton, Alan I. Marcus, and David M. Katzman. *Technical Knowledge in American Culture: Science, Technology, and Medicine Since the Early 1800s.* Tuscaloosa: University of Alabama Press, 1996.

Creese, Mary R. S., with contributions by Thomas M. Creese. *Ladies in the Labortatory? American and British Women in Science, 1800–1900.* Lanham, MD: Scarecrow Press, 1998.

Cutright, Paul Russell. *Lewis & Clark: Pioneering Naturalists.* Reprint. Lincoln, NE: University of Nebraska Press, 1989.

Daniels, George H., ed. *Nineteenth-Century American Science: A Reappraisal.* Evanston, IL: Northwestern University Press, 1972.

Daniels, George H. "The Process of Professionalization in American Science: The Emergent Period, 1820–1860." *Isis* 58, no. 2 (Summer 1967): 150–166.

———. *American Science in the Age of Jackson.* New York: Columbia University Press, 1968.

———. *Science in American Society: A Social History.* New York: Alfred A. Knopf, 1971.

———, ed. *Nineteenth-Century American Science: A Reappraisal.* Evanston, IL: Northwestern University Press, 1972.

Darwin, Charles. "Second Notebook on Transmutation of Species (February to July 1838)," ed. Gavin de Beer, *Bulletin of the British Museum (Natural History), Historical Series*, 1960, 2.

Dettelbach, Michael. "Humboldtian Science." In *Cultures of Natural History*, ed. N. Jardine, J. A. Secord, and E. C. Sparry. Cambridge: Cambridge University Press, 1996.

Dupree, A. Hunter. *Science in the Federal Government: A History of Policies and Activities to 1940.* Cambridge, MA: Belknap Press of Harvard University Press, 1957.

Elliott, Clark A. *Biographical Dictionary of American Science: The Seventeenth through the Nineteenth Centuries.* Westport, CT: Greenwood Press, 1979.

Gascoigne, Robert Mortimer. *A Chronology of the History of Science, 1450–1900.* New York: Garland Publishing, 1987.

Gould, Stephen Jay. "Art Meets Science in *The Heart of the Andes*: Church Paints, Humboldt Dies, Darwin Writes, and Nature Blinks in the Fateful Year of 1859." In *I Have Landed: The End of a Beginning in Natural History.* Cambridge, MA: The Belknap Press of Harvard University Press, 2011.

Greene, John C. "American Science Comes of Age, 1780–1820." *Journal of American History* 55, no. 4 (1965): 22–41.

————. *American Science in the Age of Jefferson*. Ames: Iowa State University Press, 1988.

Hindle, Brooke. *The Pursuit of Science in Revolutionary America*. Chapel Hill: University of North Carolina Press, 1956.

Hoffman, C. F. "The United States Coast Survey." *Literary World* 11 (Sept. 1847): 125–28.

Jaffe, Bernard. *Men of Science in America*. New York: Simon & Schuster, 1944.

Jaffe, Mark. *The Gilded Dinosaur: The Fossil War between E. D. Cope and O. C. Marsh and the Rise of American Science*. New York: Crown Publishing, 2000.

Kevles, Daniel J., Jeffrey L. Sturchio, and P. Thomas Carroll. "The Sciences in America, Circa 1880." *Science* 209, no. 4452 (4 July 1980): 26–32.

Kingsland, Sharon. *The Evolution of American Ecology, 1890–2000*. Baltimore, MD: Johns Hopkins University Press, 2005.

Kline, Robert. "Construing 'Technology' as 'Applied Science': Public Rhetoric of Scientist and Engineers in the United States, 1880–1945." *Isis* 86, no. 2 (June 1995): 194–221.

Kohlstedt, Sally G. *The Formation of the American Scientific Community: The American Association for the Advancement of Science, 1848–1860*. Urbana: University of Illinois Press, 1976.

————. "Science: The Struggle for Survival, 1880–1894." *Science* 209, no. 4452 (4 July 1980): 33–42.

Kohlstedt, Sally G. and Margaret W. Rossiter, eds. *Historical Writing on American Science: Perspectives and Prospects*. Baltimore: Johns Hopkins University Press, 1986.

Kuhn, Thomas. *The Structure of Scientific Revolutions*. Chicago: University of Chicago Press, 1962.

Levine, Lawrence W. *The Opening of the American Mind: Canons, Culture, and History*. London: Beacon Press, 1996.

Lurie, Edward. *Nature and the American Mind: Louis Agassiz and the Culture of Science*. New York: Science History Publications, 1974.

Marcus, Alan I. "Science and Technology." *A Companion to 19th-Century America*, ed. William L. Barney. Oxford: Blackwell Publishing, 2001.

Mason, S. F. *Main Currents in Scientific Thought: A History of the Sciences*. New York: Henry Schuman, 1953.

Mayr, Ernst. "Agassiz, Darwin, and Evolution." *Harvard Library Bulletin* 13 (1959): 165–194.

McColley, Grant, ed. *Literature and Science: An Anthology from English and American Literature, 1600–1900*. Chicago: Packard and Company, 1940.

Menand, Louis. *The Metaphysical Club: A Story of Ideas in America*. New York: Farrar, Straus and Giroux, 2001.

Metzger, Walter. *Academic Freedom in the Age of the University*. New York: Columbia University Press, 1965.

Miller, Howard S. *Dollars for Research: Science and Its Patrons in Nineteenth-Century America*. Seattle: University of Washington Press, 1970.

Miller, Lillian B. *The Lazzaroni.* Washington, DC: Smithsonian Institution Press (For the National Portrait Gallery), 1972.

Morison, Samuel Eliot, ed. *The Development of Harvard University since the Inauguration of President Eliot, 1869–1929.* Cambridge, MA: Harvard University Press, 1930.

National Academy of Sciences. *Biographical Memoirs.* Washington, DC: Published by the Home Secretary, 1877.

Newcomb, Simon. "Abstract Science in America, 1776–1876." *North American Review* 122, no. 250 (Jan. 1876): 88–124.

Numbers, Ronald L. *Darwinism Comes to America.* Cambridge, MA: Harvard University Press, 1998.

Oleson, A. and Brown, S. C., eds. *The Pursuit of Knowledge in the Early American Republic: American Scientific and Learned Societies from Colonial Times to the Civil War.* Baltimore, MD: Johns Hopkins University Press, 1976.

Oleson, A. and Voss, J., eds. *The Organization of Knowledge in Modern America, 1860–1920.* Baltimore, MD: Johns Hopkins University Press, 1979.

Pauly, Philip J. *Biologists and the Promise of American Life: From Meriwether Lewis to Alfred Kinsey.* Princeton, NJ: Princeton University Press, 2000.

Pfeifer, Edward J. "The Genesis of American Neo-Lamarckism." *Isis* 56, no. 2 (Summer, 1965): 156–167.

Porter, Charlotte M. *The Eagle's Nest: Natural History and American Ideas, 1812–1842.* Tuscaloosa: University of Alabama Press, 1986.

Ratner, Sidney. "Evolution and the Rise of the Scientific Spirit in America." *Philosophy of Science* 3, no. 1 (January 1936): 104–22.

Reingold, Nathan, ed. *Science in Nineteenth-Century America: A Documentary History.* New York: Hill and Wang, 1964.

Rossiter, Margaret W. *Women Scientists in America: Struggles and Strategies to 1940.* Baltimore, MD: Johns Hopkins University Press, 1982.

Rowland, Henry A. "A Plea for Pure Science." *Science* 2, no. 29 (24 Aug. 1883): 242–50.

———. "The Highest Aim of the Physicist." Presidential address in *Bulletin of the American Physical Society* 1, no. 1 (1899): 4–16.

Schneer, Cecil J. "The Great Taconic Controversy." *Isis* 69, no. 2 (June 1978): 173–191.

Shryock, Richard. "American Indifference to Basic Science in the Nineteenth Century." *Archives internationales d'histoire des sciences,* no. 5 (October 1948): 50–65.

Slotten, Hugh Richard. *Patronage, Practice, and the Culture of American Science: Alexander Dallas Bache and the U.S. Coast Survey.* New York: Cambridge University Press, 1994.

Smallwood, W. M. "The Agassiz-Rogers Debate on Evolution." *The Quarterly Review of Biology* 16, no. 1 (March 1941): 1–12.

Struik, Dirk J. *Yankee Science in the Making: Science and Engineering in New England from Colonial Times to the Civil War.* New York: Collier Books, 1962.

Theberge, Albert E. "The Coastal Survey and Army Operation During the Civil War." *The Coastal Survey 1807–1867* (Volume 1). National Oceanic and Atmospheric

Administration, 9 September 1998. http://www.lib.noaa.gov/noaainfo/heritage/coastsurveyvol1/CONTENTS.html (accessed 7 February 2012).

Timmons, Todd. *Science and Technology in Nineteenth-Century America*. Westport, CT: Greenwood Press, 2005.

Tocqueville, Alexis de. "Why the Americans Are More Addicted to Practical than to Theoretical Science." In *Democracy in America*, vol. 2, trans. Henry Reeve and ed. Francis Bowen. Cambridge, MA: Sever and Francis, 1862.

Tolley, Kim. *The Science Education of American Girls: A Historical Perspective*. New York: Routledge Falmer, 2003.

Van Tassel, David D., and Michael G. Hall. *Science and Society in the United States*. Homewood, IL: Dorsey Press, 1966.

Wallace, David Rains. *The Bonehunter's Revenge: Dinosaurs, Greed, and the Greatest Scientific Feud of the Gilded Age*. New York: Houghton Mifflin Co., 1999.

Walls, Laura Dassow. *Seeing New Worlds: Henry David Thoreau and Nineteenth Century Natural Science*. Madison: University of Wisconsin Press, 1995.

Webb, George E. *The Evolution Controversy in America*. Lexington: University Press of Kentucky, 1994.

Williams, David B. "A Wrangle over Darwin: How Evolution Evolved in America." *Harvard Magazine*, September–October 1998.

Wilson, Daniel J. *Science, Community, and the Transformation of American Philosophy, 1860–1930*. Chicago: University of Chicago Press, 1990.

Winchell, N. H. "The Taconic Controversy in a Nutshell." *Science* 7, no. 153 (8 Jan. 1886): 34.

Wright, Chauncey. "Limits of Natural Selection." *Philosophical Discussions by Chauncey Wright*, ed. C. E. Norton. New York: Lennox Hill, 1877.

———. *The Evolutionary Philosophy of Chauncey Wright*, 3 vols, ed. Frank X. Ryan and introduced by Frank X. Ryan and Edward H. Madden. Bristol, UK: Thoemmes Press, 2000.

Zeller, Suzanne. *Inventing Canada: Early Victorian Science and the Idea of a Transcontinental Nation*. Toronto: McGill-Queen's University Press, 1987.

———. "The Spirit of Bacon: Science and Self-Perception in the Hudson's Bay Company, 1830–1870." *Scientia Canadensis: Canadian Journal of the History of Science, Technology, and Medicine* 13, no. 2 (37) 1989: 79–101.